T0234029

The Prior Consultation of Indigenous Peoples in Latin America

This book delves into the reasons behind and the consequences of the implementation gap regarding the right to prior consultation and the Free, Prior and Informed Consent (FPIC) of Indigenous Peoples in Latin America.

In recent years, the economic and political projects of Latin American States have become increasingly dependent on the extractive industries. This has resulted in conflicts when governments and international firms have made considerable investments in those lands that have been traditionally inhabited and used by Indigenous Peoples, who seek to defend their rights against exploitative practices. After decades of intense mobilisation, important gains have been made at international level regarding the opportunity for Indigenous Peoples to have a say on these matters. Notwithstanding this, the right to prior consultation and the FPIC of Indigenous Peoples on the ground are far from being fully applied and guaranteed. And, even when prior consultation processes are carried out, the outcomes remain uncertain.

This volume rigorously investigates the causes of this implementation gap and its consequences for the protection of Indigenous Peoples' rights, lands, identities and ways of life in the Latin American region.

Claire Wright (PhD in Contemporary Political Processes) is a Research Professor at the Universidad de Monterrey, Mexico.

Alexandra Tomaselli (PhD in Law) is a Senior Researcher at Eurac Research, Italy.

Routledge Studies in Development and Society

For more information about this series, please visit: www.routledge.com/
Routledge-Studies-in-Development-and-Society/book-series/SE0317

The Prior Consultation of Indigenous Peoples in Latin America

Inside the Implementation Gap

Edited by
Claire Wright and Alexandra Tomaselli

Routledge
Taylor & Francis Group
LONDON AND NEW YORK

First published 2019
by Routledge
2 Park Square, Milton Park, Abingdon, Oxon OX14 4RN

and by Routledge
605 Third Avenue, New York, NY 10017

First issued in paperback 2021

Routledge is an imprint of the Taylor & Francis Group, an informa business

British Library Cataloguing-in-Publication Data
A catalogue record for this book is available from the British Library

Library of Congress Cataloging-in-Publication Data
A catalog record has been requested for this book

ISBN 13: 978-0-367-78437-9 (pbk)
ISBN 13: 978-1-138-48806-9 (hbk)

DOI: 10.4324/9781351042109

Typeset in Times New Roman
by Taylor & Francis Books

The Open Access version of Chapters 8 and 18 was funded by Eurac Research.

To B and S from Claire, and to F and L from Alexandra, with love.

Contents

Illustrations

Contributors

Irati N. Barreña is a Human Rights Analyst at the United Nations Office in Honduras and a PhD Candidate in Human Rights at the Universidad de Alcalá (Spain). She is a social worker and social anthropologist by training, and holds a Master of Laws (LLM) in the International Protection of Human Rights. She has complementary training in gender and development. Irati has worked in different NGOs, United Nations agencies, and State offices in the field of human rights in different countries. She has specialised in the analysis of human rights from different cultural and religious perspectives.

Laura Calle Alzate is an Adjunct Professor and Researcher at the Department of Social Anthropology and Social Psychology at the Universidad Complutense de Madrid (Spain). Laura is an anthropologist and a human rights activist, thus her research and teaching methodology emphasises a community-oriented approach. She has worked with Indigenous Peoples in Colombia since 2004. Her research ranges from the analysis of the phenomena of internal colonisation and the dispossession of Indigenous Peoples' lands, to the effects of extractivism on their territories and its implications for the exercise of Indigenous autonomy and governance.

Humberto Cantú Rivera is Professor at the School of Law and Social Sciences of the Universidad de Monterrey (Mexico), and Executive Director of its Business and Human Rights Institute. He holds a PhD from the Université Panthéon-Assas Paris II (France). Humberto has advised the Government of Mexico in the business and human rights treaty negotiations at the UN Human Rights Council, as well as in the development of national policies on this topic. He currently serves as the General Director remedy of the Latin American Branch of the Global Business and Human Rights Scholars Association.

Andrés Del Castillo is a Senior Legal Advisor at the Indigenous Peoples' Center for Documentation, Research and Information (DOCIP), Geneva (Switzerland), where he provides legal advice for Indigenous Peoples at the United Nations, the European Union and the OECD. He is also the Focal

Point before Multilateral Institutions and Guest Speaker on Responsible Business Conduct and the Creation of Shared Value at the University of Geneva and other institutions. Andrés is a Colombian lawyer with studies in International Economic Law in Brazil, and the European Union. He holds a Master's degree in International Administration Law from the Université Panthéon-Assas Paris II (France). Besides working in law firms in the private sector, Andrés has worked for UNESCO in Paris (France), ADECOM Network in Pondicherry (India), and the Office of the High Commissioner of Human Rights in Geneva.

Cathal M. Doyle is a Leverhulme Early Career Fellow at the School of Law of Middlesex University, London, UK. He is a founding member of the European Network on Indigenous Peoples, a member of the Committee on the Rights of Indigenous Peoples of the International Law Association, and a member of the boards of the Forest Peoples Programme (FPP) and the International Work Group for Indigenous Affairs (IWGIA). He specialises in the rights of Indigenous Peoples under international and national law, and the interface with business and human rights and sustainable development, with a particular focus on the right to consultation and Free, Prior and Informed Consent. He has published numerous books, reports, articles and book chapters addressing Indigenous Peoples' consultation and consent rights, including *Indigenous Peoples, Title to Territory, Rights and Resources: The Transformative Role of Free Prior and Informed Consent* (Routledge, 2015).

Riccarda Flemmer is a Research Fellow at the German Institute of Global and Area Studies (GIGA), and a PhD candidate in International Relations at the University of Hamburg (Germany). Her dissertation addresses the contested meaning of prior consultation in struggles over natural resources in Peru. Riccarda works at the intersection of International Relations, Legal Anthropology, and Decolonial Studies. She has authored various research articles about Indigenous Peoples' rights, socioenvironmental conflicts and development projects which have been published in journals such as *World Development*, the *Journal of Latin American and Caribbean Anthropology* and *Third World Quarterly*.

Juan C. Herrera is a Postdoctoral Visiting Research Fellow at the Max Planck Institute for Comparative Public Law and International Law in Heidelberg, Germany. He is a former Law Clerk of the Constitutional Court of Colombia. Juan's research experience and areas of practice encompass national and supranational issues related to fundamental rights, democracy, and regional integration in Latin America.

Julia Mello Neiva is a Senior Researcher on Brazil, Portugal and Portuguese-speaking Africa, a Representative for the Business and Human Rights Resource Centre, and a PhD candidate at the University of São Paulo (Brazil). She has been a human rights lawyer and researcher for over 15 years. Julia has studied at the Law Schools of Columbia University (USA), University of São Paulo, and the Pontifical Catholic University of São Paulo. She is a member of the research group "RSE et métamorphoses du droit" at the Institut des Sciences Juridique et Philosophique, Université Panthéon-Sorbonne Paris I. Julia is also a founding member of Conectas Human Rights, the Human Rights and Business Center, and the Fundação Getulio Vargas.

Anavel Monterrubio Redonda is Professor and Researcher at the Universidad Autónoma Metropolitana-Azcapotzalco (Mexico), and a Postdoctoral researcher at the National Centre of Competence in Research North-South (NCCR-NS) of the Swiss National Science Foundation. She was awarded a PhD in Sociology and a Master in Metropolitan Policies and Planning from the UAM-Azcapotzalco. Anavel has previously worked as researcher at the Centro de Estudios Sociales y de Opinión Pública of Mexico's Chamber of Deputies of Mexico, and in the public administration of the Federal District (Mexico City). Her research areas include urban habitat production, urban renovation, urban planning, and urban conflicts, social movements, housing policies, participative planning, and citizen participation.

Martin Papillon is Associate Professor in the Department of Political Science, Université de Montréal and Director of the Centre de Recherche sur les Politiques et le Development Social (CPDS) at the same university (Canada). He is the author of a number of articles on Indigenous-settlers relations in Canada, including works co-authored with Thierry Rodon on the implementation of the principle of Free, Prior and Informed Consent, the role of Impact and Benefits Agreements, as well as community-based impact assessments. His current project looks comparatively at the politics of Indigenous consent in the context of extractive projects in Canada, Brazil, Bolivia and Norway.

Thierry Rodon is an Associate Professor in the Political Science Department at Université Laval (Canada) and holds a Research Chair in Northern Sustainable Development. He is also the director of the Interuniversity Centre for Aboriginal Studies and Research (CIERA). He specialises in northern policies and Arctic governance and leads MinErAL, an international research network on extractive industries and Indigenous livelihood. He has authored several publications on Indigenous policies and governance in relation to the extractive industry in Canada.

Marzia Rosti is Associate Professor of History and Institutions of the Americas at the Università degli Studi di Milano, Italy. She holds a PhD in the Sociology of Law from the same university. Over the last years, her research has focused on the history of relations between Indigenous Peoples and national institutions in Argentina. Her recent publications include: Indigenous rights and extractivism in Argentina. *Federalismi.it*

(3), 2016; El 'modelo extractivista' y los derechos de los pueblos indígenas a los recursos naturales y al territorio en la Argentina de hoy. *Diritto Pubblico Comparato ed Europeo* (4), 2016; Neocolonialismo e a 'Outra Face' da Nação: as Exibições de Primitivos do sul do continente latino-americano no Chile (1873) e Argentina (1898). *Revista Latino-Americana de História* 6(17), 2017.

Malka San Lucas Ceballos is a Postdoctoral Researcher at the Department of Public Law of the Universitat Rovira i Virgili (URV) in Tarragona (Spain), and a Research Collaborator at the Centre D'Estudis de Dret Ambiental de Tarragona (CEDAT). She has been a visiting researcher and an external consultant at the Environmental Law Centre of the International Union for Conservation of Nature – IUCN. Malka has conducted research and published on the subjects of Indigenous Peoples' rights and extractive industries, environmental conflicts and environmental defenders. She is currently working on a research project on land grabbing and companies, through a legal analysis of these trends and their consequences.

Almut Schilling-Vacaflor is a sociologist and anthropologist, who is currently a Postdoctoral Research Fellow at Osnabrück University, Germany. Almut has headed two research projects on prior consultation and Free, Prior and Informed Consent (FPIC) in Bolivia and Peru funded by the German Foundation for Peace Research (DSF) and the Research Council of Norway (NRC). She has published extensively on participation, FPIC, environmental governance, Indigenous Peoples and socio-environmental conflicts in Latin America. Almut has published her research findings in *World Development, Development and Change*, the *Journal of Peasant Studies, Third World Quarterly* and in the co-edited anthology *New Mechanisms of Participation in Extractive Governance* (Routledge).

Alexandra Tomaselli is a Senior Researcher at the Institute for Minority Rights of Eurac Research, Bolzano-Bozen (Italy). She holds a PhD in Law from the Goethe University of Frankfurt a.M. (Germany). Since 2004, she has worked in a number of EU-funded research and international cooperation projects in Europe, Latin America, and South Asia. She has published several articles and delivered a number of lectures and presentations on the subjects of Minority and Indigenous rights, Latin American constitutionalism, legal anthropology, and other subjects related to international law, human rights law, and international climate change law both in Europe and overseas. Her recent publications include the single-authored book *Indigenous Peoples and their Right to Political Participation: International Law Standards and their Application in Latin America* (Nomos, Baden-Baden, 2016) and the co-edited bilingual book *Challenges to Indigenous Political and Socio-economic Participation: Natural Resources, Gender, Education and Intellectual Property (Desafíos de los pueblos indígenas en su participación política y socio-económica:*

Recursos Naturales, Género, Educación y Propiedad Intelectual) (Eurac E-Book, 2017). Alexandra is also a member of the International Law Association (ILA) and of its Committee on the Implementation of the Rights of Indigenous Peoples.

William Vega is a Costa Rican lawyer. He holds a Law degree from the Universidad de Costa Rica and a Master's Degree in Human Rights, Democracy and International Justice from the Universidad de Valencia (Spain), where he is currently a PhD candidate in Human Rights. He has worked for international organisations, civil society organisations, and the Government of Costa Rica. Over the past two years, he has worked as a consultant for various agencies of the United Nations, such as the International Labour Organisation, the UN Development Programme, and the Office of the High Commissioner for Human Rights.

Claire Wright is a Research Professor at the School of Law and Social Sciences of the University of Monterrey (Mexico), and Member of its Business and Human Rights Institute. She holds a PhD in Contemporary Political Processes from the University of Salamanca (Spain). Since 2006, Claire has carried out empirical research on the issues of Indigenous participation, conflicts over natural resources, social movements, participatory institutions, and emergency powers in the region. She has coordinated several research projects on these topics, financed by the Mexican Council of Science and Technology (CONACYT) and the Secretary of Public Education (SEP). Claire has also collaborated with the United Nations Development Programme (UNDP) and Spanish Cooperation (*Cooperación Española*) as an individual consultant. The results of her research have been published in both English and Spanish, in authored and edited volumes, journal articles, and book chapters. Claire's most recent publications include a co-authored book in Spanish on the right of Indigenous Peoples to consultation in electoral affairs (*El derecho a la consulta en materia electoral de los pueblos y comunidades indígenas*), published by the Electoral Institute of the State of Mexico (IEEM, 2018). Claire is a member of CONACYT's National Researchers' System, Level II.

Sara Mabel Villalba Portillo is full Professor at the Universidad Católica Nuestra Señora de la Asunción (UCA) and contract lecturer at the Universidad Nacional de Asunción (UNA) in the under- and postgraduate courses on Political and Communication Science and Political Science and Sociology. She is also a Researcher of the Consejo Nacional de Ciencia y Tecnología (CONACYT) of Paraguay. She was awarded a PhD in Contemporary Political Processes and a Master's Degree in Political Science at the Universidad de Salamanca (Spain). Her main research fields are participation and electoral systems, with a particular focus on the participation of Indigenous Peoples. Sara Mabel has also published on collective action and electoral processes. She has previously worked as Political Advisor at IDEA International and other organisations in Spain, Germany and Paraguay.

Lucía Xiloj is a Maya K'iche' lawyer. She holds a Master's Degree in Constitutional and Criminal Law from the Universidad de San Carlos (Guatemala). She was awarded a law degree in Environmental law by Tulane University (USA). For the past ten years, Lucía has worked with Indigenous Maya and Xinka communities. She has focused on their rights to land, natural resources, consultation and Free, Prior and Informed Consent (FPIC). Lucía has published a number of articles on the collective rights of Indigenous Peoples and the ILO Convention No.169 Concerning Indigenous and Tribal Peoples in Independent Countries of 1989, addressed to Indigenous communities as well as law and court practitioners.

Preface

As many researchers will know, the most interesting spaces for planning future projects with colleagues are not necessarily those offered during formal Conferences or event sessions, but rather in coffee breaks, hotel lobbies, mealtimes or – in fact – any occasion in which we can relax and let our creativity and imagination flow, beyond the shackles of academic convention and formality.

This is the case of this book. We – the editors – were comfortably sitting on Claire's sofa in Monterrey after an intense seminar on Indigenous Peoples and their rights, in late September 2016. Not really watching the TV news in front of us, we were discussing why Indigenous rights, and particularly the application of their rights to prior consultation and Free, Prior and Informed Consent (FPIC) still continue to fail. While speculating on the potential reasons for this in the region of Latin America we are familiar with, Claire suddenly launched the idea of writing a book dealing with the so-called "implementation gap". Just moments later, we had mentally prepared the first draft table of contents and identified the potential authors. Having fine-tuned our proposal, we decided to approach Routledge a few months after, and – after receiving the excellent news that our proposal had been accepted – we embarked on this joint venture in mid-2017.

Hence, the book you have in your hands is the result of over 18 months' work by two editors and a total of 19 authors, reflecting a great diversity of nationalities, mother tongues, experiences, professions, and disciplines. Nevertheless, both the editors and the authors are united in our aim of opening up and looking inside the implementation gap regarding prior consultation rights and FPIC in Latin America. The volume has survived the respective maternity leaves of the two editors and we hope – incidentally – that it is a testament to the capacity of working mothers to successfully complete our professional projects, particularly by supporting one another at crucial moments.

We feel that it is important to highlight several limitations of the volume which we feel are inherent to this sort of endeavour – and which do not take away any of the merit – but which nonetheless should be mentioned. First, is the restricted scope of our analysis of consultation and FPIC: in this volume we are interested in what these rights and mechanisms imply foremost for

Indigenous Peoples, although we are aware that the norms can be applied to other peoples including, but not limited to, Afro-Descendants. Second, although we have aimed to cover the countries in the region with the most considerable developments on this issue, there are some notable exceptions, particularly Panama (which adopted a Law on Consultation in 2016) and Nicaragua (where consultation was invoked in conflicts over the planned Inter-Oceanic canal). And, finally, given the nature of the topic, new legal and political developments are occurring constantly in the different countries, meaning that the analyses represent a moment in time rather than a fixed ending point. In any case, all major, relevant developments up to December 2018 have been included in the different chapters.

The authors of this volume have had total freedom in their analysis, although each chapter has been subject to rigorous double-blind peer review by experts. It is important to note that the authors do not necessarily speak on behalf of the institutions they represent and, likewise, this volume cannot claim to speak for, or on behalf of, the Indigenous Peoples with regard to the crucial issue of prior consultation and FPIC. Nevertheless, the normative backbone of this study is that the successful implementation of Indigenous rights at national and local levels is key to their empowerment as Peoples and, indeed, nothing short of a moral imperative for both State and private actors. We (humbly) hope that this volume will add some clarity on this issue and (why not?) add some fuel to the debate.

Monterrey (Mexico) and Bolzano-Bozen (Italy),
Claire Wright and Alexandra Tomaselli
March 2019

Acknowledgements

Just like any adventure, this volume would have been impossible without the invaluable contributions of its protagonists, i.e., the authors of the thought-provoking chapters that are included in this edited collection. Hence, as editors, our most sincere gratitude goes first and foremost to them. This volume would never have seen the light of day without the enthusiasm they showed from when we first shared a few rough ideas until they handed in the very final versions of their analysis of different issues and cases. We owe much of our analytical perspective to their input and appreciate all of their feedback and constructive criticism throughout what can only be described as a team effort.

Second, we are equally grateful to all of the fantastic Routledge staff who have promptly and efficiently followed us during the months that we took to prepare this volume. We are particularly thankful to our editorial assistants Ruth Anderson and Nonita Saha, as well as to our commissioning editor Faye Leerink and production editor Cathy Hurren.

Third, we are indebted to all of the efficient and committed colleagues who have served as peer reviewers for the chapters included in this book (in alphabetical order per surname): Guilherme Assis de Almeida, Carlos A. Baquero, Frederica Barclay, Marcelo Bogado, Nicolás Carrillo-Santarelli, Jorge Contesse, Jahir Dabroy, José Del Carmen-Ortega, Marvin Carvajal Pérez, Emilie Dupuits, Guillermina Espósito, Amelia Fiske, Marjorie Herrera-Castro, Alke Jenss, Quelvin O. Jimenez-Villalta, Sheryl Lightfoot, Carlos Mamani, Jennifer Matamoros-Pineda, Joaquín A. Mejía-Rivera, Jorge Mercado Mondragón, Roger Merino, Maritza Paredes, José Parra, Roberta Rice, Luis Sánchez-Vázquez, Charlotte Schumann, Claudia Talavera Reyes, Haydeé Valey, Víctor Tokichen Tricot, Irene Vélez-Torres, Rodrigo Villagra-Carron, Matthias vom Hau, Viviane Weitzner. Clearly, the final chapters published here are the entire responsibility of the editors and authors. Besides being all double-blind peer-reviewed, some of the chapters in this volume were also presented at the 56th International Conference of Americanists, which was held in Salamanca from 15 to 20 July 2018.

For her part, Claire expresses her gratitude to her colleagues at the Universidad de Monterrey (UDEM) for all of their financial and moral support for her research endeavours, including this volume: particularly Arturo Azuara Flores, Osvaldo Tello Rodríguez, Rafael Ibarra Garza, Carlos

Basurto Meza, Jacobo Tijerina Aguilera, Jorge Lozoya Santos, Maribell García, and María de la Luz Martínez Quezada. Specifically, she would like to thank the authorities of the UDEM for financing a considerable part of the language revision and edition undertaken in this volume, which was carried out by ENAGO, and special mention must be made of Alyson Culmer who carried out further revisions of several chapters. Claire would also like to thank Mexico's *Consejo Nacional de Ciencia y Tecnología* (CONACYT) for supporting her research on the political participation of Indigenous Peoples. Finally, she extends her gratitude to José Fredman Mendoza Ibarra for his tireless efforts to harmonise the reference system used throughout the text.

Alexandra wishes to sincerely thank her Directors and colleagues at Eurac Research for their encouragement and collegiality during the preparation and the editing of this volume. Among them, she is particularly grateful to Stephan Ortner, Roberta Bottarin and Roland Psenner for their advanced vision; and to Ivana Spirovska for her kind editing. Alexandra extends her gratitude also to Silvia Ordoñez-Ganoza, Elvira Dominguez-Redondo, Joshua Castellino, Jessika Eichler, and Philipp Altmann for their constant support and their priceless friendship. In addition, both of us are most indebted to our fantastic European and Latin American friends and families, our caring husbands, and, most of all, to our cheerful little boys who were both born while preparing this volume: without their energy and joyfulness we would have had a much harder time in preparing this book.

The authors also express their gratitude for the scientific and financial support of the following people and institutions, as follows:

Andrés Del Castillo is grateful to the Aymara historian Carlos Mamani and to José Del Carmen Ortega for their invaluable comments and suggestions regarding his chapter.

Cathal M. Doyle would like to thank Frederica Barclay and Tami Okamoto for their helpful suggestions. He would also like to thank the *Federaciones* of the four river basins (Federación de Indígenas Quechuas del Pastaza-FEDIQUEP, Federación de Comunidades Nativas del Corrientes-FECONACO, Organización de Pueblos Indígenas Kichuas, Amazónicos Fronterizos del Perú y Ecuador-OPIKAFPE, and Asociación Cocama de Desarrollo y Conservación San Pablo de Tipishca-ACODECOSPAT), and Wampis, Awajun and Aymara Indigenous representatives, and their support organisations, Peru Equidad, Derechos Humanos y Medio Ambiente (DHUMA), Pueblos Indígenas Amazónicos Unidos en Defensa de sus Territorios (PUINAMUDT) and CooperAccion, who have shared their insights with him.

Both Riccarda Flemmer and Almut Schilling-Vacaflor thank the German Foundation for Peace Research-DSF for supporting their researchers within the project: "Prior Consultation, Participation and Conflict Transformation. A Comparative Study of Resource Extraction in Bolivia and Peru". Almut is also grateful to the funding kindly provided by the Norwegian Research Council (Project: "Extracting Justice?"), and Riccarda to the DAAD Bilateral SDG Graduate Schools "trAndeS" for their scholarship.

Juan C. Herrera expresses his gratitude to Aida Torres, Ander Errasti, Andres Del Castillo, Carlos Baquero, César Carvajal, Kehinde Balogun, Manuel Góngora, Oscar Parra, Pedro Villarreal and Roy Lee for their accurate and insightful critiques. He is similarly grateful for the comments and ideas expressed at the seminars and conferences where previous versions of the research were discussed between 2016 and 2017: the ICON·S Annual Conference in Copenhagen; the 3rd International Conference on Public Policy ICCP3 in Singapore; the MPIL Agora in Heidelberg; the UB Law and Society International Colloquium in Barcelona; and the UPF Constitutional Law seminars in Barcelona, which was financed by Colciencias. His chapter in this volume – "Binding Consent of Indigenous Peoples in Colombia: An Example of Transformative Constitutionalism" – is a revised version of his article "Judicial Dialogue and Transformative Constitutionalism in Latin America: the case of Indigenous Peoples and Afro-descendants", published in the Revista Derecho del Estado, 43 (2019). The author and editors would like to thank the editors of the journal for their permission to publish the revised version of the article.

Julia Mello Neiva would like to thank Marinilzes Moradillo Mello, Guilherme de Almeida, and Amanda Romero for their invaluable support and their critical content review of her chapter.

Martin Papillon and Thierry Rodon thank the Social Sciences and Humanities Research Council of Canada for their funding contribution for writing their chapter.

Malka San Lucas Ceballos is grateful to the Centre D'Estudis de Dret Ambiental de Tarragona (CEDAT) at the Universitat Rovira i Virgili (URV) of Tarragona for their support.

William Vega expresses his gratitude to the Maleku expert Geyner Blanco (to whom he dedicates his chapter) as well as to Marjorie Herrera-Castro and Marvin Carvajal for their support.

And, Sara Mabel Villalba Portillo thanks *Programa Nacional de Incentivo a Investigadores* – PRONII organised by Paraguay's *Consejo Nacional de Ciencia y Tecnología* (CONACYT).

Last but not least, we owe a profound debt to all of the inspiring Indigenous Peoples both of us, as well as all the authors, have had the fortune to meet and talk to throughout Latin America over the years. We sincerely hope that this book will make a contribution not only at a scientific level but that it will also help to denounce and better identify why and how their rights are consistently denied.

1 Introduction

Alexandra Tomaselli and Claire Wright

Over the last decade, the debate surrounding the right to consultation of Indigenous Peoples[1] and the duty to obtain their Free, Prior and Informed Consent (FPIC) in the case of measures likely to directly affect them has intensified at international level, and particularly in the region of Latin America.[2] Likewise, in recent years there has been considerable interest among scholars and practitioners in how prior consultation and FPIC are currently – and could more usefully be – implemented through political-administrative processes at national level.[3]

Almost all of the States in this subcontinent have ratified ILO Convention No.169 Concerning Indigenous and Tribal Peoples in Independent Countries of 1989 (ILO 169); the latest was Nicaragua in 2010. During the so-called wave of "multicultural constitutionalism" (Lee Van Cott, 2000, p. 17) of the late 1980s and early 1990s, a number of Latin American States finally started to recognise (or expand the protection already provided to) Indigenous rights in their Constitutions. Only a handful of them, however, have included the right to consultation and FPIC in their *Magna Cartas*, many years after ratifying ILO 169, i.e., Ecuador in 2008 and Bolivia in 2009.

ILO 169 is the initial source of international law for Indigenous Peoples' rights to consultation (see mainly its art.6, but also articles 15.2, 17.2, 22.3, 27.3, and 28.1) and FPIC (although in its embryonic form of FIC – free and informed consent – in its article 16.2 with regard to forced relocations). This means that these rights and duties continue to be binding for the ratifying countries, since this treaty entered into force within their respective legal orders. However, the States' praxis has shown that only a few countries have made progress via-à-vis the application of these rights following the ratification of ILO 169. In some cases, these fundamental rights have been (at least partially) safeguarded thanks to norm contestation by Indigenous Peoples and the active role of the national courts, such as in the emblematic case of Colombia and its Constitutional Court (see further in the chapter by Herrera).[4]

In 2007, the adoption of the United Nations Declaration on the Rights of Indigenous Peoples (UNDRIP) gave further impetus to the rethinking of the requirements of consultation and FPIC (see its articles 10, 11, 19, 28, 29.2, 30.2, 32.2, 36.2, and 38). Indeed, UNDRIP has expanded on the requirement

DOI: 10.4324/9781351042109-1

for FPIC, and, although – in a legal *stricto sensu* – it has no direct, binding effects upon States, it has an important role in interpreting other international human rights treaties ratified by them (Doyle & Whitmore, 2014). Again, almost all Latin American countries voted in favour of this instrument (notwithstanding the notorious abstention of Colombia, at that time under the mandate of President Álvaro Uribe Vélez).

With regard to international case law, the Inter-American Court of Human Rights (henceforth, IACtHR or the Court) consecrated the right to consultation of Indigenous Peoples in the well-known cases of *Saramaka People v Suriname* (2007) and *Kichwa Indigenous People of Sarayaku v Ecuador* (2012). *Inter alia*, in these decisions, the IACtHR has established that three safeguards apply if large-scale projects affect Indigenous Peoples, namely: (1) their effective participation in the project plans; (2) a reasonable benefit from such plans; and (3) no issuing of concessions or licenses until an independent body carries out an environmental and social impact assessment (IACtHR, 2007, para. 129). Moreover, the Court ruled that States are responsible for the correct application of the right to and the processes of consultation with Indigenous Peoples, and not the companies that are working in the concerned area (IACtHR, 2012, para. 187); and States bear the duty, and not the affected Indigenous Peoples, to demonstrate in an effective way that the consultation process was fairly guaranteed (IACtHR, 2012, para. 179).

In addition, the IACtHR defined that due processes of consultation require the acceptance and dissemination of appropriate information, including environmental and health risks; continuous communication between the parties; good faith; culturally appropriate procedures aimed at reaching an agreement; fair timing, which requires the realisation of the consultation process at the early stages of a development or investment plan, and not when there is no time to obtain consent from the Indigenous Peoples involved; and the use of traditional methods of decision-making (IACtHR, 2007, para. 133; IACtHR, 2012, para. 167). Last, but not least, the Court openly affirmed that the right to consultation of Indigenous Peoples – besides being a treaty-based right – is a "general principle of international law" (IACtHR, 2012, para.164). The Court has reiterated these precedents and how crucial consultation and FPIC are for Indigenous Peoples in the following two cases: *Garifuna Community of Punta Piedra and its members v Honduras* (IACtHR, 2015a, para.182); and, *Kaliña y Lokono Peoples v Suriname* (IACtHR, 2015b, para. 206 and ff.).

The resonance of this jurisprudence has also reached other parts of the world. In 2010, the African Commission on Human and Peoples' Rights, in its landmark decision *Centre for Minority Rights Development (Kenya) and MRG on behalf of Endorois Welfare Council v Kenya*, besides recognising the right of the Endorois to ownership of their ancestral land, also reiterated the State duty to both consult and obtain the FPIC of the affected Indigenous Peoples (African Commission on Human and Peoples Rights, 2010, para. 291). In this same vein, the UN Human Rights Committee affirmed that "[...] not mere consultation [of Indigenous Peoples is required] but the free, prior and informed consent of the members of the community [...]" in the case *Ángela Poma Poma v Peru* of 2009 (HRC, 2009, para. 7.6).

Other international documents proclaim the right to consultation and FPIC of Indigenous Peoples.[5] Lately, the American Declaration on the Rights of Indigenous Peoples included the rights to consultation and FPIC of Indigenous Peoples in a number of its provisions (see in particular its articles 13.2, 18.3 and 4, 20.4, 21, 23.2, 28.3, and 29.4). Moreover, the Expert Mechanism on the Rights of Indigenous Peoples (EMRIP), after its thematic reports on participation in decision-making (EMRIP, 2010; 2011), has decided to dedicate its penultimate thematic report to FPIC (EMRIP, 2018).

The intensifying international and regional concern and debate surrounding these crucial rights, to which Indigenous leaders have significantly contributed (see the chapter by Del Castillo in this volume, and Charters and Stavenhagen, 2009, for a general overview), have thus influenced a number of States, few of which have felt compelled to adopt ad hoc legislation, most probably in order to limit the discussions at domestic level.

After the notorious conflict of Bagua, in which over 30 people died in clashes between Indigenous protestors and members of the police, Peru has led this new wave of laws on consultation by enacting Law No. 29785 in 2011 (which was later regulated by Supreme Decree No. 001–2012-MC in 2012; see the chapters by Doyle and Flemmer in this volume). Chile followed in 2013 with an ad hoc regulation (although it had started to carry out a number of consultations already in early 2009; see the chapter by Tomaselli), and so did Paraguay in late 2018 (see the chapter by Villalba), while Panama adopted a consultation law in 2016 (see the chapter by Wright and Tomaselli). In recent years, Costa Rica, Honduras, and Mexico have also discussed whether and how to regulate, mainly, consultation (but not FPIC) (see the chapters by Vega, Barreña, and Monterrubio). Other countries rely on national jurisprudence that has set standards to be applied internally (e.g., Colombia, since 1997) or legislation that refers to the right to consultation in specific branches (as in the case of Bolivia's Hydrocarbons Law (Law No. 3,058 of 2005) and its regulation (Decree No. 29,033 of 2007)).

There have also been important developments regarding consultations in practice in the region. For instance, several countries have created institutions or departments in charge of administrating prior consultation processes, including the Direction of Prior Consultation within the Ministry of the Interior in Colombia, the Direction of Prior Consultation within the Ministry of Culture in Peru, and the Direction of Participation and Consultation within the National Commission for the Development of Indigenous Peoples in Mexico. Processes given the nomenclature "prior" or "indigenous" consultation have been carried out by many governments throughout Latin America, most notably – but not exclusively – Colombia (which has carried out over 5,000 processes; see further the chapters by Herrera and Calle in this volume), Bolivia (see the chapter by Schilling-Vacaflor in this volume), and Peru (see further the chapter by Flemmer in this volume). However, as a general rule these processes fall far short of international standards, and are fraught with a series of technical and political difficulties. In many cases – despite international and national legal obligations – consultations are systematically avoided or are merely an act of window dressing in order to give the go ahead to lucrative projects. This situation was thus worthy of further conceptual, empirical, and comparative analysis.

Against this background, this volume takes inspiration from the concept of the "implementation gap" coined by the first Special Rapporteur on Indigenous Rights, Rodolfo Stavenhagen (Economic and Social Council, 2006, para. 5), to explore the current status of consultation and FPIC throughout Latin America, both in terms of its legal recognition and its practical administration. The "implementation gap" refers to the different obstacles and problems that hinder the application of Indigenous rights and thus create a lacuna between the legislation actually in force and the reality at grassroots level (Economic and Social Council, 2006, para. 5). These impediments include inconsistencies of the laws, delays in the adoption of secondary or regulatory laws, and the vacuum between legislation and real administrative, legal, and political practices (Economic and Social Council, 2006, para. 82 and ff.).

More specifically, the volume looks inside this implementation gap, asking several fundamental questions, including the following: what does the application of international standards on consultation and FPIC mean for different actors at international, national, and local levels? What success stories can be told and what factors lie behind these successes? Conversely, what happens when these standards remain unfulfilled or are only partially applied? What lies behind these failures? And finally, even when consultation processes fall short of international standards, may they have other, positive impacts? As a result of their absence or limitations, can they contribute to the empowerment of Indigenous Peoples?

Hence, this volume offers a series of country-level case studies and draws comparative lessons regarding the dimensions and nature of the above-mentioned implementation gap regarding the right to consultation of Indigenous Peoples and their FPIC, together with its multiple causes and consequences for the protection of these peoples' rights, lands, identities, and ways of life.

Having said this, this book deliberatively uses the combination of the terms "prior" and "consultation". We are well aware that international treaties and documents refer to the right of consultation and to FPIC separately. However, we follow the assumption according to which the right to consultation of Indigenous Peoples and the duty to obtain their FPIC can hardly be exercised if disjoint. The former implies seeking an agreement or the consent of the concerned Indigenous Peoples in accordance with art.6.2 of ILO 169. The latter may be effectively expressed via a fair consultation process. In particular, FPIC has been denoted both as a "principle" and a "right". The second Special Rapporteur on the Rights of Indigenous Peoples, James Anaya, defined FPIC as a "principle" (Anaya, 2013, para. 26 and ff.), while other authors argue that it is a "right" and deal with it as such (Gilbert, Tugendhat, Couillard, & Doyle 2009; Carmen, 2010; Yrigoyen Fajardo, 2011). The formulation of FPIC in UNDRIP does not indeed refer to it as a right. Nevertheless, it is undoubtedly a State duty. This implies a right-holder, thus Indigenous Peoples. And, notwithstanding the *sui generis* legal status of the UNDRIP, FPIC is considered by both the HRC and the CESCR, as well as academics, to be part of the right to self-determination of Indigenous Peoples (Doyle, 2015; Tomaselli, 2016).

Furthermore, this volume brings together a group of experts in the dynamics of consultation and FPIC that include political scientists, lawyers, anthropologists, and sociologists with grassroots experiences. All of the authors have conducted desk and field research since the mid- or late 2000s, and they hold first-hand information regarding the implementation of consultation in the country under their inquiry. This implies the observation and reporting of different types of terminology both for the laws and for those processes that are identified as the exercise of the right to consultation and FPIC.[6] These different conceptualisations are reflected in the chapters.

The book is thus organised as follows:

Part 1 "Defining prior consultation" offers an overview of the scope and nature of the right to consultation of Indigenous Peoples and their FPIC at the international, regional, and national levels. In Chapter 1, Del Castillo explores the incidence of Indigenous organisations in international fora, emphasising their role in discussions on prior consultation and FPIC within UNDRIP. In Chapter 2, Cantú Rivera reviews existing international standards at the international and regional levels and proposes an approach from the Business and Human Rights Perspective. In Chapter 3, Herrera identifies the role of the Colombian Constitutional Court as a national and regional trendsetter regarding prior consultation and, particularly, FPIC. This first section ends with Doyle (Chapter 4), who highlights the experiences of Indigenous resistance, participation, and autonomy in the pioneering national consultation law and the role of the courts in Peru, despite persistent difficulties regarding implementation.

Part 2 "Administering prior consultation" includes a series of analyses of the difficulties faced when carrying out consultation processes in practice. In Chapter 5, Schilling-Vacaflor identifies Environmental Impact Assessments (EIA) as a key variable in our understanding of prior consultation, with evidence from Bolivia. For her part, Calle discusses asymmetries of power and consultation as a space of dispute in the staging of prior consultation in Colombia's Orinoquía (Chapter 6). Next, in Chapter 7, Flemmer analyses the deficiencies of consultations in the hydrocarbon and mining sectors in Peru, while she notes their importance in terms of opening up the public debate. In Chapter 8, Tomaselli reviews recent consultations carried out on institutional and legal reforms in Chile, highlighting a series of obstacles to their success. The limits of prior consultation in practice in the Mexican case are then discussed by Monterrubio in Chapter 9.

Part 3 "Institutionalising prior consultation" considers the current or latest efforts by States to create laws, regulations, or mechanisms to institutionalise prior consultation. In Chapter 10, Vega offers a personal account of the process by which the Costa Rican mechanism was recently created, high-lighting the importance of the standard of "consultation on consultation". Next, in Chapter 11, Barreña reviews the ongoing process to create a law of consultation in Honduras, including internal and external pressures, key moments in the process, and the difficulties that it has entailed. Early efforts to institutionalise prior consultation in the case of Paraguay are reviewed by Villalba in Chapter 12.

Part 4 "Avoiding prior consultation" collects a series of studies that analyse how and why States continue to avoid prior consultation in practice. In Chapter 13, San Lucas deals with the case of the Yasuní-ITT Initiative in Ecuador and its implications in terms of public opinion and participation. For her part, in Chapter 14, Rosti discusses the implications of the lack of consultation in the context of conflicts over Indigenous lands in Argentina. Then, in Chapter 15, Mello Neiva analyses the absence of prior consultation within the broader context of threats to Indigenous rights in Brazil.

Finally, Part 5 "Re-thinking consultation" offers two experiences that invite us to re-think the relationship between Indigenous consultation, FPIC, and the law. In Chapter 16, Xiloj analyses the importance of community con-sultations in Guatemala. Finally, Canada is included as a contrasting case from beyond Latin America (Chapter 17), in which Papillon and Rodon highlight the State's sceptical attitude vis-à-vis international Indigenous rights law but which nevertheless has a rather rich domestic practice, mainly thanks to mobilisation and norm contestation by Indigenous Peoples themselves. In the concluding analysis (Chapter 18) Wright and Tomaselli bring together a series of lessons learned with relation to the nature of and reasons for the implementation gap over consultation and FPIC in Latin America.

In conclusion, we hope that the critical and insightful analyses of the case studies that are offered in this volume will help us to unpack the "implementa-tion gap" regarding the rights to consultation and FPIC of Indigenous Peoples. Specifically, the volume tries to understand why this gap occurs, and it explores the complexities and the implications that follow when the rights-holders (Indi-genous Peoples) are denied the exercise of their rights, or the duty-bearers (the States) do not fulfil their obligations. Indeed, as flagged by the UN Special Rapporteur on Business and Human Rights, Mr John Ruggie, in 2011, the missed or poor application and implementation of these crucial rights also has a financial burden. It has been estimated that – due to missed opportunities for dialogue and negotiation with the concerned Indigenous Peoples – a firm may face losses of up to $30 million per week in order to reimburse the disruptions of its works or face lawsuits (Connor, 2011; see also First Peoples Worldwide, 2013). In other words, not carrying out fair consultation processes and obtaining the FPIC of Indigenous Peoples vis-à-vis a measure likely to directly affect them has an economic cost. Furthermore – and we would argue, more importantly – if States do not implement Indigenous rights, they also risk neglecting their inter-national obligations, violating the rights of Peoples who are still in a vulnerable, a non-dominant position, fuelling discontent and social tensions.

Notes

1 This volume uses the term Indigenous People or Peoples to refer to those peoples that self-identify as such pursuant to art.1.2 of the ILO Convention No.169 concerning Indigenous and Tribal Peoples in Independent Countries of 1989 and art.33.2 of the United Nations Declaration on the Rights of Indigenous Peoples (UNDRIP) of 2007.

2 The literature on these standards has expanded considerably lately. Apart from Doyle (2015), other recent and thought-provoking contributions on the international standards or their applications in Latin America are those by MacInnes, Colchester, and Whitmore (2017) on corporate responsibility and Indigenous Peoples' consent in the extractive industry; by Tomaselli (2017) on the need for a holistic approach to Indigenous participation, consultation, FPIC, and self-government systems; by Merino (2018) on the limits of prior consultation and FPIC to truly influence the policies on which Indigenous Peoples were consulted; by Leydet (2019) on the implications of FPIC as a right to veto and the creation of a new asymmetry in the relation between Indigenous Peoples and the States (on this, see also the chapter by Herrera in this volume); by Porsani and Lalander (2018) on the failure of consultation over land acquisition as an instrument for communities' deliberation in the case of Mozambique; or by Rollo (2018) on the legitimacy of Indigenous FPIC in Canada; by Zaremberg and Torres Wong (2018) on the implications of prior consultation as a means to avoid extraction on Indigenous lands, economic redistribution, and as an instrument to decrease State repression in Bolivia, Peru, and Mexico; and by Wright, Aguirre Sotelo, and Rodríguez Cruz (2018) on the right to consultation of Indigenous Peoples on electoral affairs in Mexico.

3 In general terms, debates on prior consultation and FPIC have focused on legal standards, particularly at the international level, but also at local levels. Nevertheless, in recent years there has been significant scholarly research on the scope of prior consultations from an empirical perspective. Studies that adopt a more political approach have considered processes carried out in Bolivia (Eichler, 2018; Schilling-Vacaflor, 2017), Brazil (Schumann, 2018), Colombia (Amparo Rodríguez, 2014), Mexico (Dunlap, 2017) and Peru (Flemmer, 2018), among others. For their part, Falleti and Riofrancos (2018) have established an interesting debate on the importance of approaching prior consultation from the perspective of institutional theory, and Zaremberg and Torres Wong (2018) have carried out a macro-level study on the results of prior consultations over extractive industries, in light of participation theories. Likewise, international and regional organisations (such as IIDH, 2016), think tanks (Hartling, 2017) and NGOs (DPLF and Oxfam, 2015) have published reports on prior consultation in different Latin American countries, from a policy perspective.

4 The main instrument for norm contestation by Indigenous Peoples has been the use of the writ of *Amparo*, which may be called also *Recurso de Protección*. This refers to those proceedings that are common to a number of Latin American countries, as well as the USA and Spain, and are held before domestic Courts for alleged violations of constitutional fundamental rights (Brewer-Carías, 2009).

5 These are, *inter alia*: the Leticia Declaration of the International Meeting of Indigenous Peoples and Other Forest-dependent Peoples adopted in 1996 (para. 2, clause No. 3, and para. 6); "informed consent" by the General Recommendation No. 23 of the Committee on the Elimination of Racial Discrimination (CERD) in 1997 (UN Doc. A/52/18, annex V, 1997, para. 4d); the "Akwé: Kon guidelines" of 2004 on the interpretation of benefit-sharing (art. 8j) of the Convention on Biological Diversity of 1992; and, the "Nagoya Protocol on Access to Genetic Resources and the Fair and Equitable Sharing of Benefits Arising from their Utilization to the Convention on Biological Diversity" of 2010 (arts. 6 and 7). See further Doyle and Cariño (2013, pp. 7–10).

6 On the issue of terminology, please note that any translations of the documents from Spanish into English are the author's own and that all names for national institutions or organisations have been translated directly into English, although acronyms have been maintained in Spanish.

References

African Commission on Human and Peoples Rights. (2010). Centre for Minority Rights Development ("CEMIRIDE") (Kenya) and Minority Rights Group International on behalf of *Endorois Welfare Council v Kenya* (Endorois case), Communication No. 276/2003, 4 February 2010.

Amparo Rodríguez, G. (2014). *De la consulta previa al consentimiento libre, previo e informado a pueblos indígenas en Colombia.* Bogotá: Cooperación Alemana, Giz, Universidad del Rosario.

Anaya, J. (2013). *Report of the Special Rapporteur on the Rights of Indigenous Peoples: Extractive Industries and Indigenous Peoples.* UN Doc. A/HRC/24/41, 1 July 2013.

Brewer-Carías, A. R. (2009). *Constitutional Protection of Human Rights in Latin America: A Comparative Study of Amparo Proceedings.* Cambridge: Cambridge University Press.

Carmen, A. (2010). The Right to Free, Prior and Informed Consent: A Framework for Harmonious Relations and New Processes of Redress. In J. Hartley, P. Joffe and J. Preston (Eds.), *Realizing the UN Declaration on the Rights of Indigenous Peoples: Triumph, Hope and Action* (pp. 120–134). Saskatoon: Purich Publishing.

Charters, C. & Stavenhagen, R. (Eds.) (2009). *Making the Declaration Work: The United Nations Declaration on the Rights of Indigenous Peoples.* Copenhagen: International Work Group for Indigenous Affairs (IWGIA).

Connor, M. (2011). Business and Human Rights: Interview with John Ruggie, *Business Ethics,* 30 October. Retrieved from https://business-ethics.com/2011/10/30/8127-un-p rinciples-on-business-and-human-rights-interview-with-john-ruggie.

Doyle, C. M. & Cariño, J. (2013). *Making Free, Prior & Informed Consent a Reality. Indigenous Peoples and the Extractive Sector.* Indigenous Peoples Links (PIPLinks), Middlesex University School of Law, The Ecumenical Council for Corporate Responsibility, London. Retrieved from http://www.piplinks.org/system/files/Con sortium+FPIC+report+-+May+2103+-+web+version.pdf.

Doyle, C. M. (2015). *Indigenous Peoples, Title to Territory, Rights, and Resources. The Transformative Role of Free, Prior and Informed Consent.* London: Routledge.

Doyle, C. M. & Whitmore, A. (2014). *Indigenous Peoples and the Extractive Sector: Towards a Rights Respecting Engagement.* Baguio City: Tebtebba Foundation, Indigenous Peoples Links (PIPLinks) and Middlesex University. Retrieved from http://www.tebtebba.org/index.php/content/322-indigenous-peoples-a-the-extractive-sector-towards-a-rights-respecting-engagement.

DPLF and Oxfam. (2015). *Derecho a la consulta y al consentimiento previo, libre e informado en América Latina.* Washington, DC: DPLF. Retrieved from http://dplf. org/sites/default/files/informe_consulta_previa_2015_web-2.pdf.

Dunlap, A. (2018). "A Bureaucratic Trap": Free, Prior and Informed Consent (FPIC) and Wind Energy Development in Juchitán, Mexico. *Capitalism Nature Socialism,* 29(4), 88–108.

Economic and Social Council. (2006). *Human Rights and Indigenous Issues Report of the Special Rapporteur on the Situation of Human Rights and Fundamental Freedoms of Indigenous People, Mr. Rodolfo Stavenhagen.* UN Doc. E/CN.4/2006/78, 16 February 2006.

Eichler, J. (2018). Indigenous Intermediaries in Prior Consultation Processes: Bridge Builders or Silenced Voices? *The Journal of Latin American and Caribbean Anthropology*, 23, 560–578.

EMRIP. (2010). *Progress Report on the Study on Indigenous Peoples and the Right to Participate in Decision-Making.* UN Doc. A/HRC/15/35, 23 August 2010.

EMRIP. (2011). *Final Study on Indigenous Peoples and the Right to Participate in Decision-Making.* UN Doc. A/HRC/18/42, 17 August 2011.

EMRIP. (2018). *Free, Prior and Informed Consent: A Human Rights-Based Approach.* UN Doc. A/HRC/39/62, 10 August 2018.

Falleti, T. & Riofrancos, T. (2018). Endogenous Participation: Strengthening Prior Consultation in Extractive Economies. *World Politics*, 70(1), 86–121.

First Peoples Worldwide. (2013). *Indigenous Rights Risk Report for the Extractive Industry (U.S.). Preliminary Findings.* Retrieved from www.firstpeoples.org/images/uploads/R1KReport2.pdf.

Flemmer, R. (2018). Stuck in the Middle: Indigenous Interpreters and the Politics of Vernacularizing Prior Consultation in Peru. *The Journal of Latin American and Caribbean Anthropology*, 23, 521–540.

Gilbert, J., Tugendhat, H., Couillard, V., & Doyle, C. M. (2009). Business, Human Rights and Indigenous Peoples: The Right to Free, Prior and Informed Consent. Forest Peoples Programme, Working Paper Series, April 2009.

Hartling, J. (2017). *Guía de buenas prácticas para la Consulta Previa en las Américas.* La Paz: Fundación Konrad Adenauer. Retrieved from https://www.kas.de/c/document_library/get_file?uuid=d4104b19-01cd-7685-fb29-21d7f6dceda4&groupId=252038.

HRC. (2009). *Ángela Poma Poma v Peru*, Communication No. 1457/2006. UN Doc. CCPR/C/95/D/1457/2006, 27 March 2009.

IACtHR. (2007). *Saramaka People v Suriname,* Judgment of November 28, 2007 (Preliminary Objections, Merits, Reparations, and Costs). Inter-Am. Ct. H.R., (Ser. C) No. 172(2007).

IACtHR. (2012). *Kichwa Indigenous People of Sarayaku v Ecuador,* Judgment of June 27, 2012 (Merits and reparations). Inter-Am. Ct. H.R., (Ser. C) No. 245(2012).

IACtHR. (2015a). *Caso Comunidad Garífuna de Punta Piedra y sus miembros v Honduras,* Judgment of October 8, 2015 (Merits, Reparations, and Costs). Inter-Am. Ct. H.R., (Ser. C) No. 305(2015).

IACtHR. (2015b). *Caso Pueblos Kaliña y Lokono v Surinam,* Judgment of November 25, 2015 (Merits, Reparations, and Costs). Inter-Am. Ct. H.R., (Ser. C) No. 309(2015).

IIDH. (2016). *El derecho a la consulta previa, libre e informada: Una mirada crítica desde los pueblos indígenas.* San José: IIDH.

Lee Van Cott, D. (2000). *The Friendly Liquidation of the Past: The Politics of Diversity in Latin America.* Pittsburgh: University of Pittsburgh Press.

Leydet, D. (2019). The Power to Consent: Indigenous Peoples, States, and Development Projects. *University of Toronto Law Journal*, 69(3), 371–403.

MacInnes, A., Colchester, M., & Whitmore, A. (2017). Free, Prior and Informed Consent: How to Rectify the Devastating Consequences of Harmful Mining for Indigenous Peoples. *Perspectives in Ecology and Conservation*, 15(3), 152–160.

Merino, R. (2018). Re-Politicizing Participation or Reframing Environmental Governance? Beyond Indigenous' Prior Consultation and Citizen Participation. *World Development*, 111, 75–83.

Porsani, J. & Lalander, R. (2018). Why Does Deliberative Community Consultation in Large-Scale Land Acquisitions Fail? A Critical Analysis of Mozambican Experiences. *Iberoamerican Journal of Development Studies*, 7(2), 164–193.

Rollo, T. (2018). Imperious Temptations: Democratic Legitimacy and Indigenous Consent in Canada. *Canadian Journal of Political Science*, 1–19. doi:10.1017/S0008423918000343.

Schilling-Vacaflor, A. (2017) Who Controls the Territory and the Resources? Free, Prior and Informed Consent (FPIC) as a Contested Human Rights Practice in Bolivia. *Third World Quarterly*, 38(5), 1058–1074.

Schumann, C. (2018). Competing Meanings: Negotiating Prior Consultation in Brazil. *The Journal of Latin American and Caribbean Anthropology*, 23, 541–559.

Tomaselli, A. (2016). *Indigenous Peoples and their Right to Political Participation. International Law Standards and their Application in Latin America*. Baden-Baden: Nomos.

Tomaselli, A. (2017). The Right to Political Participation of Indigenous Peoples: A Holistic Approach. *International Journal on Minority and Group Rights*, 24(4), 390–427.

Wright, C., Aguirre Sotelo, V. N., & Rodríguez Cruz, L. A. (2018). *El derecho a la consulta en materia electoral de los pueblos y las comunidades indígenas*. Toluca: Instituto Electoral del Estado de México, Centro de Formación y Documentación Electoral.

Yrigoyen Fajardo, R. Z. (2011). El Derecho a la Libre Determinación del Desarrollo. Participación, Consulta y Consentimiento. In M.A. Wilhelmi (Ed.), *Los Derechos de los Pueblos Indígenas a los Recursos Naturales y al Territorio. Conflicto y Desafíos en América Latina* (pp. 103–146). Barcelona: Icaria.

Zaremberg, G. & Torres Wong, M. (2018). Participation on the Edge: Prior Consultation and Extractivism in Latin America. *Journal of Politics in Latin America*, 10(3), 29–58.

Part I
Defining prior consultation

1 Indigenous Peoples' contributions to multilateral negotiations on their rights to participation, consultation, and free, prior and informed consent[1]

Andrés Del Castillo

Introduction

The degree of involvement of inhabitants or groups in matters that directly affect them or that are in their interests varies considerably. In any case, the right to participation is recognised as an individual right and as a collective right in the International Bill of Human Rights and other regional legal instruments.[2]

When the rights-holders are Indigenous Peoples, their inclusion translates into a greater degree of autonomy and competencies from a collective point of view, which is known today as "the right of Indigenous Peoples to self-determination, conceived as the right to decide their political condition and determine what their future will be" (International Law Association, 2012, p. 2). This right was defined in articles 3 and 4 of the 2007 UN Declaration on the Rights of Indigenous Peoples (UNDRIP).

In order for the collective rights of self-government and autonomy to materialise, other rights of Indigenous Peoples are invoked, such as the right to participate in decision-making on matters concerning them, the right to be consulted on any project that may affect them, and the right to Free, Prior and Informed Consent (FPIC) regarding projects that have a significant impact on their rights and/or ways of life.

The manner in which these rights relate to each other may vary. For some Indigenous representatives, consent is the rule, and consultation and participation are the exceptions to the rule. In accordance with a 2018 study, the Expert Mechanism on the Rights of Indigenous Peoples (EMRIP) stated that:

> Free, prior and informed consent is a manifestation of the right of Indigenous peoples to determine for themselves their cultural, social and economic priorities. These are three interrelated and cumulative rights: the right to be consulted, the right to participate and the right to their land, territory and resources. According to the declaration, there cannot be free, prior and informed consent if one of those components is missing.
>
> (EMRIP, 2018, p. 5)

DOI: 10.4324/9781351042109-3

Indeed, classifying the three rights into participation, consultation, and consent, in determining the degree to which those rights should be obligatory, has been a common practice in multilateral negotiations on international legal instruments that contain Indigenous rights. For example, during negotiations over ILO Convention No.169 Concerning Indigenous and Tribal Peoples in Independent Countries of 1989 (ILO 169), the parties preferred the terms "participation" and "consultation" instead of "self-determination" and "consent" as required by some Indigenous Peoples (Dahl, 2009).

Hence, by bearing in mind that the origin of those rights precedes their compilation in multilateral regulatory texts, this article offers an overview of how participation, consultation, and FPIC have been repeatedly affirmed by the Indigenous delegates who participated in the preparatory work for the adoption of the international legal instruments on their rights.

Having said that, this chapter starts from the assumption that FPIC is an internationally recognised right, and is part of customary international law (Sambo, 2013, p. 149). Indeed, in addition to the practice of international organisations on Indigenous Peoples' issues, there is abundant State practice in adopting resolutions related to Indigenous Peoples which can contribute to the crystallisation of the status of this right as customary international law. A 2018 report by the International Law Commission of the UN, adopted by the General Assembly, warns that while only the practice of States can create or express customary international law, in certain cases, the practice of international organisations also contributes to the formation, expression, and identification of rules of customary international law (General Assembly, 2018, p. 130). According to the same report, although the behaviour of entities that are not States or international organisations does not contribute to the formation or expression of regulations of customary international law, "[it may] have an indirect role in the identification of customary international law, by stimulating or recording the practice and acceptance as law (*opinio juris*) of States and international organizations" (General Assembly, 2018, p. 131).

Therefore, this chapter begins with a short outline of the first affirmations of Indigenous rights to participation, consultation, and consent. Next, it highlights the discussions that shaped the articles that refer to such rights in the ILO Convention No.107 concerning the Protection and Integration of Indigenous and Other Tribal and Semi-Tribal Populations in Independent Countries of 1957 (ILO 107) and ILO 169. Finally, it focuses on the role of the international Indigenous movement in the development of FPIC in the preparation and adoption process of the UNDRIP.

Early affirmations of Indigenous rights to participation, consultation, and consent[3]

According to the theory of the origin of public international law – which claims to exist only after the emergence of Nation States – States are the exclusive subjects of international law. However, throughout different periods of colonial rule, there were also rules that regulated relations between States and other political entities, including Indigenous Peoples. For example, the agreements reached between colonial powers and local political entities were some of the means by which rights, territorial sovereignty, and property were acquired or recognised (Hébié, 2015).

During decolonisation processes, new Nation States were created with borders that sometimes united but often divided several of the Peoples that lived there prior to the arrival of colonial powers. The principle of "inherent sovereignty" in US jurisprudence establishes that the powers which are legally conferred on a People are those which predate the colonisation of what is now known as the Americas and which have never been revoked (Green & Work, 1976). These are known as ancestral rights, prior to the formation of modern States, under the principle of *prior in tempore, potior in iure*. Other principles of international law such as the Blue Water Thesis and the *Uti possidetis* were used against Indigenous Peoples to assert States' inherent sovereignty and territorial dominion (Gilbert, 2006).

Nevertheless, examples of FPIC were present, for example in several sections of the *Popol Vuh*, one of the sacred texts of the Quiche Maya (Goetz, Morley Griswold, Weil Kaufman, & Jackson, 1954) where two of the deities use consultation.[4] Colonial practices, mainly in the Americas, were consistently and uniformly based on treaties and agreements between colonial powers and native peoples. Finally, the Spanish Requirement (*Requerimiento*) of 1513, which was a declaration by the Spanish monarchy of Castile, included a divinely ordained right to take possession of the territories of the Indigenous Peoples, including what could also be construed as a reference to consultation.

Rights are derived from these sovereign exercises between Nation States and local authorities, including Indigenous Peoples, such as participation, representation, consultation, and consent on decisions that affect the rights holders, which are, in essence, a means to guarantee other rights. While important, treaties are not the only proof of sovereign power; however, they are evidence of the recognition of Indigenous Peoples' sovereignty. A former UN Special Rapporteur on the Rights of Indigenous Peoples reaffirmed "[the] widespread recognition of 'overseas peoples' – including Indigenous Peoples [...] – as sovereign entities by European powers [...]" (Commission on Human Rights, 1999, p. 17).

In the twentieth century, the League of Nations was created. It emerged from the 1919 Treaty of Versailles, which ended World War I. The inclusion of the doctrine of guardianship in article 23.b of the League of Nations Covenant shaped the doctrine's definition during the interwar period (Rodríguez-Piñero Royo, 2004, p. 60).

In 1923, Chief Hoyaneh Deskaheh was the first Indigenous representative to come before the League of Nations to try – albeit unsuccessfully – to persuade the international community to recognise the Iroquois Confederacy as an independent state, in accordance with Article 17 of the League of Nations Covenant. His request was based on the treaties and agreements made between his people and the colonial powers. According to the request, consent was a right derived from the sovereignty recognised by the colonial authority, which was later known as the self-determination of the peoples (Deskaheh, 1923). In 1924, a Maori delegation headed by Pita Te Turuki Tamati Moko – secretary of Tahupōtiki Wiremu Rātana, a Maori religious leader – travelled to Geneva to speak to the League of Nations about the Treaty of Waitangi of 1840 (Orange, 2004, p. 417). After the creation of the UN in 1945, a group of Indigenous delegates from Ecuador went to the UN headquarters in New York, and met with Benjamín Cohen Gallerstein, Under-Secretary-General for public information to thank the UN for the support they received after the 1949 earthquake.

Only after many decades, the late independence of many colonies, the emergence of national civil liberties and human rights movements, and the adoption of ILO 107 on Indigenous and Tribal Peoples, were other Indigenous Peoples able to go to the UN with a robust agenda that included the recognition and establishment of their rights.

In September 1977, several Indigenous delegates participated in the first Conference on Discrimination against Indigenous Peoples in the Americas (Dunbar-Ortiz, 2015). The economic commission of this conference recommended "support [for] the right of self-determination of aboriginal people in the development of their land and resources according to their own values and social structures and laws" (International Indian Treaty Council, 1977, p. 15). For its part, the legal commission of the conference brought with it the issue of self-determination, understood as "the inherent legal right of Indigenous Peoples to control and regulate their own affairs" (International Indian Treaty Council, 1977, p. 21), including the principle in the draft Declaration of Principles for the Defence of the Indigenous Nations and Peoples of the Western Hemisphere. In addition, the commission concluded that "[...] lands, land rights and natural resources [...] should not be taken, and their land rights should not be terminated or extinguished without their full and informed consent" (International Indian Treaty Council, 1977, p. 23).[5]

Participation, consultation, and consent in the ILO Conventions

In 1957, the ILO adopted Convention No.107 (ILO 107) as well as Recommendation No.104 regarding Indigenous and Tribal Peoples. For its part, ILO 107 recognised the obligations of States with respect to Indigenous Peoples and considered it essential to adopt general international standards for their protection. There is no record of the participation of Indigenous Peoples in the negotiation and adoption of the Convention. ILO 107 is still in force for those countries that have ratified it, except for those that subsequently ratified ILO 169, because that ratification involved *ipso jure* the immediate denunciation of ILO 107, in accordance with its Article 36.

Regarding the right to participation, ILO 107 had a protectionist and integrationist perspective, and both the preamble and Article 5.C stipulate the State duty of collaboration, which represents a nuanced tone of the right to participation, both individually and collectively. Nothing is mentioned regarding consultation, but in Articles 4[6] and 12, the right to free consent is specifically recognised in two cases: when applying the provisions of the Convention which relate to integration, and when transferring Indigenous Peoples from their habitual territories, albeit with the exceptions provided for by domestic legislation regarding national security, the economic development of the country, or the health of these populations.

The revision process of ILO 107 was carried out in response to repeated requests from several organisations, particularly the UN. It should be noted that in 1982, a Working Group on Indigenous Peoples (WGIP) was established at the UN level and that there is a correlation between the revision of the Convention and the international Indigenous movement (ILO, 1987). For the commission of ILO 107, there was consensus on the fact that the integrationist doctrine was no longer acceptable, and that Indigenous and Tribal Peoples should be genuinely associated with any decision that affects them.

> The essence of the revision [of Convention 107] was procedural and the principle was the participation of Indigenous and tribal peoples as social partners with the right to organise – an approach which was entirely consistent with the work of the ILO.
>
> (International Labour Conference, 1989a, pp. 25–26)

For its part, ILO 169 refers to States' obligations, considering participation in article 2.5.C, 6.B, 7.1 and 2, 15.1, 22.1 and 2, 23.1, 27.2, and 29. The right to consultation is framed as a duty of the State in article 6.1 and 2, 15.2, 17.2, 22.3, 27.3, and 28.1; and the right to informed consent in its articles 6.2, and 16.2. In this case, there is evidence of indigenous delegates' participation in the negotiation and adoption of the Convention. Even government advisers highlighted the valuable contribution they made (International Labour Conference, 1989b). As mentioned previously, this is in contrast to ILO 107. Indeed, once the report of the Committee on ILO 107 was submitted, adopting the text of the Convention, Indigenous organisations took the floor and said in their final interventions: "We did not come here so that the ILO could tell the world it had consulted Indigenous Peoples during the revision of ILO 107. Because in point of fact we have not been consulted" (International Labour Conference, 1989c, p. 31.1).

The principles of participation, consultation, and consent were repeatedly presented by Indigenous representatives during the revision and adoption process of ILO 107 (Huaco Palomino, 2015). For example,

> Several representative non-governmental organisations of Indigenous and Tribal Peoples argued that it was contradictory to speak of respect for the cultures, institutions and life systems of these peoples while reserving the right of governments entitling them to undertake actions to which the former are opposed.
>
> (International Labour Conference, 1989d)

The UNDRIP drafting process

UNDRIP sets forth participation as a collective right in articles 5, 13, 18, 27 and 41. For its part, consultation is contained in seven articles (15.2, 17.2, 19 [together with FPIC], 30.2, 32.2, 36.2 and 38) as a duty of the State. Finally, FPIC is enshrined in articles 10, 11.2, 19, 28.1, 29.2 and 32.2.

As mentioned above, the WGIP was established in 1982 (ECOSOC, 1982). At the first meeting, it proposed to start working on a declaration – and later even on a convention – and suggested the model of open negotiations for Indigenous delegates who did not have consultative status before the Economic and Social Council (ECOSOC). In that same year, the Indian Law Resource Center (1982) submitted a series of principles to guide the deliberations of the WGIP, including the imperative that

> [I]ndigenous Peoples shall not be deprived of their rights or claims to land, property, or natural resources, without their free and informed consent. No state shall claim or retain, by right of discovery or otherwise, the territories of Indigenous Peoples, except such land as may have been lawfully acquired by valid treaty or other freely-made cession. Under no circumstances shall [...] Indigenous Peoples or groups [be] subject to discrimination with respect to their rights or claims to land, property, or natural resources.
>
> (p. 1)

Since 1982, it has become clear that the subject of FPIC has not only been limited to land. Indeed, "[A]ny family planning programme or programme for the adoption or foster care of Indigenous children must be approved only after prior consultation and close collaboration with, and with the active participation and control of, the Indigenous communities and groups concerned" (WGIP, 1982, p. 2).

In 1983, with regard to land and natural resources, Indigenous delegates highlighted the need for the participation of Indigenous Peoples in all decision-making processes regarding development projects in their territories or the land they lived on or regarding projects that could have an impact on their lives, thereby underlining the need for consultation and consent (Commission on Human Rights, 1983; Anti-Slavery Society for the Protection of Human Rights, 1983). In 1985, the World Council of Indigenous Peoples (1984) submitted to the WGIP a document which outlined 17 fundamental principles of Indigenous Peoples. These included principles 9 and 12 on free and informed consent in cases related to traditional land and its resources. They specified the scope of free and informed consent as follows:

> Principle 9. Indigenous People shall have exclusive rights to their traditional lands and its resources, where the lands and resources of the Indigenous Peoples have been taken away without their free and informed consent such

lands and resources shall be returned (...) Principle 12. No action or course of conduct may be undertaken which, directly or indirectly, may result in the destruction of land, air, water, sea ice, wildlife, habitat or natural resources without the free and informed consent of the Indigenous Peoples affected.

(World Council of Indigenous People, 1984, p. 11)

That same year, another declaration of principles was submitted to the WGIP by a special assembly of Indigenous organisations, which reviewed the drafted principles by the World Council of Indigenous Peoples (1985), and included the issue of granting usufruct of the land with reference to FPIC: "Rights to share and use land, subject to the underlying and inalienable title of the Indigenous nation or people, may be granted by their free and informed consent, as evidenced in a valid treaty or agreement" (p. 2).

For its part, the study on discrimination by José Martínez Cobo, the first Special Rapporteur on the Rights of Indigenous Peoples, began in 1972 and was completed in 1986, based on 37 specialist works, making it the largest study of its kind (Commission on Human Rights, 1986; 1993). However, this study did not include information on Africa and Asia, which are home to many Indigenous Peoples. The WGIP session in 1990 was very important because identification criteria were extended, going from groups that had been subjected to European colonialism to other groups in which the historical process was not the only relevant factor, thanks to the participation of Indigenous delegates from those regions in the 1980s. For example, lifestyles connected to land use and self-identification as Indigenous became relevant (Eide, 2009).

In 1993, coinciding with the International Year of the World's Indigenous Peoples, the World Conference on Human Rights was celebrated, thanks to the participation of Indigenous delegates, including Rigoberta Menchú, who won the Nobel Peace Prize in 1992. Consequently, the Vienna Declaration and Programme of Action (1993) was adopted, which stated that "[...] [S]tates must guarantee the total and free participation of indigenous populations in all aspects of society, particularly in matters that concern them" (p. 8), and called for the early finalisation of the draft declaration.

In August of the same year, the WGIP also finalised the draft of a declaration on the rights of Indigenous Peoples, which included free and informed consent. The draft was presented to the Sub-Commission on the Promotion and Protection of Human Rights, which, after approving the text, sent it on to the Commission on Human Rights. In 1995, they established the Working Group on the Draft Declaration on the Rights of Indigenous Peoples (WGDD) as a subsidiary body (Commission for Human Rights, 1993b). The WGDD met for over a decade in Geneva with the aim of preparing a draft declaration that would address the staunch disagreements of opposing States. In their meetings, participation was not limited to member States, Indigenous organisations, and other NGOs in consultative status with ECOSOC, but rather was extended to other relevant Indigenous organisations with no ECOSOC consultative status (ECOSOC, 1995).

Although Indigenous Peoples have endorsed the draft contained in the annex to resolution 1994/45 of 26 August 1994 as representing the minimum standard, some States proposed substantial changes to the text, including removing all reference to self-determination from the Declaration and replacing it with the term self-management. As proof of the position of Indigenous delegates, in February 2003, the International Indian Treaty Council (2003) submitted a written statement before the Commission on Human Rights, mentioning that: "[o]ur right to self-determination is not up for negotiation. It must be made clear that Indigenous Peoples must be recognised as peoples with the same fundamental rights as all other peoples" (p. 4). The same year, the WGIP decided that it would initiate the preparation of a legal commentary on the principle of FPIC from Indigenous Peoples in relation to developments affecting their land and natural resources, and included it in its 2004 work programme, when various Indigenous statements were made on the subject. For example:

> The Indigenous Peoples [...] need [...] to promote, to encourage our [Indigenous] participation in economic and social development and decision-making, mainly those that concern our communities. This minimum framework would allow us to express our free, prior and informed consent on the positions and decisions that are made when development projects or inherent policies affect our lives, land or territory and natural resources.
> (Organización OTM de los Niños Mayas de Guatemala, 2004, p. 1)

Many Indigenous Peoples' organisations supported the 2004 Preliminary Working Paper on the principle of FPIC in relation to development affecting their lands and natural resources, which was inspired by varied statements from Indigenous delegates and updated in 2005 (Commission on Human Rights, 2004; 2005). Demonstrating the value of the aforementioned working document, the Saami Council and the Inuit Circumpolar Conference affirmed: "the right to free, prior and informed consent is truly a right, and any attempt to deviate from such standards constitutes a violation of international, and, as additionally shown, often also of domestic law" (Saami Council & the Inuit Circumpolar Conference, 2004, p. 2).

During the same session, an Indigenous representative of the *Dayak-iba* group in the State of Sarawak, Malaysia, clarified the meaning of FPIC:

> Free, prior and informed consent means: 1. All members of the communities, who are affected, consent to the decision. 2. Consent is determined in accordance with customary laws, rights and practices. 3. Freedom from external manipulation, interference or coercion. 4. Full disclosure of the intent and scope of the activity. 5. Decisions are made in a language and process understandable to the communities. 6. Indigenous Peoples' customary institutions and representative organisations must be involved at

all stages of the consent process. 7. Respect for the right of Indigenous Peoples to say NO.
(Dayak Peoples, Orang Asii Peoples, & Kadazan-Dusun Peoples, 2004, p. 2)

This subtlety is important because it was contained in a request from Indigenous Peoples themselves on the basis of the 1994 draft declaration. Such consensus on the concept implied that these were the minimum standards that any declaration should contain. For instance, other Indigenous organisations stated that: "[t]he principles of prior and informed consent and full collaboration with the affected indigenous peoples must be applied for the effective implementation by the States of the provisions throughout the declaration" (International Indian Treaty Council et al., 2004, p. 8).

With the risk of the deterioration of the minimum standards of the draft declaration, proposals were included to replace the verb "obtain" with "seek" with respect to FPIC (WGDD, 2004, pp. 10–13). This occurred in the same year that Indigenous Peoples held a four-day hunger strike and spiritual fast at the UN headquarters in Geneva to draw the world's attention to the continued attempts of some States, as well as the UN process itself, to weaken and undermine the draft declaration (Carmen, 2009).

Since the first sessions of the UN Permanent Forum on Indigenous Issues, the need to clarify the principle of FPIC was noted, and after a seminar in 2005, a report clarified the methodological approach of the principle, concluding that various international instruments, as well as pronouncements, provide a normative basis for FPIC. The report also highlighted that FPIC, as a principle, is based on an approach to development from a human rights perspective, and as a right, is related to the right to self-determination as well as the treaties and rights regarding land, territory, and natural resources, and that consultation and participation are crucial components of a consent process (Permanent Forum on Indigenous Issues, 2005).

In 2007, the General Assembly adopted the UNDRIP after negotiations between the Indigenous caucuses, friendly States that supported the UNDRIP, and the African states. Another bloc of States, including New Zealand, the USA, Australia, and Canada opposed the draft because of their concerns, for instance, that UNDRIP went too far in expanding FPIC as a right (General Assembly, 2007). Later on, in 2014, the General Assembly itself adopted a resolution reaffirming UNDRIP agreements on consent before adopting and applying legislative or administrative measures that affect Indigenous Peoples (General Assembly, 2014).

Even after the adoption of the UNDRIP, there has been a significant presence of Indigenous delegates at instances at different levels, which has allowed them to illustrate the dynamic nature of the right to FPIC. For example, the latest report of the EMRIP flagged that "[f]ree, prior and informed consent operates fundamentally as a safeguard for the collective rights of Indigenous Peoples" (EMRIP, 2018, p. 4).

Conclusions

This chapter has argued that Indigenous delegates made a significant contribution to multilateral negotiations on international instruments that proclaim Indigenous Peoples' rights, particularly to what is understood by participation, consultation, and FPIC. However, these contributions were often not reflected in the final texts.

It has been shown that, since the first intervention of Indigenous delegates from many regions of the world, participation, consultation, and FPIC have been placed at the centre of multilateral discussions concerning Indigenous rights, grounding them on and aligning them with the principles/rights of self-determination and equality or non-discrimination, cultural rights, and the principle of permanent sovereignty over natural resources.

The dynamic presence of Indigenous representatives at the international level has also redefined the practice of international organisations.[7] It has raised a debate on how international organisations should work with Indigenous Peoples, illustrating and developing institutional policy frameworks through which the principle of FPIC has also been solidified.

Finally, it is worth questioning whether, in addition to being an exercise of their right to self-determination, the participation of Indigenous Peoples in international organisations with the aim of seeking the recognition of their rights could contribute directly to the formation of regulations of customary international law, or whether, on the contrary, this contribution is indirect.

Notes

1 The opinions expressed in this document are the sole responsibility of the author and do not represent the official position of the Indigenous Peoples' Centre for Documentation, Research & Information (Docip).
2 See article1 and 25 to the International Covenant on Economic, Social and Cultural Rights (1966), articles 1, 16, 20, 25, 27, and 47 the International Covenant on Civil and Political Rights (1966), and article 27 the Universal Declaration of Human Rights (1948).
3 See Doyle (2015) for a detailed analysis of the genesis of FPIC.
4 "Then came the word. Tepeu and Gucumatz came together in the darkness, in the night, and Tepeu and Gucumatz talked together. They talked then, discussing (*consultando*) and deliberating; they agreed, they united their words and their thoughts" (Goetz, Morley Griswold, Weil Kaufman, & Jackson, 1954, p. 4).
5 See Akaitcho Dene and Hupacasath First Nation (2004) on the 1975 Advisory Opinion of the ICJ submitted to the General Assembly on the case of Western Sahara.
6 According to article 37 of the Convention 107, the English and French versions of the text are equally authoritative. However, in article 4, the English version replace the French verb "*consentement*" by "willing", so it is necessary to follow the rules of articles 31, 32, and 33 of the Vienna Convention on the Law of Treaties. Under article 32, the record of proceedings of article 4 was adopted unanimously, without discussion, and it was reflected by the text of the "Proposed Conclusions concerning the Protection and Integration of Indigenous and Other Tribal and Semi-Tribal Populations in Independent Countries" (International Labour Conference, 1956, p. 748).
7 On the international participation of Indigenous Peoples see Charters (2010).

References

Carmen, A. (2009). International Indian Treaty Council Report from the Battle Field: The Struggle for the Declaration. In C. Charters & R. Stavenhagen (Eds.), *Making the Declaration Work: The United Nations Declaration on the Rights of Indigenous Peoples* (pp. 86–95). Copenhagen: IWGIA.

Charters, C. (2010). A Self-Determination Approach to Justifying Indigenous Peoples' Participation. *International Journal on Minority and Group Rights*, 17(2), 215–240.

Commission on Human Rights. (1983). Study on the Problem of Discrimination Against Indigenous Populations. Report of the WGIP on its second session. UN Doc. E/CN.4/Sub.2/1983/22.

Commission on Human Rights. (1986). Study of the Problem of Discrimination Against Indigenous Populations. Final reports submitted by the Special Rapporteur, Mr José Martínez Cobo. UN Doc. E/CN.4/Sub.2/1986/7.

Commission on Human Rights. (1993a). Study of the Problem of Discrimination Against Indigenous Populations. Final reports submitted by the Special Rapporteur, Mr José Martínez Cobo. UN Doc. E/CN.4/Sub.2/1993/2.

Commission on Human Rights. (1993b). Report of the WGIP on its Eleventh Session: Draft Declaration on the Rights of Indigenous Peoples. UN Doc. E/CN.4/Sub.2/1993/29.

Commission on Human Rights. (1999). Study on Treaties, Conventions and Other Constructive Agreements Between States and Indigenous Peoples. Special Rapporteur Mr José Martínez Cobo. UN Doc. E/CN.4/Sub.2/1999/20.

Commission on Human Rights. (2004). Preliminary Working Paper on the Principle of Free, Prior and Informed Consent of Indigenous Peoples in Relation to Development Affecting their Lands and Natural Resources that would serve as a Framework for the Drafting of a Legal Commentary by the WGIP on this Concept. Submitted by Antoanella-Iulia Motoc and the Tebtebba Foundation, 22nd session of the WGIP. UN Doc. E/CN.4/Sub.2/AC.4/2004/4.

Commission on Human Rights. (2005). Expanded Working Paper on the Principle of Free, Prior and Informed Consent of Indigenous Peoples in Relation to Development Affecting their Lands and Natural Resources that would serve as a Framework for the Drafting of a Legal Commentary by the WGIP on this Concept. Submitted by Antoanella-Iulia Motoc and the Tebtebba Foundation, 23rd session of the WGIP. UN Doc. E/CN.4/Sub.2/AC.4/2005/WP.1.

Dahl, J. (2009). *IWGIA: A History.* Copenhagen: IWGIA.

Dayak Peoples, Orang Asii Peoples, & Kadazan-Dusun Peoples. (2004). Statement on Item: Indigenous People & Conflict Resolution for the 22nd session of the WGIP, 19–24 July 2004. Retrieved from https://cendoc.docip.org/collect/cendocdo/index/assoc/HASHfa3f/5336b246.dir/008_as.pdf.

Doyle, C. M. (2015). *Indigenous Peoples, Title to Territory, Rights and Resources: The Transformative Role of Free Prior and Informed Consent.* London: Routledge.

Dunbar-Ortiz, R. (2014). *An Indigenous Peoples' History of the United States.* Bostol: Peacon Press.

ECOSOC. (1982). Study of the Problem of Discrimination Against Indigenous Populations, Resolution 1982/34. UN Doc. E/RES/1982/34.

ECOSOC. (1995). Establishment of a Working Group of the Commission on Human Rights to Elaborate a Draft Declaration in Accordance with Paragraph 5 of General Assembly Resolution 49/214. UN Doc. E/RES/1995/32.

Eide, A. (2009). The Indigenous Peoples, the Working Group on Indigenous Populations and the Adoption of the UN Declaration on the Rights of Indigenous Peoples. In C. Charters & R. Stavenhagen (Eds.), *Making the Declaration Work: The United Nations Declaration on the Rights of Indigenous Peoples* (pp. 32–47). Copenhagen: IWGIA.

EMRIP. (2018). Free, Prior and Informed Consent: An Approach Based on Human Rights. UN Doc. A/HRC/39/62.

General Assembly. (2007). Report of the Human Rights Council: Explanations of Vote Australia, Canada, and Record of Voting of Draft Resolution A/61/L.67.

General Assembly. (2014). Outcome Document of the High-Level Plenary Meeting of the General Assembly known as the World Conference on Indigenous Peoples. UN Doc. A/RES/69/2.

General Assembly. (2018). Adoption of the Conclusions on the Identification of Customary International. Drafted by the International Law Commission at its Seventieth Session, A/73/10. UN Doc. A/RES/73/203.

Gilbert, J. (2006). *Indigenous Peoples' Land Rights under International Law: From Victims to Actors*. New York: Transnational Publisher.

Goetz, D., Morley Griswold, S., Weil Kaufman, L., & Jackson, E. G. (1954). *The Book of the People = Popol vuh: The National Book of the Ancient Quiché Maya*. Los Angeles: Plantin Press.

Green, J. & Work, S. (1976). Comment: Inherent Indian Sovereignty. *American Indian Law Review*, 4(2), 311–342.

Hébié, M. (2015). *Souveraineté territoriale par traité: Une étude des accords entre puissances coloniales et entités politiques locales*, Geneva: Presses Universitaires de France P.U.F. / Publications de l'Institut universitaire des Hautes Etudes Internationales.

Huaco Palomino, M. (2015). *Los trabajos preparatorios del Convenio N° 169 de la Organización Internacional del Trabajo (OIT) sobre Pueblos Indígenas y tribales en países independientes*. Lima: Editora Presencia.

Indian Law Resource Center. (1982). Principles for Guiding the Deliberations of the WGIP. Submitted to the Division of Human Rights Unit of the UN. Retrieved from http://cendoc.docip.org/collect/cendocdo/index/assoc/HASH010e/71eec333.dir/Wgip82Indianlawressourcecenter_07.pdf.

International Indian Treaty Council. (1977). Official Report of the International NGO Conference on Discrimination Against Indigenous Populations. Retrieved from https://www.ohchr.org/Documents/Issues/Business/WGSubmissions/2013/International%20Indian%20Treaty%20Council.pdf.

International Indian Treaty Council. (2003). Written statement submitted. UN Doc. E/CN.4/2003/NGO/126.

International Indian Treaty Council et al. (2004). Written statement submitted. Report Addendum of the WGDD.

International Law Association. (2012). Rights of Indigenous Peoples: Conclusions and Recommendations of the Committee on the Rights of Indigenous Peoples, Resolution No. 5/2012.

International Labour Conference. (1956). Proposed Conclusions Concerning the Protection and Integration of Indigenous and Other Tribal and Semi-Tribal Populations in Independent Countries. Committee on Indigenous Populations, Geneva.

International Labour Office. (1987). Partial Revision of the Indigenous and Tribal Populations Convention, 1957 (No. 107), Report VI (1), Geneva.

International Labour Conference. (1989a). Partial Revision of the Indigenous and Tribal Populations Convention, 1957 (No. 107). Intervention of the Representative of the Four Directions Council. Meeting of the International Labour Conference, Geneva.

International Labour Conference. (1989b). Intervention of Government Adviser from Denmark, and Reporter of the Committee on Convention No. 107, Geneva.

International Labour Conference. (1989c). Intervention of the Representative of the International Organisation of Indigenous Resource Development. In Provisional Record No. 31, Geneva.

International Labour Conference. (1989d). Partial Revision of the Convention on Indigenous and Tribal Peoples in Report IV (2A). Responses and comments from the member states on Report IV(1) of the 76th Meeting which proposed the new text of the revised Convention, Geneva.

Nishnawbe-Aski Nation, Canada. (1982). Cover letter submitted to the Division of Human Rights Unit of the UN. Retrieved from http://cendoc.docip.org/col lect/cendocdo/index/assoc/HASH0155/92b023af.dir/Wgip82Nishnawbe-AskiNa tion_03.pdf.

Orange, C. (2004). *An Illustrated History of the Treaty of Waitangi*. Wellington: Bridget Williams Books.

Organización OTM de los Niños Mayas de Guatemala. (2004). Statement for the Item: Legal Aspects of the Concept of FPIC During the 22 Session of the WGIP. Retrieved from http://cendoc.docip.org/collect/cendocdo/index/assoc/HASHfa40/ bf69136b.dir/053_es.pdf

Permanent Forum on Indigenous Issues. (2005). Report of the International Workshop on Methodologies Regarding Free, Prior and Informed Consent and Indigenous Peoples. UN Doc. E/C.19/2005/3.

Rodríguez-Piñero Royo, L. (2004). La OIT y los pueblos indígenas en el derecho internacional. Del colonialismo al multiculturalismo. *TRACE*, 46, 59–81.

Sambo, D. (2013). Evaluation of the Situation of the Indigenous Peoples in the World and the Role of the UN: Accomplishments, Failures, and Examples of Successful Practices. Geneva.

Sami Council and the Inuit Circumpolar Conference. (2004). Standard Setting: Legal Commentary on Free, Prior and Informed Consent. Retrieved from http://cendoc. docip.org/collect/cendocdo/index/assoc/HASHfa88/70edb246.dir/100_as.pdf.

Secretary-General of the League of Nations. (1923). Letter to Sir J.E. Drummond. Retrieved from http://cendoc.docip.org/collect/deskaheh/index/assoc/HASH0102/ 5e23c4be.dir/R612-11-28075-30626-8.pdf.

United Nations. (1983). Anti-Slavery Society for the Protection of Human Rights. Retrieved from http://cendoc.docip.org/collect/cendocdo/index/assoc/HASH0165/ 5994d046.dir/Wgip83_WorkingGroup_02.pdf.

United Nations. (2004). Provisional Agenda Item 5: Free, Prior and Informed Consent. Retrieved from http://cendoc.docip.org/collect/cendocdo/index/assoc/HASHfa 41/9bed9246.dir/087_as.pdf.

Vienna Declaration and Programme of Action. (1993). World Conference on Human Rights. Un Doc. A/CONF.157/23.

WGDD. (2004). Amended Text of the Draft Declaration on the Rights of Indigenous Peoples, Information Provided by Denmark, Finland, Iceland, New Zealand, Norway, Sweden, and Switzerland. UN Doc. E/CN.4/2004/WG.15/CRP.1.

WGIP. (1982). Study on the Problem of Discrimination Against Indigenous Populations. Note to the WGIP, by the Special Rapporteur: Martinez Cobo. UN Doc. E/ CN.4/Sub.2/AC.4/1982/CPR.1.

World Council of Indigenous Peoples. (1984). Review of Developments Pertaining to the Promotion and Protection of Human Rights and Fundamental Freedoms of Indigenous Populations. Retrieved from http://cendoc.docip.org/collect/cendocdo/index/assoc/HASHe46d/5595a0eb.dir/210457.pdf.

World Council of Indigenous Peoples. (1985). Declaration of Principles, submitted at the Fourth Session of the WGIP. Retrieved from http://cendoc.docip.org/collect/cendocdo/index/assoc/HASH0143/b349f4e0.dir/210481.pdf.

2 Towards a global framework on business and human rights, Indigenous Peoples, and their right to consultation and free, prior and informed consent

Humberto Cantú Rivera

Introduction

Over the past decades, the right to Consultation and Free, Prior and Informed Consent (FPIC) has been one of the cornerstones that protects the integrity and survival of Indigenous Peoples, especially in regions of the world that are attractive sites for resource exploration undertaken by extractive industries or development projects. These industries have regularly embarked on projects that create jobs and allow for a certain level of social and economic development in the areas surrounding them; however, their most usual feature is the extensive damage caused by their activities, which directly or indirectly impacts Indigenous Peoples' rights. As a result of this common and recurrent scenario, numerous parties have started to call for different normative and regulatory measures to prevent such negative impacts on the territories and livelihoods of Indigenous Peoples. Such measures, under international human rights law, have traditionally been addressed to States, who have a duty to protect, respect, and fulfil the human rights of those in their territories or under their jurisdiction. However, recent attempts have focused on the activities of business enterprises, which are often materially responsible for the damage caused to Indigenous Peoples' rights. Nevertheless, there is an important tension between what States have agreed to in multilateral settings – such as the International Labour Organization (ILO), the United Nations (UN) or the Organization of American States (OAS) – and the expectations of civil society organisations and Indigenous Peoples themselves. This tension serves to illustrate the difficulties in ensuring an adequate protection of certain vulnerable groups vis-à-vis economic interests. As will be argued, this tension has only deepened through the diverging interpretations of the content of the right to Consultation and FPIC, a situation that calls for a renewed focus on the obligation of States to strike a balance between the respect for and protection of human rights, on the one hand, and the need to ensure economic and social development, on the other.

The aim of this chapter is to analyse the different areas of international human rights law that intersect with business activities within Indigenous Peoples' territories, with the intention of discerning and discussing the scope

DOI: 10.4324/9781351042109-4

of Indigenous Peoples' rights, as well as the human rights obligations of States and the responsibilities of business enterprises, in accordance with the UN Guiding Principles on Business and Human Rights ("UNGPs" or "Guiding Principles"). Thus, this chapter will take into consideration normative standards adopted by different multilateral organisations, as well as jurisprudential developments – especially from the Inter-American Court of Human Rights (IACtHR) – at the international level, in order to analyse the progress achieved so far in the determination of the specific contents and meaning of the right to Consultation and Free, Prior and Informed Consent. To that end, a first section will address the global business and human rights framework, and the manner in which it has started to address Indigenous Peoples' rights; the following section will analyse the existing international standards in relation to Consultation and Free, Prior and Informed Consent, before turning to the role that the Inter-American Human Rights System has played in the clarification of existing international human rights obligations for States, and in the development of a regional *corpus juris* that is slowly addressing the responsibilities of businesses to respect human rights. Finally, a concluding section will briefly examine the role that businesses have to play in relation to Consultation and FPIC, and how this relates to the larger issue of corporate sustainability.

The global framework on business and human rights and its focus on Indigenous Peoples

Following a six-year mandate, the UN Special Representative of the Secretary-General on the issue of human rights and transnational corporations and other business enterprises (the "SRSG" or the "Special Representative") presented its final report to the Human Rights Council (2011): the UN Guiding Principles on Business and Human Rights. The Guiding Principles are divided into three main pillars, revolving around the State duty to protect human rights, the corporate responsibility to respect these rights, and the need to ensure effective remedy for victims of business-related human rights abuses. The report was well-received by the Human Rights Council, which decided to endorse it, a decision that eventually had an impact in numerous other international standard-setting processes and organisations, thereby creating a sort of "ecosystem" of corporate responsibility (Cantú Rivera, 2017, p. 48).

The first pillar of the UNGPs, addressing the State duty to protect, is manifestly inspired by existing international human rights law. Under international human rights law, the State is the main duty-bearer in terms of protecting, respecting, and fulfilling human rights for people within its territory or under its jurisdiction (De Schutter, 2014, p. 280). This entails different obligations for States; for example, they have a duty to respect human rights, which means avoiding human rights violations through their actions or omissions. In terms of protection, the State has a legal responsibility to ensure that third parties do not threaten or abuse human rights, and, when this may be the case, to prevent, investigate, sanction, and ensure adequate reparation. Fulfilling human rights implies a different type of responsibility, which involves the need to ensure adequate access to the different rights enshrined in international human rights conventions (Tomuschat, 2014, p. 141). As a result, States find that their duty to protect against business-related human rights abuses is an existing feature within international human rights law, and the UNGPs merely "restate" the approach to apply to situations involving business impacts on human rights.

On the other hand, the second pillar of the Guiding Principles is the main contribution to international human rights law resulting from this process. Corporate responsibility in the field of human rights had been a much-questioned idea in past UN efforts on this matter, especially as a result of the lack of international legal personality of business enterprises, and the idea that human rights obligations would be transferred to corporations (Vagts, 2003). However, as a result of a multi-stakeholder approach throughout his mandate, the SRSG achieved consensus that businesses have a core responsibility to respect human rights throughout their operations and supply chains. This responsibility is not only based on societal expectations (Deva, 2012), but derives also from the corporation's own interest in risk management and, more recently, several national legislations. Thus, as a result of this consensus, the second pillar was modelled upon three important aspects: first, the need for business enterprises to have an explicit commitment to respect human rights; second, the need for action to implement that commitment – namely via human rights due diligence, a process whereby enterprises should identify, address, and mitigate risks posed to human rights as a result of their activities or business relationships; and third, to participate actively in legitimate processes to ensure adequate remedy, either through judicial, non-judicial, or non-State based mechanisms. Thus, the Guiding Principles provide an explicit roadmap of what business enterprises can and should do to avoid harm to all human rights resulting from their operations, activities, or business relations.

Finally, the third pillar of the UNGPs revolves around the need for adequate access to effective remedy. The SRSG decided that the focus of this pillar should move beyond the traditional judicial approach that is enshrined in most international human rights instruments, particularly because other forms of remedy may be available and more advantageous to victims than judicial proceedings (Ruggie, 2013, p. 103), which tend to become lengthy and expensive processes that discourage or create new barriers for victims instead of ensuring adequate access to reparation. Thus, he suggested that non-judicial mechanisms (ranging from regulatory measures by the State and the complaints-handling function of NHRIs, to other conciliation or mediation-based mechanisms), as well as non-State based mechanisms (such as operational-level grievance mechanisms or multi-stakeholder platforms), could be a further layer of protection for victims in their pursuit for reparation. The expansion of the different remedies available would mean that victims should be able to choose which mechanism is more appropriate for the kind of remedy sought, depending on the type of human rights violation and the severity of the damage, without necessarily having to rely on judicial remedies as the only available avenue.

The adoption of the UNGPs by the Human Rights Council was accompanied by the creation of a follow-up mechanism, the UN Working Group on Business and Human Rights ("the UNWG"), a Special Procedures mandate (Cantú Rivera, 2015, p. 7) with the task of overseeing the adequate dissemination and implementation of the UNGPs (Addo, 2014). While the SRSG did not devote specific attention to the challenges that corporate activities and operations may pose for specific vulnerable groups, the second report of the UNWG to the General Assembly (2013), focused precisely on the issue of Indigenous Peoples' rights in connection with corporate activities. As will be addressed in the following section, the rights of Indigenous Peoples vis-à-vis development or extractive projects had already been considered by the first UN Special Rapporteur on the rights of Indigenous Peoples, Rodolfo Stavenhagen, in a report to the Commission on Human Rights (2003), as well as by James Anaya in his stint as Special Rapporteur.

In any case, the UNWG's focus on Indigenous Peoples' rights was particularly significant, given that it was an early contribution to the mandate on this particular issue. In this regard, it is relevant that a specific section of the UNWG report devoted attention to these rights, highlighting that FPIC is both an indicator of the fulfilment of the State duty to protect, on the one hand, and an instrument to prevent negative impacts on Indigenous Peoples' rights (United Nations Working Group on business and human rights [UNWG], 2013), on the other. A crucial aspect that is addressed in the report is the need to ensure legitimate consultation with Indigenous Peoples when assessing the possibility of undertaking a development project that may pose a threat to their livelihood or their rights, which includes the need to ensure adequate consultation not just with representatives of the group, but also of those that are normally underrepresented – most notably Indigenous women and children. As a starting point, the UNWG report refers to the UN Declaration on the Rights of Indigenous Peoples ("UNDRIP"), and to ILO Convention 169 ("ILO 169"), which – as was the case until then – provided most of the existing legal framework in relation to Indigenous Peoples' rights. However, one aspect that appears to be a glaring omission by the UNWG in this regard is the lack of proper consideration of the IACtHR's case law or *corpus juris*, given that in numerous judgments it has defined the precise meaning of ILO 169 and UNDRIP in the context of the American Convention on Human Rights, under the umbrella of a *pro persona* interpretation (Rodiles, 2016). Indeed, only a handful of these cases were referred to in the report. As is well known, the international legal framework regulating Indigenous Peoples' rights vis-à-vis development or extractive projects continues to expand, an issue that shall be addressed in the following section.

International standards on the right to consultation and Free, Prior and Informed Consent

The rights to consultation and FPIC are key aspects that differentiate Indigenous Peoples' rights from the human rights of other vulnerable groups (Errico, 2007, p. 745), and yet they have become a harbinger for the promotion and protection of other interrelated civil, political, economic, social, and cultural rights. The current normative framework regarding Indigenous Peoples' rights, however, started to develop well before the discussions surrounding corporate human rights responsibilities. In this regard, different instruments have set forth the obligations of States – and the correlative rights of individuals and groups – to promote Indigenous Peoples' subsistence, and to ensure the preservation of their traditions and culture.

The most relevant instrument is ILO 169 of 1989, which prescribed what would become the basis for many other regional and international developments in this area. The Convention addresses different elements of Indigenous Peoples' rights, but most of the focus has revolved around consultation, on the one hand, and consent, on the other hand. Article 6 of the Convention sets forth that governments shall consult the peoples concerned whenever legislative or administrative measures that may affect them are being considered; establish means to allow free participation of the peoples in decision-making; and ensure that consultations are carried out in good faith and in a form appropriate to the circumstances, specifically mentioning that the aim of such consultations shall be to achieve agreement or consent on the proposed measures. The aspect of consultation, as can be observed, is not merely a procedural obligation for States, but an autonomous substantive right of Indigenous Peoples on its own, aimed specifically at ensuring their participation in decision-making and policies that may affect their livelihood, culture, and survival. Thus, this right seeks to empower Indigenous Peoples and ensure that governments avoid discriminatory practices and policies against them, by taking all necessary measures to ensure that the special character of several of their rights is considered, such as their relationship with the land and with their own heritage and customs.

A relatively different basis can be perceived regarding Free, Prior and Informed Consent (FPIC). Article 16.2 of ILO 169 specifically sets forth that FPIC will be required whenever relocation of peoples is considered necessary, and that these peoples retain the right to return to their original territories or land whenever possible, and as soon as the conditions that led to their relocation cease to exist. Thus, ILO 169 establishes a dual system whereby Indigenous Peoples shall always have the right to be consulted in relation to any project, legislation, or administrative decision that may affect them or their traditional livelihood, but their consent will only be required if they must be relocated. In this regard, it will only be applicable in very specific contexts, therefore having a limited scope – as opposed to consultation, which has a wider reach and a more general application in any context of extractive or development projects. For some, this differentiation between both objectives can lead to the fragmentation of those rights, especially if achieving consent is not the main goal of any consultation, which can thus render it meaningless (Gonza, 2014, p. 528); others, however, highlight the importance of a complementary approach to conciliate State interests in their own national development with the rights of Indigenous Peoples to their own forms of governance and development (Errico, 2007, p. 752).

Several of these aspects were taken up again during the negotiation of the UNDRIP (on this, see the chapter by Del Castillo in this volume). Article 10, for instance, sets forth that FPIC is required to relocate Indigenous Peoples from their lands or territories; article 19 stipulates that consultations in good faith must be conducted before adopting or implementing legislative or administrative measures that may affect them, with the specific intent of achieving consent. Finally, article 32.2 provides for a State duty to consult with the aim of achieving consent prior to the approval of any project that may impact Indigenous Peoples' rights, including their lands and territories. As can be observed, UNDRIP basically replicates the standard set forth in ILO 169, in an attempt to consolidate an international standard that guarantees the participation of Indigenous Peoples in projects that may pose a risk to their survival.

Another recently adopted document is the American Declaration on the Rights of Indigenous Peoples, negotiated within the Organization of American States. While the instrument addresses different substantive and procedural rights that have also been recognised in other international instruments, the issues of consultation and consent are included in a section devoted to Indigenous rights to development. Thus, in section 4 of article XXIX, the Declaration provides that "States shall consult and cooperate in good faith with the indigenous peoples concerned through their own representative institutions in order to obtain their free and informed consent prior to the approval of any project affecting their lands or territories and other resources, particularly in connection with the development, utilization or exploitation of mineral, water, or other resources." Despite this important similarity with UNDRIP (Hohmann & Weller, 2018, p. 2), some have argued that the American Declaration falls short from the standards set forth in UNDRIP, especially with regard to compensation and restitution (Errico, 2017).

An analysis of the three aforementioned international standards reveals an important emphasis in the declaratory instruments on the apparent existence of a "right to veto" bestowed upon Indigenous Peoples, which is not as apparent in ILO 169, which only requires consent whenever relocation must take place in order for the development project to continue. In a more restrained manner, ILO 169 makes an effort to conciliate national development interests with Indigenous Peoples' rights but stops short of creating an international veto power for Indigenous Peoples vis-à-vis development projects. However, as will be analysed below, this aspect has been interpreted in a broader manner by different organs, who have construed ILO 169 and UNDRIP as establishing a considerable influence vis-à-vis development projects that may threaten their existence, culture, or livelihood (Barelli, 2018, p. 253).

Several United Nations bodies have tried to promote the agenda of Indigenous Peoples' rights at the international level, by interpreting the different declaratory and conventional provisions on Indigenous Peoples' rights in relation to development or extractive projects; however, this section will focus exclusively on those Special Procedures mandates that have specifically contributed to this issue. Within UN Special Procedures, the clearest link has been explored by the different UN Special Rapporteurs on Indigenous Peoples' rights (UNSR). For example, in 2003 former UNSR Rodolfo Stavenhagen clearly identified the potentially negative impacts of major development projects on Indigenous Peoples' rights:

> Wherever such developments occur in areas occupied by indigenous peoples it is likely that their communities will undergo profound social and economic changes that are frequently not well understood, much less foreseen, by the authorities in charge of promoting them. Large-scale development projects will inevitably affect the conditions of living of indigenous peoples. Sometimes the impact will be beneficial, very often it is devastating, but it is never negligible.
>
> (UNSR, 2003, p. 2)

It was also in this report that the distinction between consultation and consent began to blur, as the UNSR (2003) appeared to indicate that there was a right to consent in relation to any large-scale development projects in indigenous territories, which actually contradicts the provisions in ILO 169.

This was partially corrected by the next Special Rapporteur, James Anaya, who – in relation to the right to consultation – stated that "[t]his provision of the Declaration should not be regarded as according indigenous peoples a general 'veto power' over decisions that may affect them, but rather as establishing consent as the objective of consultations with indigenous peoples" (Human Rights Council, 2009, par. 46). Rather, he emphasised that the provisions found in ILO 169 and UNDRIP to that end focused particularly on the importance of ensuring that consultation procedures genuinely tried to build consensus among the different stakeholders (Human Rights Council, 2009, par. 48). In his 2013 report, the UNSR reaffirmed the right of Indigenous Peoples to express consent in order for extractive or development projects that might affect their rights to proceed, although with an important caveat:

> In the view of the Special Rapporteur, however, when States make such efforts to consult about projects and, for their part, the Indigenous Peoples concerned unambiguously oppose the proposed projects and decline to engage in consultations, as has happened in several countries, the States' obligation to consult is discharged.
>
> (UNSR, 2013, par. 25)

He further elaborated that if a State decides to go ahead with a project despite the absence of consent from Indigenous Peoples, it shall still have the obligation to respect and protect their rights, and especially to ensure that impact assessments are undertaken in order to minimise limitations of their rights, to adopt mitigation measures, as well as to ensure adequate compensation and benefit sharing (UNSR, 2013, par. 38).

This caveat is important, especially if consideration is given to the fact that any type of development or extractive project may pose a threat to numerous rights of Indigenous and non-Indigenous Peoples. However, despite the fact that consultation is mainly a procedural obligation that serves to protect many other substantive rights, it is clear that States will need to balance their efforts in order to ensure adequate protection for Indigenous Peoples' rights without necessarily limiting their own national development agenda, which for many developing and less-developed countries still relies primarily on the extraction of natural resources. In that regard, Inter-American case law has been instrumental in developing a clearer image of the procedural requirements that States must fulfil in order to discharge their obligation, a welcome addition that should, in theory, clarify the expectations regarding State and business conduct.

The Inter-American Human Rights System and Indigenous Peoples' rights

The Inter-American Human Rights System, which includes the Inter-American Human Rights Commission (IACHR or the Commission) and the Court, has been a key player in the development of an international *corpus juris* on Indigenous Peoples' rights. These developments have not been limited to the scope of the American Convention on Human Rights: on the contrary, they have expanded their reach to take advantage of numerous other international and regional instruments, reports, and jurisprudence. Especially through the Court's case law, the rights of Indigenous Peoples to consultation and FPIC in situations or contexts that may affect their traditional livelihood and territories have been developed, thereby defining the contours of State obligations to protect the collective rights of Indigenous Peoples. In addition, the Commission has had an important role in advancing the understanding of State obligations in relation to this group, especially through country and thematic reports (including, *inter alia*, on Indigenous Peoples' rights over their ancestral lands and natural resources, or on Indigenous Peoples in voluntary isolation and initial contact).

In this regard, one of the earliest country reports issued by the Commission to recognise the reality of Latin America as a hub for natural resource extraction was its 1997 report on the human rights situation of Ecuador (IACHR, 1997), which, in its chapter VIII, highlights the issue of oil companies and the effects of their activities on the environment – and as a result, on numerous economic, social, cultural, civil, and political rights – (IACHR, 1997, chapter VIII). This report highlighted the role that States should have vis-à-vis private actors, in one of the early manifestations of the Commission in relation to this issue. In its 2015 report "Indigenous Peoples, Afro-Descendent Communities, and Natural Resources: Human Rights Protection in the Context of Extraction, Exploitation, and Development Activities" (IACHR, 2015), the Commission addressed this issue and divided the human rights obligations of States into two different groups: first, obligations in the context of extractive, exploitation, and development activities; and second, obligations and guarantees in relation to Indigenous and Tribal Peoples and Afro-Descendent communities.

With regard to the context of development projects, the Commission determined that most State obligations revolve around aspects contained in articles 1.1 and 2 of the American Convention on Human Rights, which stipulate State duties to respect and ensure the rights enshrined in it, and to adopt the necessary legislative or other measures to implement the provisions of the Convention. To that effect, the Commission explicitly underlined the State obligation to have an adequate legal framework, a prerequisite to enforce legal obligations in relation to development projects, and, especially, to ensure the adequate prevention, mitigation, and eradication of negative impacts on human rights. Furthermore, it should be able to adequately supervise those activities, to ensure effective participation and access to information – a vital element for Indigenous Peoples who may be affected by development projects –, and to guarantee access to justice whenever a human rights violation occurs. The second set of State obligations deals specifically with the topic of this chapter, and refers to the different elements and rights that have been identified by the Inter-American Human Rights System as belonging to Indigenous Peoples, in order to avoid denial of their physical and cultural survival (which has regularly been the case through assimilation processes in different countries). Some of these elements include important procedural rights and correlative State duties, such as the right to effective participation; to undertake impact studies and assessments to identify potential human rights violations (although the case law of the Inter-American Court has not yet advanced in that explicit direction, focusing so far on environmental and social impact assessments); to have access to shared benefits; and generally, to consultation and FPIC. As such, the 2015 report by the Inter-American Commission (IACHR, 2015) is a sort of repository of the law and practice developed by both organs of the Inter-American Human Rights System, that served especially to highlight the importance of the issue of business and human rights, and to underline the role that the State has in ensuring the effective protection of the human rights of Indigenous Peoples.

While the case law of the IACtHR on Indigenous Peoples' rights has been largely examined elsewhere (Pasqualucci, 2006; Burgorgue-Larsen & Úbeda de Torres, 2011), this chapter will address succinctly some of the most relevant aspects to have been dealt with by the Court in recent judgments, especially in relation to impact assessments, on the one hand, and consultation and consent, on the other hand. In relation to impact assessments, the Court has slowly progressed in recognising that they are essential elements to ensure informed consultations with Indigenous Peoples. While under international law they are not elements that will determine whether a project can – or cannot – proceed, in the opinion of Gillespie, "the important outcome is that a participatory and robust process is undertaken and reliable information is gathered for decision makers before a decision is made that may have significant environmental impacts" (Gillespie, 2008, p. 233). The Inter-American Court has gradually required that States undertake different types of impact assessments – so far, environmental and social impact assessments – prior to the start of any type of work or project that may affect, directly or indirectly, the rights of Indigenous Peoples. This trend commenced in the *Saramaka People v Suriname* judgment in 2007 (IACtHR, 2007), where the Court established different types of safeguards to ensure that the State complies with its duty to protect human rights through positive obligations in relation to the environment, namely that the assessments are undertaken by autonomous and technically-capable entities, are supervised by State authorities, and that they take place before any concession is granted. This was reiterated in the *Kichwa Indigenous People of Sarayaku v Ecuador* judgment of 2012 (IACtHR, 2012), where the Court further refined the parameters established in *Saramaka* in relation to the implicit links between impact assessments and the different elements of consultation that should be taken into account.

Business activities in the recent case law of the Inter-American Court of Human Rights

The case-law of the IACtHR regarding the impacts of business operations or extractive or development projects on the rights of Indigenous Peoples would find new ground in its 2015 judgment in the case of the *Kaliña and Lokono Peoples v Suriname*, in particular as a result of the fact that the Court explicitly quoted the UNGPs (discussed above) when addressing the issue of environmental and social impact assessments. The Court noted that the actions affecting the environment were undertaken by private companies, and thus, the State should have supervised their activities in order to ensure that actions were taken to prevent, mitigate, and eventually repair the damages inflicted directly or indirectly upon the Kaliña and Lokono Peoples. One perplexing aspect, however, is the reason why the Court fell short of requiring that human rights impact assessments are undertaken by companies or required by States, in accordance with the UNGPs, especially if the Guiding Principles were explicitly quoted as establishing corporate "duties" (reflected in the word "must" used by the Court, which departs, however, from the explicit text of the UNGPs, which speaks about responsibilities) to "protect" and respect human rights. Despite the general progressiveness of the IACtHR, this appears to be a missed opportunity to advance and consolidate the purpose of impact assessments for the explicit protection of human rights, and especially of Indigenous Peoples' rights vis-à-vis corporate activities.

At the end of the day, human rights impact assessments go beyond the environmental and social aspects that may affect a community, to establish a standard-based measurement that can potentially identify impacts that the other two assessments cannot, due to their restrictive focus on social practices – which may deeply embed within them important human rights violations – or the environment. Despite this, the fact that the Court quoted the UNGPs is in itself an important achievement, by integrating them into its *corpus juris* and thus making them legally relevant for States under its jurisdiction. This decision may also affect the manner in which the Court interprets the question of consultation in the future, particularly if further integration of the different elements of the UNGPs takes place. It may well be the case for the Court to require States to adopt legislative and other measures to effectively require business enterprises to undertake human rights impact assessments and adopt human rights due diligence processes, which coupled with the existing requirements in relation to environmental and social impact assessments, could potentially pave the way for an integrated impact assessment standard (Cantú Rivera, 2018).

It is clear that the Inter-American Human Rights System, through its reports and case law, has directly contributed to the refinement and evolution of international standards in relation to FPIC, on the one hand, and to its distinctive elements, most notably the obligation to undertake impact assessments, on the other. In particular, while the progressive interpretation made by the Court may contrast with political choices made by States in multilateral settings, the conciliation of both sets of interests may provide for more effective implementation by States of their human rights obligations, which would theoretically contribute to the realisation and fulfillment of Indigenous Peoples' rights.

Final thoughts: what role for businesses vis-à-vis consultation and Free, Prior and Informed Consent?

Having devoted this chapter mostly to the issue of the obligations of States under international human rights law to undertake prior consultation processes with Indigenous Peoples in relation to extractive or development projects that may affect them, it is not only appropriate but necessary to consider the other side of the spectrum: what does FPIC mean for businesses, and what role can they play in this regard while considering the responsibilities prescribed in the Guiding Principles? In general terms, two aspects can be considered to be of relevance for business enterprises which decide to invest in areas that may be linked to Indigenous Peoples: first, the aspect of the "social licence" they require to make an appropriate investment; and, second (and in direct connection with the first aspect), the promise of the sustainability of their operations and activities.

To begin with, the "social licence" refers to a concept whereby, for any business project to succeed, companies must obtain the "permission" of any community that may be affected by its activities or operations, a situation that requires constant communication, dialogue, and understanding from both parties about their needs and conditions (Morrison, 2014). This social licence goes well beyond the issue of legal requirements and permits, and in theory would allow any project to run more smoothly, helping all parties to prevent undesired conflict that may result in financial or other losses while deepening the connections that would allow for positive developments. This concept, so well-known in the context of extractive industries, can, however, be applied to any type of business project which may involve risks to local communities, and is especially relevant for operations that may pose a risk to Indigenous Peoples' rights. In a sense, the concept of the social license to operate is closely related to the issue of consultation, which has been identified as a requirement that must be complied with whenever any project may pose a risk to the livelihood or territories of Indigenous Peoples' rights. In legal and theoretical terms, this obligation corresponds exclusively to the State, but in practice is often executed directly by business enterprises themselves. Thus, it can also relate to some of the elements of corporate human rights due diligence as prescribed by the Guiding Principles, which largely address and underscore the importance of adequate communication with all relevant stakeholders throughout the life of any given project. In this sense, compliance with the second pillar of the UNGPs may well be reflected in the obtention of a social license to operate, as long as a genuine and honest dialogue between Indigenous Peoples, company representatives, independent experts, and State authorities takes place, which would imply better prospects for any specific business project being developed within Indigenous Peoples' territories.

The issue of communication and the social licence to operate is also directly related to a widespread corporate practice in the twenty-first century, that of the "sustainability agenda". Many companies operating worldwide frequently claim that they are "sustainable" or that their practices are so, with sustainability relating to several components, including economic, social, and environmental performance. This has been accompanied by numerous calls for corporate sustainability, including those made in relation to the Sustainable Development Goals adopted in 2015 by the United Nations. However, would it be possible to call "sustainable" a development project that did not integrate the concerns of any given society in relation to their traditions or customs, or a project that did not have adequate mitigation measures to ensure environmental preservation? Of course, the economic viability of a project will necessarily be the aspect that helps to make any project a reality, an aspect that could bring about important results in terms of improving the general standards of living of any given community; but when business enterprises unilaterally declare that they are sustainable, those claims need to be based on actual corporate practices that seriously take into consideration their impact on the human rights of those communities that may be affected by their projects. While this is not the appropriate space to delve deeper into the issue of what sustainability means in terms of corporate operations and activities, it is important to highlight that the sustainability of a project that may pose a threat to the survival or livelihood of Indigenous Peoples necessarily considers the issues raised before, and most notably those of dialogue and consultation in a constructive manner that allows Indigenous Peoples to participate directly in making decisions that may ultimately affect them positively or negatively, as well as enhance or diminish their prospects of survival. To that extent, the issue of FPIC must be considered by business enterprises as not just a practice or step that needs to be fulfilled in order to obtain a permit to start operations; rather, it is precisely the component of the project that may ensure its economic, social, and political viability, while reaffirming a genuine commitment to respecting the human rights of peoples that may be affected by it – as well as the sustainable character of its operations.

References

Addo, M.K. (2014). The Reality of the United Nations Guiding Principles on Business and Human Rights. *Human Rights Law Review*, 14(1), 133–147.

Barelli, M. (2018). Free, Prior, and Informed Consent in the UNDRIP: Articles 10, 19, 29(2), and 32(2). In J. Hohmann & M. Weller. (Eds.), *The UN Declaration on the Rights of Indigenous Peoples: A Commentary* (pp. 247–271). Oxford: Oxford University Press.

Burgorgue-Larsen, L. & Úbeda de Torres, A. (2011). *The Inter-American Court of Human Rights: Case Law and Commentary*. Oxford: Oxford University Press.

Cantú Rivera, H. (2015). The United Nations Human Rights Council: Remarks on its History, Procedures, Challenges and Perspectives. In H. Cantú Rivera (Ed.), *The Special Procedures of the Human Rights Council* (pp. 1–24). Cambridge: Intersentia.

Cantú Rivera, H. (2017). Los desafíos de la globalización: reflexiones sobre la responsabilidad empresarial en materia de derechos humanos. In H. Cantú Rivera (Ed.), *Derechos humanos y empresas: reflexiones desde América Latina* (pp. 37–83). San José: IIDH.

Cantú Rivera, H. (2018). Extractive Industries, Indigenous Peoples' Rights, and the Need for a Cohesive Doctrine on Human Rights Due Diligence. *Revista Internacional de Derecho y Ciencias Sociales*, 29.

Commission on Human Rights. (2003). Report of the Special Rapporteur on the Situation of Human Rights and Fundamental Freedoms of Indigenous People, Rodolfo Stavenhagen, submitted in accordance with Commission resolution 2001/65. UN Doc. E/CN.4/2003/90.

De Schutter, O. (2014). *International Human Rights Law*. Cambridge: Cambridge University Press.

Deva, S. (2012). *Regulating Corporate Human Rights Violations: Humanizing Business*. London: Routledge.

Errico, S. (2007). The Draft UN Declaration on the Rights of Indigenous Peoples: An Overview. *Human Rights Law Review*, 7(4), 741–755.

Errico, S. (2017). The American Declaration on the Rights of Indigenous Peoples. *ASIL Insights*, 21(7). Retrieved from https://www.asil.org/insights/volume/21/issue/7/american-declaration-rights-indigenous-peoples.

General Assembly. (2013). Report of the Working Group on the Issue of Human Rights and Transnational Corporations and Other Business Enterprises. UN Doc. A/68/279.

Gillespie, A. (2008). Environmental Impact Assessments in International Law. *Review of European Community and International Environmental Law*, 17(2), 221–233.

Gonza, A. (2014). Derecho a la propiedad privada. In C. Steiner & P. Uribe (Eds.), *Convención Americana sobre Derechos Humanos: Comentario* (pp. 503–530). Bogotá: Konrad Adenauer Stiftung.

Hohmann, J. & Weller, M. (2018). Introduction. In J. Hohmann & M. Weller (Eds.), *The UN Declaration on the Rights of Indigenous Peoples: A Commentary* (pp. 1–6). Oxford: Oxford University Press.

Human Rights Council. (2009). Report of the Special Rapporteur on the Situation of Human Rights and Fundamental Freedoms of Indigenous People, James Anaya. UN Doc. A/HRC/12/34.

Human Rights Council. (2013). Report of the Special Rapporteur on the Situation of Human Rights and Fundamental Freedoms of Indigenous People, James Anaya: Extractive Industries and Indigenous Peoples. UN Doc. A/HRC/24/41.

Human Rights Council. (2011). Report of the Special Representative of the Secretary-General on the Issue of Human Rights and Transnational Corporations and Other Business Enterprises, John Ruggie: Guiding Principles on Business and Human Rights: Implementing the United Nations "Protect, Respect and Remedy" Framework. UN Doc. A/HRC/17/31.

IACHR. (2009). Indigenous and Tribal Peoples' Rights over their Ancestral Lands and Natural Resources. OEA Doc. OEA/Ser.L/V/II. Doc. 56/09.

IACHR. (1997). Report on the Situation of Human Rights in Ecuador. OEA Doc. OEA/Ser.L/V/II.96 Doc. 10 rev. 1.

IACHR. (2015). Indigenous Peoples, Afro-Descendent Communities, and Natural Resources: Human Rights Protection in the Context of Extraction, Exploitation, and Development Activities. OEA Doc. OEA/Ser.L/V/II. Doc. 47/15.

IACtHR. (2007). *Saramaka People v Suriname*, Judgment of November 28, 2007 (Preliminary Objections, Merits, Reparations, and Costs). Inter-Am. Ct. H.R., (Ser. C) No. 172(2007).

IACtHR. (2012). *Kichwa Indigenous People of Sarayaku v Ecuador*, Judgment of June 27, 2012 (Merits and reparations). Inter-Am. Ct. H.R., (Ser. C) No. 245(2012).

Morrison, J. (2014). *The Social License*. London: Palgrave MacMillan.

Pasqualucci, J. (2006). The Evolution of International Indigenous Rights in the Inter-American Human Rights System. *Human Rights Law Review*, 6(2), 281–322.

Rodiles, A. (2016). The Law and Politics of the "Pro Persona" Principle in Latin America. In H.P. Aust & G. Nolte (Eds.), *The Interpretation of International Law by Domestic Courts* (pp. 153–174). Oxford: Oxford University Press.

Ruggie, J.G. (2013). *Just Business: Multinational Corporations and Human Rights*. New York: W.W. Norton.

Tomuschat, C. (2014). *Human Rights: Between Idealism and Realism*. Oxford: Oxford University Press.

Vagts, D.F. (2003). The UN Norms for Transnational Corporations. *Leiden Journal of International Law*, 16(4), 795–802.

3 Binding consent of Indigenous Peoples in Colombia

An example of transformative constitutionalism[1]

Juan C. Herrera

Introduction

The Free, Prior and Informed Consent (FPIC) of Indigenous Peoples on matters that have the potential to affect their interests and territories has become one of the most powerful tools that positive and jurisprudential law has created in recent decades to protect the collective rights of these peoples.[2] The Constitutional Court of Colombia (CCC) is one of the most relevant and engaged bodies in the construction of transformative constitutionalism in Latin America. "Transformative constitutionalism seeks to remake a country's (supposedly deficient) political and social institutions by moving them closer to the sets of principles, values, and practices found in the constitutional text" (Landau, 2005, p. 1535).[3]

Together with other Courts, especially the Inter-American Court of Human Rights (IACtHR), the CCC is part of a group of Latin American courts that have found a creative way to embrace the main characteristics and principles of the recent constitutions issued in the region. This is a unique version of judicial case law that links transformative and dialogic constitutionalism to plausible results.

Regarding the Indigenous question in the Latin American constitutional context, Gargarella has asked: "how should we solve, then, the questions posed by the emerging tensions between the rights and interests of indigenous groups and rights and interests of the rest of the population?" (Gargarella, 2013, p. 180).

This chapter offers an answer to that question by studying Colombian constitutional case law. It thus complements and critically annotates the jurisprudential and legal grounds of the right to prior consultation of Indigenous Peoples by focusing on the issue of binding consent. For this reason, this chapter does not provide an empirical review of how the right to prior consultation itself is implemented. Instead, it explains how the court introduced one of the region's most transformative case laws regarding FPIC. Furthermore, it provides strong reasons for not linking prior consultation to a "veto right" but rather to a contextualised "binding consent". Finally, it explains why the outcomes of the precedents are at risk. One way or another, our comprehension of a good example of judicial practice may help to reduce the "implementation gap".

DOI: 10.4324/9781351042109-5

Transformative case law of the Constitutional Court of Colombia

Contextualisation

Constitutional scholars have shown increased interest in the jurisprudential "proactivism" of the CCC (Landau, 2005, p. 736) and the transformative role of its justices (Uprimny, 2006, p. 127). In fact, prior consultation and Indigenous rights are a good example of vigorous commitment to transformative constitutionalism (Cepeda & Landau, 2017, p. 241).

In Colombia, before the Constitution of 1991, populations such as (i) Indigenous, (ii) Afro-Colombians, (iii) *Raizales* islanders, (iv) *Palenqueras*, and (v) Roma were constitutionally dismissed as independent or special groups[4] or "people outside the Constitution" (Gargarella, 2014, p.1). After Independence, these peoples were voluntarily and/or involuntarily compelled to reside in peripheral areas such as the Amazon region in the South, the Pacific coast in the West, and the Caribbean in the North. In addition, the "development" process enforced and demanded by major cities, such as Bogotá, with more than eight million people in the centre of the country, Medellin with 3.7 million in the North-West, Cali with 2.9 million in the South, and Barranquilla with 1.8 million in the North, represents a direct threat for the protected peoples.[5]

Moreover, the past 50 years have seen increasingly rapid advances and developments in those centres of industrial production and the rest of the country. Overpopulation and other factors linked with "development" have created a two-pronged problem. On the one hand, national and transnational actors are exploring or exploiting natural resources in areas where these communities have been living for centuries. On the other, the cyclical violence of the country in some cases and the western influence in others are pushing some groups or families from small rural communities to urban areas.[6] These dramatic changes often result in a very inadequate quality of life and in most cases a quick push towards homelessness for people who are barely surviving.[7]

Constitutional protection[8]

Colombia's 1991 Constitution has provisions that directly protect Indigenous Peoples and creates the perfect platform for their participatory rights, which have, historically, been subjected to discrimination. The most relevant are: article 7 (ethnic and cultural diversity), article 10 (language), article 40 (rights of citizen participation), article 171 (two Senators are elected in a special national constituency for Indigenous communities) and articles 246, 286, 287, 329, and 330 (Indigenous territories as administrative entities, faculties within local authorities, management, judicial jurisdiction, and the development of policies). The constitutional ranking of these particularities was tantamount to a small revolution and strategically developed relevant aspects of the ILO Convention No.169 Concerning Indigenous and Tribal Peoples in Independent

Countries of 1989 (ILO 169). According to the doctrine of the "constitutional block", international instruments such as ILO 169 are integrated into the constitution with a special status. Indeed, "[...] a block of constitutionality exists, comprising national constitutional norms and regional/universal human rights instruments" (Góngora-Mera, 2017, p. 236).

For its part, article 330 is the main source of the connection between prior consultation and the rights of Indigenous Peoples. According to the Constitution, councils formed by and regulated according to their traditions shall govern native territories. As a result, councils should do the following:

(i) implement regulations of land uses and settlement of their territories; (ii) design the policies, plans, and programmes of economic and social development within their territory, in accordance with the National Development Plan; (iii) promote public investments in their territories and ensure their proper implementation; (iv) collect and distribute their resources; (v) ensure the preservation of natural resources; (vi) coordinate the programmes and projects promoted by the different communities in their territory; (vii) collaborate with the maintenance of public order within their territory in accordance with the instructions and provisions of the national government; (viii) represent the territories before the national government and other entities to which they are members; (ix) attend to matters stipulated by the Constitution and the law; and (x) exploit natural resources without harm to the cultural, social and economic integrity of indigenous communities.

(Constitution of Colombia, 1991, article 330)

Article 330 of the Colombian Constitution is at the heart of jurisprudential and constitutional understandings of the right to consultation. Indeed, it states that:

[...] [E]xploitation of natural resources in indigenous territories shall be done without harming the cultural, social and economic integrity of indigenous communities. The decisions taken with respect to such types of exploitation in their territories must be encouraged by the government and coordinated with the representatives of the respective communities.

(Constitution of Colombia, 1991, article 330)

The constitutionalisation of the extensive guarantees described above introduced an important instrument for the protection and legitimatisation of Indigenous Peoples, to some extent recognising the centuries of systematic violations of their rights and lands. For the very first time, Colombia recognised its own national and regional reality as well as its status as a multicultural State that takes "seriously" the right of these peoples to self-determination. "These ideas challenge previously dominant western conceptions of the cultural homogeneous and legally monolithic state" (Anaya, 2001, p. 61).

Regarding legal reforms, the most relevant debate about prior consultation is its regulation as a fundamental right and the determination of whether a so-called veto power exists. Thus, from that perspective, the Statutory Act (*Ley Estatutaria*) deserves a mention. This Act constitutes a special law that requires an absolute majority for legislation to be passed in Congress and a legislative period of one year. Subsequently, legislation passed is subject to mandatory review before the CCC.

The Colombian government, the private sector, and several civil society and leadership organisations are trying to discuss a draft law, which has paradoxically stalled due to the lack of prior consultation. However, the solution is not *per se* the regulation. In fact, some observations made on the Peruvian case are worrying: "[w]e have found that the actual influence exerted by the groups consulted on the content of the law and the decree was very limited" (Schilling-Vacaflor, & Flemmer, 2015, p. 835).[9]

Review of legislation (Abstract Control)

The following sections analyse the basic elements or standards developed by the CCC regarding the review of legislation (Abstract Control) and the specific review of individual petitions related to violations of human rights (Concrete Control).[10]

Abstract Control is a complex tool of the CCC due to the difficulties in explaining and determining what exactly should be consulted with the communities regarding regulations. Decision C-030/08 stands out as clearly the most important and comprehensive precedent related to the Abstract Control of legislation and the fundamental right to prior consultation of Indigenous Peoples.[11] In one of its decisions of 2013, the Court affirmed that judgment C-030 of 2008 re-conceptualises the jurisprudence of the court and makes progress in establishing the requirements and characteristics of prior consultation. That is, it adheres to the main precedent in this area, as it consolidates the standard in implementing prior consultation in the legislative process, to guarantee the realisation of the right of such communities to participate in decisions that directly affect them (Constitutional Court of Colombia, 2013, para. 6.3.3).

To interpret the content and scope of the legislative measures in specific cases, Decision C-030/2008 of the CCC distinguished between the direct impact on Indigenous Peoples and the impact on Colombian society as a whole. For instance, mining regulation affects all Colombian citizens. However, if there is a chapter or article that mentions or exclusively affects Indigenous Peoples, the CCC reviews whether the protected communities were consulted or not during the legislative process.

Figure 3.1 shows the distribution of 41 decisions analysed in this research during the 25-year period from 1992 until 2017.[12] There were only two cases out of 41 (5%) where the CCC set aside parts of the legislation which directly affects the right to prior consultation, and merely five cases (12%) in which

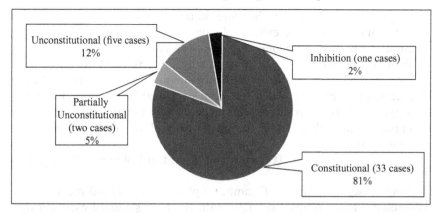

Figure 3.1 Abstract control or review of legislation in the case law of the Constitutional Court of Colombia related to prior consultation of Indigenous Peoples in the last 25 years.
Source: Herrera (2019)

the whole regulation was reviewed. This means that the Court declared the lack of prior consultation during the law-making of this regulation to be unconstitutional. Contrary to the various critics who consider prior consultation a problematic issue for the mainstream development model and the "general interest" (El Espectador, 2016), in more than 80% of the cases reviewed (33 cases), the Court has found the regulation made by the Congress to be constitutional. On just one occasion, inhibition or dismissal was the mechanism for rejecting the study of legal action due to the mediocre quality of the litigation (2%).

Consequently, the CCC has applied several hermeneutical methods: (i) textual interpretation of the regulatory body; (ii) systematic interpretation; (iii) historical interpretation; (iv) contextual interpretation including the precedents and the controversies surrounding the ruling process; and (v) teleological interpretation. In short, the key element here is the direct impact of the regulation on the interests of the community. In those cases where it is possible to prove or find a lack of prior consultation, the jurisprudence has declared some articles or the whole law to be unconstitutional, once again, only in 17% of the cases studied in 25 years. In the remaining more than 80%, it has declared that the regulation was constitutional.

Decision C-389/16, the last of the case law in the period studied here, is an excellent example of how it is possible to solidify binding consent. This decision recapitulates the leading case of Abstract Control and links constitutionality to the considerations of the leading case of concrete control, T-129/11, and the other decisions related to direct interventions in the territories of the communities. In the holding of Decision C-389/16, the Grand Chamber of the Court makes the following important remark regarding measures to be considered during direct interventions in the protected territories:

The Grand Chamber will therefore declare Articles 122, 124 and 133 of the Mining Code to be enforceable, on the understanding that it is constitutionally admissible only if it is considered that ethnic communities shall be consulted in relation to mining projects likely to affect them directly (C-371/14, T-129/11, T-769/09, among many others). And if, in accordance with that which has been explained in the preceding paragraphs, decisions that directly and intensely affect their rights can only be implemented if they obtain the prior, free and informed consent of the communities.

(Constitutional Court of Colombia, 2016, p. 152)

To summarise, the Grand Chamber upholds the commitment to the Colombian ethnic groups, especially with the high standard established in Decisions T-769/09, T-129/11 and the many others that reinforce and solidify the case law of the CCC or what Bonilla (2013) has properly coined the "pluralistic multicultural model", i.e., "[t]his model appeals to a pluralistic structure of the State as well as to intercultural equality, corrective justice, self-government rights, and cultural integrity in order to justify and give content to the right to prior consultation" (p. 35).

The Concrete Control regarding the protection of constitutional rights (Acciones de Tutela)

In decisions regarding the protection of constitutional rights, on several occasions, the CCC has studied problems where Indigenous Peoples have had to address the violation of their territories or the exploitation of natural resources, both by private and/or public actors.

82 cases ruled on by the Constitutional Court related to such concrete control (see Figure 3.2).[13] In the five rejected cases (6%), the jurisprudence pointed out that such *Acción de Tutela* was not the proper mechanism for the specific case. Finally, in just one case, the court annulled the decision, due to the inappropriate configuration of the parties in conflict. In short, 24% of the cases were in some way negative to the claim of protection and in 76% of the judgments, the justices found that the violation of the right to prior consultation existed, and ordered protection. According to these results, it is possible to affirm that once the Court decides to review a case, there is a high probability that it will rule in favour of protecting Indigenous Peoples' rights. See Figure 3.2.

Of equal importance is separating the macro and micro interventions to show the different scales of situations that peoples protected by ILO 169, the Inter American Convention and the national constitution are facing in Colombia (see Figures 3.3 and 3.4). Likewise, it is of importance to demonstrate the type of cases that have been selected and ruled on by the CCC. The dividing line between the categories comes from the scale or impact of the interventions.

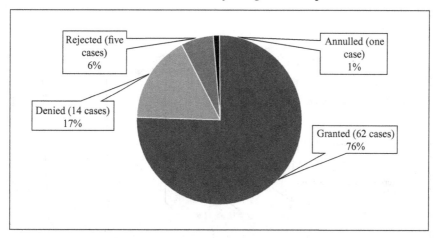

Figure 3.2 Concrete control or *"tutelas"* in the case law of the Constitutional Court of Colombia related to prior consultation of Indigenous Peoples in the last 25 years.
Source: Herrera (2019)

Infrastructure projects or interventions related to roads, dams, ports, and large-scale project interferences aroused the attention of the Court 23 times (see Figure 3.3). Extractive industries were the reason for the Court to intervene on 17 occasions. And five cases concerned the construction of a military base and towers or the use of *Acción de Tutela* to stop a legislative process. In total, 45 macro intervention cases were ruled upon.

Furthermore, the Court studied 37 micro interventions on issues related to territorial integrity and ruled on seven cases. It also reviewed small infrastructure

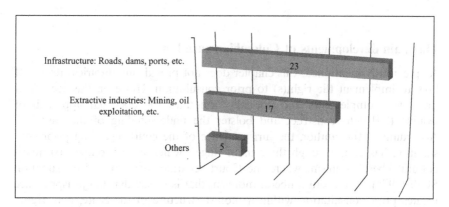

Figure 3.3 Macro interventions in concrete control cases in the case law of the Constitutional Court of Colombia related to prior consultation of Indigenous Peoples in the last 25 years.
Source: Herrera (2019)

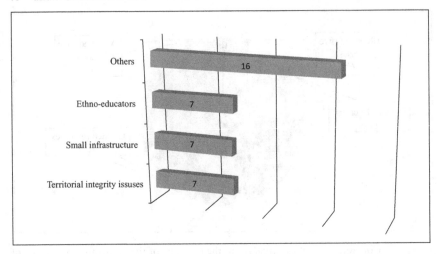

Figure 3.4 Micro interventions in concrete control cases in the case law of the Con-
        stitutional Court of Colombia related to prior consultation of Indigenous
        Peoples in the last 25 years.
Source: Herrera (2019)

constructions (see Figure 3.4) such as wastewater treatment works or a small
luxury hotel and spa on Providencia Island (seven similar cases). In addition are
another seven cases of ethnic-educators without the authorisation of the communities to teach in their territories, and other cases such as the relocation of an
informal salesperson on a private beach and a mobile radio station in Indigenous
territories (16 different cases). This division of macro and micro interventions
shows as well that the FPIC of the communities should be checked without
regard for the scale of the intervention, according to the arguments that follow.

The main developments of Colombian case law

As previously mentioned, this chapter does not provide an empirical review of
how to implement the right(s) to prior consultation. However, this section of
the study complements and critically annotates the doctrinal approach of
Bonilla (2013) to construct and bolster the understanding of the case law.
According to this author, the jurisprudence of the court regarding prior consultation has gone through three main stages, which are as follows: (i) multicultural liberal monism, where the Court separates consent and consultation
SU-039/97; (ii) procedural liberal monism, that is to say that the jurisprudence
locates prior consultation within a monist structure of the State, and argues
that FPIC is not a component of this right (for instance, C-030/08, C-615/09,
C-175/09, among others); and (iii) multicultural liberal pluralism in which cultural integrity and self-governance are related to the right to veto and this right

is linked to a pluralist interpretation *pro homine* (initiated by T-769/09 and extensively explained in case T-129/11) (Bonilla, 2013, p. 248).[14]

At the time of his book's publication in 2013, Bonilla had studied 23 cases. This chapter, as mentioned, takes into consideration 82 *Acciones de Tutela* decisions and 41 reviews of legislation, for a total 123 cases. However, Bonilla's categories may still be applied with the modifications that my analysis presents and clarifies on the matter. It includes SU-133/17, which confirmed and amalgamated the Concrete and Abstract Control decisions that the CCC has developed in 25 years of case law, and the protection of Indigenous Peoples and right to prior consultation and FPIC. This case confirmed the *ratio decidendi* of cases C-389/11 and Decision T-129/11, and the other decisions that developed this landmark precedent.[15] The main message behind the unification of this type of judgment is to clarify or correct previous, contradictory judgments.[16]

Why is Decision T-129/11 transformative?

The situation surrounding Decision T-129/11 involved a mixture of problems relating to the Indigenous Emberá Katío community located near the border with Panama. The community was facing the construction of the Pan-American Highway, gold mining, the installation of electrical towers, and the systematic appropriation of their territories by settlers.[17] The CCC studied the standards developed since 1992, and, in 2011, it introduced new ones settling an impressive judicial dialogue with the Inter-American Court in the case of *Saramaka v Suriname* of 2007.[18]

For Colombian jurisprudence, the prior consultation of Indigenous Peoples constitutes a fundamental right.[19] The standards developed point out that the consultation process should be studied by the authorities using a differential approach strategy. It should be in accordance with the specific traditions of the groups involved, including the timeframe of the consultation process and the particularities of each case. Moreover, this case determined the relevance of an environmental license and archaeological management plan before intervention in protected territories.

Most importantly, the Court introduced an elemental concept on the timing of prior consultation, which cannot be understood only at the previous stage, but also "before", "during", and "after" the intervention according to each concrete case.[20] According to the CCC "[...] the process is not limited only to the prior stage of the intervention in ethnic territories, [but] a reformulation is required to balance the aims of ILO 169 and subsequent developments on this matter. [...] [In other words] [a]ffected communities can make use of the possibility to review and put forward their views on the intervention, not only in the previous stage but during and after the implementation [of it] [...]" (Corte Constitucional de Colombia, 2011, p. 75). Additionally, three situations have been elaborated on, in which it is mandatory to pursue FPIC. This is when:

- it involves the removal or displacement of communities;
- it involves the storage or dumping of toxic waste;
- it represents a high social, cultural, and environmental impact on a community that may lead to endangering its existence and continuity.[21]

When an external party cannot obtain the approval of the community in the events described, or where the destruction or disappearance of groups is inferable, authorities and the communities involved must apply or claim the principle of *pro homine* interpretation in favour of the fundamental rights of the community involved. The *pro homine* principle "relativises the absolute protection of conflicting rights [and] thereby creates an open circumstance for striking an appropriate balance most favourable to persons in terms of their substance" (Negishi, 2017, pp. 479–480).

Is there a veto power?

Is it plausible to claim that, in the situations described, the *pro homine* interpretation is a kind of veto power? Bonilla (2013) considers that the Court declared that "[t]he right to consultation includes veto power for cultural minorities in certain circumstances", i.e., according to this author, the CCC "[i]ntroduces an element that contradicts the interpretation that the Court had articulated on prior consultation in the previous two stages [...]. If no agreement is reached, the minority would have the right to veto" (p. 243). Consistent with Bonilla's opinion, Decision T-129/11 forms the cornerstone of a new so-called "pluralistic" model of approach in contrast to the "monistic" interpretation employed in previous decisions, which is referred to in this chapter as "binding consent". Other scholars believe the opposite of Bonilla in relation to binding consent. For instance: "[u]nfortunately, the last ruling being revised (T-129/11) states that prior consultation is not a veto right, which contradicts the postulates of the decision" (Abello, 2012, p. 122). Another critical author has stated that:

> [i]t is true that consultation is a fundamental right; however, it does not constitute a right to veto [...] participation does not imply veto [...] [and] the right to participation could be confused in the popular imagination with a right to veto – and this can be exercised in bad faith or intended to block or delay.
>
> (Salinas Alvarado, 2011, p. 241)

According to Yriart (2015), "[t]he no-veto doctrine is out of place with the body of the law the Court develops otherwise on the subject" (p. 502). In contrast, Vallejo (2016) affirms that: "without there being a veto right, the decision [...] must take into account the considerations made by traditional communities during the consultation process" (p. 169).

It is clear that some academic interpreters of the jurisprudence have found that a veto power exists in some cases but not in others. As a former law clerk of the CCC under the supervision and orders of Justice Jorge I. Palacio, I had the opportunity to draft Decision T-129/11. During the drafting process, the issue of "veto power" was a relevant concern. Apart from this, the new standards were widely discussed. In fact, no passage of the decision states expressly that there is a veto power. What is more, in para. 7.1, the Court asked: "Is the right to Free, Prior and Informed Consent a veto power?" (Corte Constitucional de Colombia, 2011, para.7.1). The judgment answered the question by emphasising that the problem should not be put forward in terms of "who vetoes whom" (Corte Constitucional de Colombia, 2011, para.7).

As Padilla Rubiano (2013) properly points out, the CCC suggests looking beyond the veto and considers consultation as an exercise and experience of democratic cultural formation. However, "in cases of a negative response by the Indigenous Peoples to an initiative that they consider to be seriously harmful, what is exercised is the right to self-determination and not, some form of veto power" (pp. 364–365). This means that a proper interpretation of decision T-129/11 is harmonised with the right of Indigenous Peoples to decide what to prioritise in their own understanding of the "development" and "preservation" of values and territories.

Leaving interpretations of secondary sources aside and returning to the primary sources, the Grand Chamber of the CCC has expressed nine opinions in *dicta* with sparse elaboration that a veto power does not exist due to the fact that no right is absolute.[22] Out of these, only case C-641/12 specified "the impossibility of drawing a uniform rule in this regard" (Constitutional Court of Colombia, 2012, para. 4.5). Fortunately, C-389/16, the last decision of the Grand Chamber, and, in fact, the one that ended the period of the justices from the 2009–2017 period, clarified possible doubts about the issue.[23] According to the Court, the implementation of a measure that directly and intensely affects the fundamental rights of Indigenous Peoples is inadmissible without their consent but not as a consequence of the current discussion on the existence or not of a right to a veto. Indeed, in the words of the Court,

[t]he expression "veto" creates the impression of an arbitrary barrier, which does not require reasons to impose itself against other points of view and ways of action and, therefore, does not seem to respond adequately to the meaning that inspires consultation, conceived as a dialogue in good faith, among equals, and aimed to reach agreements that take into account the environmental, social and economic impacts of a measure, in an attempt to reconcile different conceptions of development.

(Constitutional Court of Colombia, 2016, p. 150)

Hence, Bonilla's doctrinal and philosophical reconstruction still plays a key role because the leading case T-129/11, complemented by T-376/12,[24] states clearly that the interpretation and application of the Constitution must be decided according to *pro homine* and proportionality principles. In some cases, conceivable consequences may include either halting the intervention process or not. Therefore, it is important to apply these principles and standards to the specific components of each case. For example, learning from the solution of Decision T-129/11, if it is possible to relocate a road without harming the community, there is no way of considering binding consent. If the impact of a specific mining project affects the community in terms of displacement, toxic waste and/or other high impacts, or it might lead to endangering its existence and continuity, there may be room for binding consent.[25]

Indeed, binding consent is less problematic than the idea of "veto power". I understand this category as going beyond a mere expression of opinion made by the peoples concerned, but at the same time it is different from a knockout argument to end a dialogic process, such as a veto. Rather than presenting a mere yes-or-no alternative, binding consent is open to multiple options for each concrete case.

Why are the outcomes of the case law at risk?

Twice the CCC has exhorted Congress and the Presidency of the Republic to exercise their constitutional and legal powers to regulate and, through their competent bodies, materialise the fundamental right of prior consultation, taking into account its jurisprudence.[26] The answer from the executive and legislative powers is the current draft law or "Statutory Act" of 2016 and 2018[27] that may regulate the fundamental right of prior consultation and determine the development of several international instruments and CCC case law.

However, in articles 13.c, 14.e, and 22 of the draft law of 2016 and article 2.9 of the draft law of 2018, it is stated that prior consultation does not entail a veto power of the legislative or administrative measures under consultation, and the duration of the process cannot be over six months. Deciding whether the *sword* of veto power is in the hand of Colombian ethnic groups or not is a complex issue, and efforts to find a balanced solution will take time. Future regulation and the cases before the Court will have to be reviewed by a completely new body. In 2017, the CCC changed four justices: Calle, Palacio, Pretelt, and Vargas, who had openly protected the right to prior consultation of Indigenous Peoples. Four current justices, Guerrero,[28] Linares,[29] Lizarazo,[30] and Ortiz,[31] differ in varying restrictive degrees with regard to the criteria of the transformative leading cases.

If the Court, in the years to come, does not improve the above-mentioned transformative precedents, at least it should follow them under the principle of progressive realisation and non-regression. However, the pressure of the media and other actors is evident (Redacción Semana, 2016). The Colombian government has shown a restrictive approach, breaking the consensus of the American Declaration on the Rights of Indigenous Peoples at the regional level, pointing out at the national level that prior consultation of Indigenous Peoples is "a headache" (El Espectador, 2013) or an "extortion mechanism" (El Tiempo, 2016). Taking the current constitutional and jurisprudential level of protection in Colombia into consideration, it is possible to conclude that the tribunal has shown a strong commitment to the protection of Indigenous Peoples.

Nevertheless, most of the work falls on the shoulders of the CCC, and protection applied through this institution should be the *ultima ratio* or last resort scenario. Innovative solutions in the coming years should consider prevention and avoid what Gargarella (2013) denominates the "problem of translation" or the tendency to "simplify what is normally too complex; it represents an attempt to solve problems that are mainly non-juridical through juridical means" (p. 181).

Conclusion: towards binding consent?

The past two decades have seen the rapid development of prior consultation and FPIC in the case law of the Colombian Constitutional Court. This tribunal has empowered the voice and self-government of Indigenous Peoples in several forms. National and regional authorities should consider these constitutional precedents an archetypical example of transformative constitutionalism in Latin America.

In order to materialise FPIC in the cases studied, the CCC proposes applying the *pro homine* principle in particular matters and re-conceptualising the notion of prior for the "before", "during" and "after" stages. This mechanism foresees the participation of the peoples concerned during all the stages of the intervention affecting their territories. In addition, the case law of the Court points out the reasons why it is impossible to set a specific term in advance for the consultation process, and why consultation should be applied in all kind of micro or macro interventions.

Questions have been raised about the relevance of binding consent instead of veto in the exclusive hands of any actor involved in the conflict. The tension between the rights and interests of Indigenous Peoples and the rights and interests of the rest of the population is already a polarised matter. The role of the judges and actors intervening in prior consultation is to take the demands of the "minorities" and the majorities seriously, to diminish the tension and not to intensify it. Consequently, public and private actors should consider the observations that have been explained and carefully constructed by the Colombian Constitutional Court over a period of 25 years.

Notes

1 This article is a revised version of Herrera (2019).
2 About the origin and meaning of the category "FPIC", see Hanna and Vanclay (2013).
3 On this, see also Bogdandy, Ferrer Mac-Gregor, Morales Antoniazzi, & Piovesan (2017) and – about the origins of Transformative Constitutionalism – Klare (1998).

4 According to the constitutional case law, the right to consultation applies not only to Indigenous Peoples but also to Afro-Descendants (incl., the *Raizal*) and the Roma (Constitutional Court of Colombia, 2011, p. 65). This chapter focuses on Indigenous Peoples only, but all the considerations drawn here may be equally applied to all the above-mentioned groups.

5 The total population of Colombia by 2005 was 42,888,592; the 2017 projection was around 49,000,000. Indigenous peoples represent approximately 3.43% of the total population; 10.62% are Afro-Colombians; 0.01% are Roma; and 85.94% are classified as non-ethnic (Departamento Administrativo Nacional de Estadística, 2007).

6 Around 50% of Latin American Indigenous Peoples have moved to urban centres. See, World Bank (2015).

7 For instance, the case of the *Nukak Maku* community. More than 50% of the population have been displaced from their traditional nomadic territories due to the armed conflict or to the action of settlers. The population is estimated at about 450–550 survivors (Stavenhagen, 2013, pp. 151–152).

8 All the references to the Colombian Constitution and case law in general are translated by the author. References to the case law of the Constitutional Court of Colombia (or CCC), where not further specified, will be identified under the style used for the tribunal (e.g. T-129/11) and can be accessed at: www.corteconstitucional.gov.co.

9 On the Peruvian context, see also Wright (2014) and the chapters by Doyle and Flemmer in this volume.

10 To check 25 years of case law, I have applied the "dynamic analysis of precedents" method. See López (2011).

11 Some specific cases are explained in Section 4.4.3 of Decision C-196/12. For a summary in English, see Cepeda & Landau (2017, pp. 264–270).

12 The decisions are: C-169/01, C-418/02, C-891/02, C-620/03, C-245/04, C-208/07, C-921/07, C-030/08 (leading case), C-461/08, C-750/08, C-175/09, C-615/09, C-063/10, C-608/10, C-702/10, C-915/10, C-941/10, C-027/11, C-187/11, C-490–11, C-196/12, C-317/12, C-318/12, C-366/11, C-882/11, C-937/11, C-051/12, C-293/12, C-331/12, C-398/12, C-395/12, C-540/12, C-641/12, C-765/12, C-767/12, C-822/12, C-943/12, C-068/13, C-194/13, C-253/13, C-371/14, C-501/14, and C-389/16.

13 The decisions are: T-428/92, T-405/93, SU-039/97, T-652/98, T-634/99, SU-383/03, T-955/03, T-737/05, T-382/06, T-880/06, T-154/09, T-769/09, T-547/10, T-745/10, T-1045A/10, T-116/11, T-129/11 (leading case), T-235/11, T-379/11, T-474/11, T-601/11, T-693/11, T-698/11, T-376/12, T-477/12, T-513/12, T-514/12, T-680/12, T-693/12, T-823/12, T-993/12, T-1080/12, T-049/13, T-245/13, T-172/13, T-300/13, T-390/13, T-657/13, T-795/13, T-858/13, T-871/13, T-117/14, T-204/14, T-294/14, T-353/14, T-355/14, T-384A/14, T-396/14, T-461/14, T-462A/14, T-576/14, T-646/14, T-800/14, T-849/14, T-857/14, T-969/14, T-247/15, T-256/15, T-359/15, T-438/15, T-485/15, T-550/15, T-597/15, T-660/15, T-661/15, T-764/15, T-766/15, T-005/16, T-041/16, T-110/16, T-197/16, T-213/16, T-226/16, T-288A/16, T-313/16, T-436/16, T-475/16, T-530/16, T-605/16, T-704/16, T-730/16 and SU-133/17.

14 44 of 64 (Concrete Control) and 6 of 23 (Abstract Control) decisions of the CCC issued after the publication of T-129/11 have quoted or applied the standards derived from that landmark precedent until the end of 2016. These are: T-601/11, T-693/11, T-698/11, C-882/11, T-376/12, C-395/12, T-513/12, T-514/12, C-641/12, T-680/12, T-693/12, T-993/12, T-1080/12, T-049/13, C-068/13, T-245/13, T-172/13, T-300/13, T-657/13, T-858/13, T-204/14, T-294/14, T-353/14, C-371/14, T-384A/14, T-396/14, T-461/14, T-462A-14, T-646/14, T-800/14, T-849/14, T-969/14, T-247/15, T-256/15, T-438/15, T-485/15, T-550/15, T-660/15, T-661/15, T-766/15, T-005/16, T-197/16, C-389/16, T-213/16, T-226/16, T-288A/16, T-436/16, T-475/16, T-530/16, and T-704/16.

15 In terms of the "dynamic analysis of precedents" method, the above-mentioned decisions are the ones that consolidate the precedents (López, 2011, 161).

16 The "SU" (*Sentencia de Unificación*) category means unification judgment that is extraordinarily discussed in the Grand Chamber.
17 For more details about this case, see Cepeda and Landau (2017, pp. 268–270).
18 See IACtHR (2007).
19 For this reason, the regulation must undergo treatment by the special law "Ley Estatutaria" or Statutory Act.
20 "Prior" literally means previous. Nevertheless, prior has more meanings than an indication of an event or situation "coming before in time". It also means "important" and "existing" (Oxford Dictionaries, 2016, p. 1400).
21 The Court embraced the recommendations made by the second UN Special Rapporteur on the Rights of Indigenous Peoples, James Anaya, made on 26 May 2011, after the promulgation of the prior consultation law in Peru.
22 See, for instance, the CCC Decisions C-882/11, C-366/11, C-367/11, C-937/11, C-331/12, C-540/12, C-068/13, C-253/13 and C-371/14.
23 *Inter alia*, these criteria can also be found in the CCC Decisions T-530/11 and T-704/16.
24 The CCC Decision T-376/12 makes a remark in the leading case T-129/11 applying the principle of proportionality to clarify and balance the participation of the communities in the process of consultation. See considerations 25–31 of the T-376/12 judgment. This decision also applied a remarkably progressive interpretation of the United Nations Declaration on the Rights of Indigenous Peoples (UNDRIP).
25 Debates in relation to the veto power are also being addressed in international fora as well as in other jurisdictions. See also Doyle (2015, pp. 161–167).
26 See the resolution part of CCC Decision T-129/11 and C-317/12.
27 These draft laws are, respectively, the Proyecto de Ley Estatutaria de 26 de octubre de 2016 del Ministerio del Interior (2016), y Proyecto de Ley Estatutaria de 11 de septiembre de 2018 currently under discussion at the Senate as draft law No. 134/18 (Senado, 2018).
28 CCC Decision SU-133/17.
29 Dissenting opinion of CCC Decisions C-389/16 and SU-133/17.
30 Dissenting opinion of CCC Decision SU-133/17.
31 CCC Decision T-313/06.

References

Anaya, J. (2004). International Human Rights and Indigenous Peoples: The Move toward the Multicultural State. *Arizona Journal of International and Comparative Law*, 21, 13–61.

Abello, C. (2012). Consulta previa en casos de minería para comunidades indígenas y tribales. *Trans-pasando Fronteras* (2), 111–124.

Bonilla, D. (2013). *Constitutionalism of the Global South: The Activist Tribunals of India, South Africa, and Colombia*. New York: Cambridge University Press.

Cepeda, M. & Landau, D. (2017). *Colombian Constitutional Law: Leading Cases*. New York: Oxford University Press.

Corte Constitucional de Colombia. (2011). Decision No. T-129/11, 3 March 2011. Retrieved from: http://www.corteconstitucional.gov.co/sentencias/2011/T-129-11.rtf.

Corte Constitucional de Colombia. (2012). Decision No. C-641/12, 22 August 2012. Retrieved from: http://www.corteconstitucional.gov.co/relatoria/2012/C-641-12.htm.

Corte Constitucional de Colombia. (2013). Decision No. C-253/13, 25 April 2013. Retrieved from: http://www.corteconstitucional.gov.co/relatoria/2013/c%2D253%2D13.htm.

Constitutional Court of Colombia. (2016). Decision No. C-389/16, 27 July 2016. Retrieved from: http://www.corteconstitucional.gov.co/sentencias/2016/C-389-16.rtf.

Departamento Administrativo Nacional de Estadística. (2007). *Colombia, una Nación multicultural, su diversidad étnica.* Bogotá: Departamento Administrativo Nacional de Estadística.

Doyle, C.M. (2015). *Indigenous Peoples, Title to Territory, Rights and Resources: The Transformative Role of Free, Prior and Informed Consent.* London: Routledge.

El Espectador. (2013). Santos dice que consultas previas y audiencias públicas "son un dolor de cabeza". *El Espectador,* 16 August. Retrieved from: https://www.elespectador.com/noticias/politica/santos-dice-consultas-previas-y-audiencias-publicas-son-articulo-440645.

El Espectador. (2016). La satanización de la consulta previa. *El Espectador,* 25 February. Retrieved from: http://www.elespectador.com/opinion/editorial/satanizacion-de-consulta-previa-articulo-618787.

El Tiempo. (2016). Consulta previa se volvió un mecanismo extorsivo. *El Tiempo,* 24 April. Retrieved from: http://www.eltiempo.com/archivo/documento/CMS-16572247.

Gargarella, R. (2013). *Latin American Constitutionalism, 1810–2010: The Engine Room of the Constitution.* Oxford: Oxford University Press.

Gargarella, R. (2014). "We the People" Outside of the Constitution: The Dialogic Model of Constitutionalism and the System of Checks and Balances. *Current Legal Problems,* 67(1), 1–47.

Góngora-Mera, M. (2017). The Block of Constitutionality as the Doctrinal Pivot of a Ius Commune. In A. Von Bogdandy, E. Ferrer Mac-Gregor, M. Morales Antoniazzi & F. Piovesan (Eds.). *Transformative Constitutionalism in Latin America: The Emergence of a New Ius Commune* (pp. 235–253). Oxford: Oxford University Press.

Hanna, P. & Vanclay, F. (2013). Human Rights, Indigenous Peoples and the Concept of Free, Prior and Informed Consent. *Impact Assessment and Project Appraisal,* 31(2), 146–157.

Herrera, J.C. (2019). Judicial Dialogue and Transformative Constitutionalism in Latin America: the case of Indigenous Peoples and Afro-descendants. *Revista Derecho del Estado,* 43.

IACtHR. (2007). *Saramaka People v Suriname,* Judgment of November 28, 2007 (Preliminary Objections, Merits, Reparations, and Costs). Inter-Am. Ct. H.R., (Ser. C) No.172(2007).

Klare, L. (1998). Legal Culture and Transformative Constitutionalism. *South African Journal of Human Rights,* 14(1), 146–188.

Landau, D. (2005). The Two Discourses in Colombian Constitutional Jurisprudence: A New Approach to Modeling Judicial Behavior in Latin America. *George Washington International Law Review,* 37(3), 687–744.

López, D. (2011). *El derecho de los jueces.* Bogotá: Legis.

Ministerio del Interior. (2016). Proyecto de Ley Estatutaria de 26 de octubre de 2016. Retrieved from: https://www.mininterior.gov.co/sites/default/files/noticias/proyecto_de_ley.pdf.

Negishi, Y. (2017). The Pro Homine Principle's Role in Regulating the Relationship between Conventionality Control and Constitutionality Control. *European Journal of International Law,* 28(2), 457–481.

Oxford Dictionaries. (2006). *Oxford Dictionary of English.* London: Oxford University Press.

Padilla Rubiano, G. (2013). Consulta previa en Colombia y sus desarrollos jurisprudenciales. Una lectura desde los pueblos indígenas, las empresas y el Estado. In C. Steiner (Ed.). *Anuario de Derecho Constitucional Latinoamericano 19* (pp. 333–352). Bogotá: Universidad del Rosario/Konrad Adenauer-Stiftung.

Redacción Semana. (2016). La Corte Constitucional versus los empresarios. *Semana*, 15 October. Retrieved from: https://www.semana.com/nacion/articulo/fallos-del-la -corte-constitucional-impactan-el-desarrollo-economico-en-las-regiones/499115.

Relatoría de la Corte Constitucional de Colombia. (n.d.). Retrieved from: http://www. corteconstitucional.gov.co/relatoria.

Salinas Alvarado, C.E. (2011). La consulta previa como requisito obligatorio dentro de trámites administrativos cuyo contenido pueda afectar en forma directa a comunidades indígenas y tribales en Colombia. *Revista Derecho del Estado*, 27, 235–259.

Senado. (2018). Proyecto de Ley Estatutaria de 11 de septiembre de 2018. Draft law No. 134/18. Retrieved from: http://leyes.senado.gov.co/proyectos/images/docum entos/Textos%20Radicados/proyectos%20de%20ley/2018%20-%202019/PL% 20134-18%20Consulta%20Previa.pdf.

Schilling-Vacaflor, A. & Flemmer, R. (2015). Conflict Transformation through Prior Consultation? Lessons from Peru. *Journal of Latin American Studies*, 47(4), 811–839.

Stavenhagen, R. (2013). *Peasants, Culture and Indigenous Peoples: Critical Issues.* Berlin: Springer-Colegio de México.

Uprimny, R. (2006). The Enforcement of Social Rights by the Colombian Constitutional Court: Cases and Debates. In R. Gargarella & T. Roux (Eds.). *Courts and Social Transformation in New Democracies: An Institutional Voice for the Poor?* (pp. 127–152). Aldershot: Ashgate.

Vallejo Trujillo, F. (2016). Prior Consultation Process in the Judgments of the Constitutional Court of Colombia. *Revista de Estudios Constitucionales*, 14(2), 143–182.

Von Bogdandy, A., Ferrer Mac-Gregor, E., Morales Antoniazzi, M., & Piovesan, F. (2017). *Transformative Constitutionalism in Latin America: The Emergence of a New Ius Commune.* Oxford: Oxford University Press.

World Bank. (2015). *Indigenous Latin America in the Twenty-First Century: The First Decade.* Washington, DC: World Bank Group. Retrieved from: https://openknow ledge.worldbank.org/bitstream/handle/10986/23751/Indigenous0Lat0y000the0first0d ecade.pdf;sequence=1.

Wright, C. (2014). Indigenous Mobilisation and the Law of Consultation in Peru: A Boomerang Pattern? *The International Indigenous Policy Journal*, 5(4), 1–16.

Yriart, M. (2015). Jurisprudence in a Political Vortex: The Right of Indigenous Peoples to Give or Withhold Consent to Investment and Development Projects – The Implementation of *Saramaka v Suriname*. In Y. Haeck, O. Ruiz-Chiriboga, & C. Burbano (Eds.). *The Inter-American Court of Human Rights: Theory and Practice, Present and Future* (pp. 477–546). Cambridge: Intersentia.

4 Indigenous Peoples' experiences of resistance, participation, and autonomy

Consultation and free, prior and informed consent in Peru

Cathal M. Doyle

Introduction

The State duty to consult in order to obtain Indigenous Peoples' Free, Prior and Informed Consent (FPIC) has a pivotal role in establishing and developing respectful rights-based relationships between States and Indigenous Peoples and in facilitating Indigenous Peoples' self-determined development. Consultation and participation have been described as the "cornerstone" of the ILO Convention No.169 Concerning Indigenous and Tribal Peoples in Independent Countries of 1989 (ILO 169) (ILO, 2013, p. 11), and significant attention has been directed towards consultation's procedural dimensions by the ILO Supervisory bodies over the last three decades. In the decade since the UN Declaration on the Rights of Indigenous Peoples (UNDRIP) was adopted in 2007, international and regional human rights bodies have increasingly addressed the substantive dimension of the duty, invoking the self-determination-based requirement to obtain Indigenous Peoples' FPIC whenever potentially significant impacts on their rights are foreseeable in the context of natural resource exploitation projects (Doyle, 2015).

As in most other Latin American countries, Indigenous Peoples in Peru have suffered long histories of discrimination, oppression, and cultural and physical decimation. Even well-intentioned, but ultimately discriminatory, ILO programmes in the 1950s and subsequent agrarian reforms initiatives in the 1960s resulted in their assimilation, the denial of customary land rights for some and the revocation of formal status as Indigenous for others (ILO, 1988a, pp. 48–9). Peru participated in developing ILO 169 in the late 1980s, ratifying it and enacting further Constitutional reforms affording recognition to Indigenous Peoples' rights in 1994 and 1993, respectively. It also engaged in the drafting, and promoted the adoption, of the UNDRIP in 2007.

Despite these important actions, Peru waited until 2011 before taking meaningful steps towards implementing its obligations regarding consultation. It did so through a law on "prior consultation" (Consultation Law, 2011). However, certain provisions of the law and its implementing rules, and the process leading to their adoption, expose serious deficiencies in the State's conception of consultations and FPIC. The effect has been to render the consultations inconsistent with their object and purpose of safeguarding Indigenous Peoples' rights.

DOI: 10.4324/9781351042109-6

Other Latin American countries, such as Colombia, Honduras, Mexico, and Paraguay are considering enacting legislation to regulate consultations and are looking to the Peruvian experience for guidance. Given the significance of this experience, the chapter explores how the requirement for consultation and consent has evolved in the Peruvian case, both prior and subsequent to the enactment of the law on consultation. It examines the role the judiciary has played in ensuring consultations are regulated in accordance with international human rights law (IHRL) standards, and the range of positions adopted by Indigenous Peoples in Peru when faced with the State's failure to implement its consultation obligations.

Constitutional context, ILO 169 and the UNDRIP

The 1993 Peruvian Constitution was adopted one year prior to the country's ratification of ILO 169. It represented a significant step forward in terms of Indigenous Peoples' rights by affording protection to Indigenous communities' land rights and their cultural diversity and recognising the role of their customary laws. Unlike its Colombian neighbour's 1991 Constitution, it did not explicitly recognise Indigenous Peoples as "peoples" with rights to autonomy or rights in relation to natural resources, and made no reference to participation or consultation in the context of natural resource exploitation. This more restrictive and conservative approach to Indigenous rights was also evident in Peru's position during discussions on the draft provisions of ILO 169, when contrasted with that of Colombia. It was most clearly manifested in Peru's decision to abstain in the vote for the final text of the Convention based on the possible veto power in article 15(2) (ILO, 1989b, p. 32/12). According to Peru, that provision was potentially incompatible with its then (1979) Constitution, under which measures deemed to be in the public necessity or utility could overwrite the inalienability and imprescriptibly of Indigenous Peoples' lands (ILO, 1989b, p. 25/22, para. 156). Peru also expressed concern about the potential for the term "peoples" to imply a right to self-determination (ILO, 1989b, p. 25/7 para. 36). Peru did, however: highlight the importance of social and environmental impact assessments and full participation of Indigenous Peoples in development planning (ILO 1988b); insist that "[r]ecourse to force or coercion as a means of promoting integration, development, the improvement of living conditions or any other similar end should be excluded" (ILO, 1988a, p. 23); point to the "need to preserve and safeguard wherever possible the right of national or ethnic sovereignty of the indigenous people over their ancestral territories" (ILO, 1988b, p. 47) and their "right to maintain their territorial base with full sovereignty" (ILO, 1989a, p. 10); and stress that "[p]rovisions should be included which establish the obligation of governments not only to recognise [Indigenous Peoples' representative] organisations but also to encourage their formation and contribute to strengthening them" (ILO, 1988b, p. 102).

By 2006, when it came to the vote on the UNDRIP at the Human Rights Council, and subsequently at the General Assembly in 2007, the tables appeared to have turned. Unlike Colombia, which abstained from voting on the Declaration because of its FPIC provision, Peru actively promoted the Declaration's adoption, and ensured that its consultation, FPIC, and self-determined development provisions remained relatively unscathed in the final negotiations. (UNGA, 2007).

Implementation of the right to consultation in Peru

Despite its statements and position at the international level, between 1994 and 2007 at the national level the Peruvian State remained largely disengaged on the issue of consultation and indeed on Indigenous Peoples' rights in general. Consultation and consent were partially addressed in certain sector-specific laws, such as the National Protected Areas (NPA) Law of 1997 and its 2001 regulation. This enabled some Indigenous Peoples, such as the Awajun People in the context of the creation of a National Park, to engage the Peruvian State in consultations (ODECOFROC, 2009). However, implementation of this piecemeal approach to consultation was flawed, with the trust of groups such as the Awajun subsequently betrayed, and failures to adequately consult in areas such as the Parque Ichigkat Muja in the northern Amazon leading to large scale protests and conflict, most notably in Bagua in 2009 (Zúñiga & Okamoto, 2018).

The Peruvian Constitutional Court and consultation of Indigenous Peoples

In 2007, the Peruvian Constitutional Tribunal (henceforth, Constitutional Court or the Court) first engaged with the issue of consultation in the context of a challenge to oil exploitation in the Cordillera Escalera conservation area (PCT, 2007). The Constitutional Court affirmed that ILO 169 formed part of Peruvian law, having erroneously concluded that this was not the case in a 2005 decision (PCT, 2005). In a series of paragraphs addressing both the right to cultural identity and how the Convention regulated natural resource exploitation, the Court quoted the Inter-American Court of Human Rights (IACtHR) reasoning in *Saramaka v Suriname* (IACtHR, 2007) that consent, in addition to consultation, was required in the context of large-scale extractive projects (PCT, 2007, paras. 30–40). However, while noting the absence of any regulation pertaining to consultations in Peru, the Constitutional Court concluded that addressing this was the responsibility of the legislature, and did not issue any findings in relation to the implementation of the right (PCT, 2007, para. 40).

On 9 June 2010, the Constitutional Court issued its first significant decision addressing the right to consultation following a challenge made by Gonzalo Tuanama to a decree providing a temporary regime for rural land titling (PCT, 2009a).[1] The Court clarified that the failure to realise fundamental rights, such as the right to consultation, could not be justified due to the absence of domestic regulation and outlined a series of principles underpinning good faith consultations, which it framed as intercultural dialogues. This procedural aspect of the ruling was aligned with recommendations of the ILO supervisory bodies and represented a significant development. The Constitutional Court's engagement with the substantive dimension of the right, namely its consent component, was less satisfactory. The ruling stated that if, following good faith consultation, no consensus was reached, the State could implement its decision provided it addressed the demands of the concerned peoples to the maximum extent possible. The absence of any reference to the principle of proportionality and the related requirement for FPIC was not in keeping with developments in IHRL, in particular the IACtHR *Saramaka v Suriname* ruling, the jurisprudence of the Human Rights Committee in *Ángela Poma Poma v Peru* (HRC, 2009), and UNDRIP articles 3, 19 and 32. This divergence also emerged in the Constitutional Court's view that Indigenous Peoples' participation in consultation was an obligation, rather than a right which they could choose to exercise or not.

On 11 November 2009, the Constitutional Court controversially dismissed a writ of *Amparo* filed by the Amazonian Indigenous Peoples' organisation, AIDESEP, requesting the suspension of the oil exploitation in blocks 39 and 67, which AIDESEP alleged posed serious risks to Indigenous Peoples in voluntary isolation (PCT, 2009b).[2] AIDESEP also alleged that it and other institutions representative of Indigenous Peoples should have been consulted. Faced with conflicting and inconclusive evidence as to the presence of groups in voluntary isolation in the concession area, the Court deemed a writ of *Amparo* to be inappropriate to protect those groups. It also rejected the position of the lower courts that, in contexts involving Indigenous Peoples in isolation, the right to consultation was evidently not relevant. Instead, the Constitutional Court held that neighbouring communities should be consulted in order to guarantee their right to participation in decisions impacting on them. It noted that the right to self-determination of Indigenous Peoples, albeit delimited by the Constitution and the principle of territorial integrity, underpinned their right to consultation. While summarising the characteristics of consultations outlined in its 2009 decision in the Tuanama case, the Constitutional Court went on to affirm that the right to consultation should be implemented gradually by the concerned companies with oversight from the competent State authorities. This deviation from the principle of "prior" consultations was justified on the basis that the companies involved had acted in good faith and followed the guidance issued by the government and therefore should not have their investments paralysed, and that AIDESEP had not presented conclusive evidence of imminent threats to the rights of Indigenous Peoples (PCT, 2009b).

On 23 August 2010, the Constitutional Court found against the Ministry of Energy and Mines for failing to regulate the right to consultation, leading to the issuance of Supreme Decree No. 023–2011-EM in May 2011. This decree was in turn criticised for restricting the application of articles 6 and 15 of ILO 169, for failing to involve Indigenous Peoples in determining what norms should be consulted, and for non-adherence with the principles of flexibility, good faith and transparency (CNDDHH, 2013). In 2011, the Constitutional Court, correcting its view to the contrary in 2010 (PCT, 2010), clarified that the Constitutional right to consultation had existed since ILO 169 entered into force in 1995 as it formed part of the Constitutional block (PCT, 2011).

This body of jurisprudence outlined important principles embodied in the right to consultation. In general, however, it failed to address the substantive implications of respect for Indigenous Peoples' rights in terms of when consent is a required outcome of consultations. It did, however, demonstrate a degree of willingness of the judiciary to move beyond the conservative, somewhat reactive and inconsistent nature of its initial engagement with the Indigenous rights framework (Córdova Flores, 2013).

Adoption of the Peruvian Consultation Law

The increased focus on the right to consultation in Peru emerged in a context of a rapid acceleration in the issuance of mining and oil and gas concessions in or near Indigenous territories in the early 2000s. A series of decrees issued without consultation in 2008 further facilitated access to Amazonian Indigenous Peoples' territories, purportedly to meet the exigencies of the investment provision of the US-Peru FTA. Proposals by the concerned peoples were ignored or met with contempt.[3] Large scale protests ensued. The government of Alan Garcia responded in July 2009 by declaring a state of emergency and initiating a military-style operation aimed at breaking road blockades near the Amazonian town of Bagua. The ensuing confrontation left 33 dead and around 200 injured (Andrada & Gonza, 2010).

Following this tragedy and subsequent interventions by UN and ILO human rights monitoring and supervisory bodies (UNSR, 2009; CERD, 2009; ILO, 2009), the Peruvian Executive established a national coordination group on the development of Indigenous Peoples of the Amazon. This coordination group in turn established a roundtable mandated to develop a draft Consultation Law (2011) with the participation of Amazonian Indigenous Peoples (Schilling-Vacaflor & Flemmer, 2013). On 19 May 2010, the Congress approved the draft. However, the Executive returned it for modification, claiming that it granted Indigenous Peoples a veto power. This led to the introduction in its article 15 of a qualification that the government will make the final decision in cases where consent is not forthcoming. The law entered into force on 7 September 2011.

A consultation process ensued involving the six national Indigenous Peoples' organisations in order to develop the law's implementing regulations. Four of these organisations formed a Unity Pact and established a series of "minimal non-negotiable principles" (Unity Pact, 2011). Central to these was the obligation to obtain consent for large-scale extractive and energy projects in accordance with IHRL.[4] The government's inflexibility to interpret the law along those lines, together with concerns about how it was conducting the consultations on the regulation, led the Unity Pact organisations to withdraw from the process (Wright, 2014). The unfortunate outcome was that the final implementing regulations impose additional restrictions on the right to consultation and compound the law's limitations on the requirement for consent (IACmHR, 2015, pp. 97–8; MC, 2012; Ruiz Molleda, 2011; Schilling-Vacaflor & Flemmer, 2013).

The irony of a failed consultation on the regulation of a law on consultation did not bode well for the law's implementation. Regulations issued by the ministries responsible for coordinating consultations, such as the Ministry for Energy and Mines, further restrict the right. In the case of the mining sector, consultations are limited to the last steps in the project authorisation process, after all decisions pertaining to assessing, mitigating, and preventing potential project impacts have been made (Hiruelas, 2018; Leyva, 2017). In the hydrocarbon sector, consultations relate to the decree approving a concession contract, and not to the contract's content (Hiruelas, 2018). In both sectors, consultations do not afford Indigenous Peoples an opportunity to influence project design or implementation. They therefore have no say on how projects impact on their rights and well-being.

Peru continues to present the law and its implementation as exemplary to the international community. Given this and the aforementioned issues, it is important to probe the actual practice of the Peruvian State. A brief consideration of three emblematic cases – involving Indigenous Peoples' participation in consultation processes, their resistance where no consultations were held, and their proactive assertion of the self-governance rights and use of judicial avenues to affirm their right to consultation and consent – exposes some of the significant ongoing deficiencies in the law's implementation and interpretation.[5] These experiences also serve to highlight important developments in recent court decisions in relation to the requirement for consent and the centrality of Indigenous Peoples' mobilisation, resistance and self-directed organisation to the realisation of their autonomy and decision-making rights.

Experiences of Peruvian Indigenous Peoples with Consultation

Indigenous Peoples' participation in consultation – the Block 192 Experience

In 1971, Peru issued the country's largest oil concession known as Block 1AB (now called Block 192). The project caused extensive contamination, profoundly impacting on the lives of the Achuar, Kichwa, and Quechua peoples (Doyle & Campanario Baqué, 2017; Raynal, 2015).[6] The affected Amazonian Indigenous Peoples were never consulted, despite the significant changes made to the concession following the ratification of ILO 169 – including changes to the block's territorial coverage, to associated easement rights and of concession operators – and the profound historical and ongoing impacts on their territories, environment, social structures, and health.[7] To assert their rights, the Achuar, Kiwcha, and Quechua communities – some commencing in the 1990s and others in the mid-2000s – organised themselves into Federations and held large scale mobilisations. They were frequently left with no option but to shut down oil operations in order to have their concerns addressed. On occasions this was met by police force, leading to deaths and criminal cases against community representatives (Doyle & Campanario Baqué, 2017).

Their advocacy eventually resulted in legislative and administrative reforms, congressional investigations into the contamination, and, in 2013, declarations by the State of environmental and health emergencies. A series of agreements was reached with the government in March 2015. These agreements included commitments to initiate studies on remediation and health impacts and to complete land titling. Importantly, the government also agreed to hold consultations on the content of a new contract to extend the project for an additional 30 years, as opposed to merely consulting on the decree authorising the contract as required under the Consultation Law's flawed implementing regulations. This represented a significant achievement. It meant that rights-protecting provisions could be agreed with Indigenous Peoples and included in the contract, providing the peoples with a degree of juridical security. In their formal agreement to participate in consultations, the Federations identified a series of core issues to be addressed. These were endorsed by the UN Special Rapporteur (UNSR, 2014, para. 36) and included land titling, remediation, and contamination avoidance, participation in environmental assessments, independent and effective monitoring, fair and equitable benefit-sharing, and community development.

The consultation process ran from 21 May to 18 August 2015. A series of delays by the government agencies responsible for the process, exacerbated by the absence of government officials with the necessary decision-making powers in the initial meetings, and the State's inflexibility to extend the consultation timeframe – to avoid delaying the contract's initiation – meant that these core issues were inadequately addressed (Leyva, 2015). Trust was further eroded by the government's inclusion of a newly formed non-governmental organisation, consisting of three breakaway communities in the final "negotiation" meeting, thereby entrenching divisions within and between communities and their Federations and weakening their negotiating power. On the last day of negotiations, the government presented a take-it-or-leave-it offer that only addressed benefit-sharing (Inguil, 2018). The Federations were provided with a couple of hours in which to decide if they would accept the offer. Most communities, represented by three of the four Federations, refused the offer, but remained open to continued consultations. They also insisted that their decision-making processes required that they return to their communities to discuss any offer. Some communities, represented by one Federation and the newly created organisation, accepted the offer, believing there to be no alternative. Based on this partial acceptance, the benefit-sharing arrangement was included in the contract and the Ministry for Culture and the Ministry of Energy declared the consultations successfully concluded (MC, 2015; MEM, 2015).

According to the government, it was under no obligation to reach an agreement with the majority of the communities and their Federations. In response, the Federations pointed to the lack of good faith on the part of the government, noting that, in addition to the aforementioned irregularities, the government's decision to unilaterally terminate consultations when the dialogue stage had only commenced, without making any effort to obtain their consent or agreement, breached the Consultation Law (2011) and its implementing regulations (Leyva, 2015). The communities were again forced to occupy the company facilities under constant threat of military intervention. On 5 November 2015, agreements were reached addressing some of their concerns but, being outside the consultation process, these agreements are not protected by a contract and impose no obligations on the concession operator.

Falling oil prices forced the government to enter into a short-term service contract (without consulting the concerned Indigenous Peoples) in place of the intended 30-year contract which it will seek to renegotiate (Zúñiga, Barclay & Campanario, 2017). The communities have called for a new consultation process over the planned contract to ensure their core concerns are addressed. In response, the government presented a rights-denying technical interpretation of the Consultation Law (2011) rejecting this call (García Delgado, 2017; VMI, 2017). It subsequently presented a contradictory position (López Tarabochia, 2017), finally agreeing that a new consultation process will be held in relation to the renegotiated contract. This decision followed a series of public statements and reports by three thematic UN special procedures – on the rights of Indigenous Peoples, on toxic wastes, and on business and human rights – that supported the communities' position, something these special procedures have consistently done since 2013 (UNSR, 2013; UNSR, 2014; UNSRs, 2017; UNWGB & HR, 2017). Ironically, while ignoring or rejecting much of their guidance, Peru presented its engagement with these UN Special Procedures in the case of block 192 as an example of its cooperative efforts to overcome challenges in the implementation consultation procedures (Chavez Basagoitia, 2015).

Despite its promising start, rather than constituting an example of a prior consultation in good faith, the case demonstrates many of the classic characteristics of bad faith and culturally inappropriate consultations. The Indigenous Peoples concerned were willing to provide their consent and enter into long-term agreements that would have secured the conditions for stable operations, while addressing ongoing and future impacts and the communities' developmental aspirations and needs. Instead of making every effort to pursue a genuine inter-cultural dialogue towards these ends, the consultation process ended up bypassing many of the communities' core concerns and demands. It contributed to the fragmentation of their representative institutions and was incompatible with their decision-making timeframes, processes and realities. The inadequacy of the government's response to the communities' legitimate demands for a new genuine consultation process suggests that it has learned little from this experience.

Rather than building rights-based consensual relationships with communities, assisting in remedying ongoing harms and empowering them to pursue self-determined development, consultations, where the State agrees to hold them, remain largely formalities with no implications for Indigenous Peoples' rights (Inguil & Zevallos, 2018). The government has yet to understand that consultations in order to obtain consent are a mechanism to protect Indigenous Peoples' rights, and to achieve a balance of Constitutional rights and societal interests in a manner that is proportionate, reasonable, objective, and respectful (Inguil, 2018; Leyva, 2018).

The Aymara People and resistance to the Bear Creek Mine

Another disturbing trend, also evident in both the Bagua and Block 192 cases, is the State's propensity to pursue criminal cases against Indigenous representatives in cases where the absence of good faith consultation leads to large scale mobilisations. Essentially, the State generates protest by ignoring Indigenous Peoples' rights and then responds in a repressive rights-denying manner when these peoples attempt to assert those rights. Its response to the protests of the Aymara peoples to Bear Creek Mining's Santa Ana project in the south of Puno is illustrative of this approach.[8]

In 2011, while the Law on Consultation was pending, over 15,000 Aymara shut down the city of Puno for three days and forced the government to cancel the project, which they believed threatened their way of life and cultural and economic survival. The Puno Superior Court subsequently affirmed that the government had failed to consult and seek FPIC, despite its obligation to do so under ILO 169 and the Inter-American Convention on Human Rights, an obligation which existed prior to the adoption of the law on consultation (Puno Superior Court, 2017b). However, rather than address and remedy this violation of the Aymara peoples' rights, the government took criminal actions against their representatives for their role in the protests, accusing them, among other things, of extortion, public disorder, and damage to property. The extortion charges against the Aymara representatives were dismissed by the Puno Court, but one representative, Walter Aduviri, was sentenced to seven years imprisonment on the basis of command responsibility, under the absurd and discriminatory proposition, developed by the Court at its own behest, that he was the indirect perpetrator as he controlled and directed the will of the Aymara people in the Puno region, and ordered unknown persons to damage government property (Puno Superior Court, 2017a). In addition to lacking any basis in fact, this proposition runs completely contrary to the collective decision-making processes and traditional governance structures of the Aymara, something which their right to consultation and consent recognises and seeks to guarantee.

On 5 October 2018, following an appeal, the Supreme Court found a violation of Aduviri's due process rights, as the charge of indirect perpetrator had not been part of the original charges filed against him by the State (PSC,

2018). It therefore referred the case back to the Puno Superior Court for a retrial. Had the ruling stood, it would have set an extremely dangerous precedent, as it had effectively established that any representative who voices the collective will of an Indigenous People could be subject to individual criminal prosecution and conviction, without the need for any supporting evidence to demonstrate his or her role in the commission of alleged wrongs. The chilling effect on those considering acting as Indigenous spokespersons in contexts where communities resist projects that are imposed without consultation and consent are clear. While the government was prosecuting these Indigenous representatives, Bear Creek sought $600 million in damages from the State in international arbitration (ISDS, 2017). Ironically, the Peruvian government's defence of its decision to cancel the project was premised on the company's failure to consult with, and obtain the consent of, the Indigenous communities through their chosen representatives, the very people the government was prosecuting for giving voice to the communities' opinions.

The Wampis assertion of self-government and judicial challenge

Given the failure of the Congress and the Executive to implement the right to consultation and obtain FPIC in a meaningful manner, a growing number of Indigenous Peoples are adopting proactive strategies that combine the assertion of autonomy and customary law with the pursuit of judicial remedies. In this regard, the case of the Wampis and Awajun peoples is exemplary.

On 29 November 2015, the Autonomous Territorial Government of the Wampis Nation declared itself Peru's first Autonomous Indigenous Government (Wampis-Nation, 2015c). Through their pursuit of territorial planning and the formalisation and development of their internal governance structures, the Wampis aim to pursue a self-determined path (Wampis Nation, 2015b). To this end, they have developed their own governing statute applicable to the entirety of their integrated autonomous territory and have sought its formal recognition from the State (Wampis Nation, 2015a). This proactive assertion of the Wampis' development plans and priorities entails the pursuit of a self-determined alternative to State-led land and resource planning that has important ramifications for the right to consultation and FPIC, territorial recognition and self-government.[9]

Faced with historical and ongoing State efforts to impose projects in their territories, in 2014 the Wampis and the Awajun mounted a legal challenge to an oil concession, Block 116, which had been issued without consultation or consent in 2006. On 28 March 2017, the Lima Superior Court (SCL) issued a landmark decision upholding the allegation of a violation of the right to citizen's participation (consultation and consent), and declared the decrees authorising the contract and approving the Environmental and Social Impact Assessment (ESIA) to be invalid (SCL, 2017).

This decision has some potentially profound implications (Servindi, 2017). Firstly, as with the decision of the Puno Superior Court in relation to Bear Creek, it rejects the government's repeated assertion that the consultation requirement under the 2011 law is not retroactive (Ruiz Molleda, 2015). Instead, the Court held that as ILO 169 orders and provides sufficient guidance for the application of the right to consultation, a specific law was not necessary for its direct implementation. This was something which the Court noted was also affirmed in the jurisprudence of the Colombian Constitutional Court (SCL, 2017, paras. 50–61; Doyle, 2015 pp. 167–8).[10] Secondly, the Lima Superior Court ruling notes that, until 2011, the government had systematically failed to comply with its obligations under ILO 169, IHRL and the jurisprudence of the IACtHR. It also held that as ILO 169 was the legal framework for the laws that existed prior to 2006, companies could not claim to have signed their contracts in good faith in the absence of prior consultation (SCL, 2017, paras. 69–71). All activities had therefore to be suspended and the companies had to withdraw from the Indigenous territories. Thirdly, and of great importance for any future projects, the Lima Superior Court held that if the government wanted to enter into a new contract, consultation would be required in relation to the contract and the associated ESIA, and these consultation processes would have to obtain the consent of the affected communities (SCL, 2017, para. 25). As this was a context-specific requirement for consent, it did not imply a general veto power.

In addressing the basis for the right to consultation, the Superior Court affirmed that it is an expression of self-government and self-determination, aimed at facilitating decision-making within an intercultural dialogue and grounded in the right of Indigenous Peoples to freely determine their own development priorities. It referred to the 2007 Constitutional Court Cordillera Escalera decision, which, like important sections of a 2013 ruling of the Peruvian Supreme Court of Justice (PSC, 2013, paras. 6.6, 6.10), quotes the consent requirement in the IACtHR *Saramaka v Suriname* ruling when affirming that prior consultation and FPIC are required when there are potentially significant impacts on Indigenous Peoples' rights.[11] Similar proportionality-based reasoning was invoked by the UN Human Rights Committee in 2009 in relation to FPIC in its *Ángela Poma Poma v Peru* decision (HRC, 2009). While the 2017 Lima Superior Court decision does not explicitly mention *Saramaka, Poma Poma,* or the principle of proportionality when affirming the need for FPIC, its reference to the IACtHR jurisprudence, IHRL, and decisions of the Colombian Constitutional Court, leave little doubt as to its reasoning. Given its implications and the government's aversion to FPIC, an appeal by the State and the company was inevitable. The decision was upheld in 2018, signaling that a new era of *de jure* respect for Indigenous rights in Peru may be on the horizon. The Wampis for their part have decided to develop their own consultation and consent protocol to ensure that any consultation the government initiates will be in accordance with their laws and IHRL.

Conclusion

The Peruvian government's adoption of the first law on prior consultation of Indigenous Peoples in Latin America was a ground-breaking development. However, as is evident from the three cases analysed in this chapter, flawed implementing procedures which favour the interests of corporations over the rights of these peoples, and the government's dubious interpretation of the law and the absence of good faith in its implementation, render it incompatible with its object and purpose of securing Indigenous rights. The predictable backlash against it raises questions as to the potential effectiveness of a similar law in other Latin American jurisdictions.

Many of the issues in Peru appear to centre more on the context, interpretation, and implementation of the law than its content (UNSR, 2014). While the law has flaws (Ruiz Molleda, 2011; 2015), its procedural and substantive provisions must be interpreted in accordance with ILO 169 and IHRL. The Consultation Law's (2011) provision permitting the State to take the final decision where consent is not forthcoming is expressly conditioned on guaranteeing respect for Indigenous Peoples' rights. The question of proportionality consequently must be addressed and unavoidably leads to a presumption of a requirement for FPIC in the context of extractive, energy, and agribusiness projects. The 2017 Lima and Puno Superior Court decisions addressing FPIC, and the reasoning of the Supreme Court of Justice in 2013, reflect this reality and are an important step towards aligning Peruvian jurisprudence with IHRL. They should act as a baseline for the interpretation and implementation of the Consultation Law (2011), including its retrospective dimension, in a manner that is consistent with Indigenous Peoples' decision-making rights. Doing so would help transform consultations from rubber stamping exercises into genuine avenues for the protection of Indigenous Peoples' self-government, territorial, and cultural rights, and could render the Peruvian Consultation Law a genuinely functioning model for other countries to emulate.

Irrespective of developments in the law, deep-seated discrimination and powerful vested interests will inevitably continue to impose barriers to genuine intercultural dialogue. Given this reality, self-defined Indigenous institutions – such as the above-mentioned Autonomous Government of the Wampis Nation, the Federations of the Achuar, Kichwa and Quechua peoples, and the traditional governance structures and representatives of the Aymara communities – will remain the primary defence against the imposition of rights-denying projects.

Consent is only meaningful where there is freedom to choose between, and to autonomously formulate, alternative developmental options (Doyle, 2018). By facilitating and supporting Indigenous Peoples in the formulation and pursuit of their own development options the Peruvian government would take an important step towards ensuring that Indigenous Peoples and their autonomous governance institutions are finally empowered to pursue self-determined development and realise their own visions for their futures. Doing so would be nothing less than complying with its own publicly stated position that international and national law should establish the obligation of governments not only to recognise but to encourage the formation and contribute to the strengthening of Indigenous Peoples' representative organisations (ILO, 1988b, p. 102).

Notes

1 The Court concluded that the decree was not applicable to Indigenous Peoples and instead applied to peasant communities, and consequently dismissed the case in relation to a breach of the duty to consult.
2 The peoples concerned included the Waorani, Pananujuri and Aushiris or Abijiras.
3 The then President Alan Garcia infamously referring to Indigenous Peoples as the "dog in the manger" (García Pérez, 2007).
4 Other issues include the restrictions on the determination of who is Indigenous and who is "affected" by a proposed project/measure and therefore who is to be consulted; who does the consulting and the fact that the entity who implements the consultation is the entity implementing the project/measure; what impacts are to be consulted and what constitutes a direct impact; and how consultation processes are determined, including the control of the implementing entity and the rigidity of the timeframes envisaged. Some of these issues have been addressed by the ILO Supervisory bodies in response to representations in relation to Peru, see Ruiz Molleda (2015).
5 The absence of consultation or flaws in consultation processes have been documented in relation to other cases. See for example UNSR (2014) for a strong critique of consultations on Block 169.
6 In addition to documented sources, the account of the consultation process is based on interviews conducted with Indigenous representatives, civil society representatives, and government officials as part of the research for Doyle & Campanerio Baqué (2017), and the author's engagement with the concerned Indigenous representatives and their support organisations in the context of his work with the UN Special Rapporteur on Rights of Indigenous Peoples in 2016.
7 The block was originally held by Occidental Petroleum and subsequently by Pluspetrol, a dubious Argentinian company with subsidiaries in the Cayman Islands and a mailbox headquarters in the Netherlands for taxation and investment protection purposes.
8 The author visited the communities and assisted Derechos Humanos y Medio Ambiente Puno (Human Rights and the Environment Puno) (DHUMA) in submitting a communication to the Special Rapporteur on the Rights of Indigenous Peoples in relation to the case.
9 Discussion with Wampis representative Shapion Noningo Sesen.
10 See PCT (2017) footnote 38 citing Colombian Constitutional Court sentences SU-039 (1997), SU-383 (2003), C-030 (2008), C-461 (2008) and C-615 (2009). This Colombian jurisprudence establishes context-specific, self-determination, territorial and cultural rights-based, constraints on certain State actions which render them impermissible in the absence of the Indigenous rights-holders' consent. See further in chapters by Herrera and Calle in this volume.
11 The 2013 Supreme Court decision declared two decrees regulating citizen participation to be void due to a failure to comply with the right to consultation and consent.

References

Andrada, F. & Gonza, A. (2010). *El derecho a la consulta de los pueblos indígenas en Perú*. Seattle: DPLF.

CERD. (20.09). *Concluding Observations to Peru*. UN Doc. CERD/C/PER/CO/14–17, 31 August 2009.

Chavez Basagoitia, L. (2015). Presentation of the Report of the Special Rapporteur on the Rights of Indigenous Peoples Intervention of the Ambassador Luis Enrique Chavez Basagoitia, Permanent Representative of Peru. Geneva.

CNDDHH. (2013). *Reglamento de MINEM desnaturaliza derecho a la consulta*. Lima: Grupo de Pueblos Indígenas Coordinadora Nacional de Derechos Humanos. Retrieved from http://www.ibcperu.org/uncategorized/reglamento_de_minem_desna turaliza_derecho_a_la_consulta.

Consultation Law. (2011). Congreso De La Republica. Ley del derecho a la consulta previa a los pueblos indígenas u originarios, Reconocido en el Convenio 169 de la Organización Internacional del Trabajo (OIT). Ley N°29785. El Peruano. Lima miércoles 7 de septiembre de 2011. Normas Legales.

Córdova Flores, A. (2013). *The Right of Indigenous Self-Determination and the Right to Consultation in the Peruvian Constitutional Tribunal Jurisprudence (2005–2011)*, LLM Thesis. Victoria BC: University of Victoria.

Doyle, C.M. & Campanario Baqué, Y. (2017). *El Daño No Se Olvida*. Lima: Equidad.

Doyle, C.M. (2015). *Indigenous Peoples, Title to Territory, Rights and Resources: The Transformative Role of Free Prior and Informed Consent*. London: Routledge.

Doyle, C.M. (2018). *Free Prior and Informed Consent, Development and Mining on Bougainville*. In C. Hill & L. Fletcher (Eds.). *Growing Bouganville's Future* (pp. 4–12). Sydney: Jubilee Australia Research Centre.

García, F. (2017). Lote 192: Gobierno rechaza pedido de nuevo proceso de consulta previa, *El Comercio*, 16 September. Retrieved from https://elcomercio.pe/peru/loreto/lote-192-m inisterio-cultura-rechaza-solicitud-consulta-previa-comunidades-noticia-458486.

García Pérez, A. (2007). El síndrome del perro del hortelano. *El Comercio*, 28 October. Retrieved from http://peruesmas.com/biblioteca-jorge/Alan-Garcia-Perez-y-el-p erro-del-hortelano.pdf.

Hiruelas, N. (2018). En el Perú no se está consultando medidas que pueden afectar derechos colectivos de los pueblos indígenas. *El Comercio*, 1 February. Retrieved from http://elgranangular.com/blog/entrevista/en-el-peru-no-se-esta-consultando-m edidas-que-pueden-afectar-derechos-colectivos-de-los-pueblos-indigenas.

HRC. (2009). *Ángela Poma Poma v Peru*. Communication No. 1457/2006. Un Doc. CCPR/C/95/D/1457/2009.

IACmHR. (2015). *Indigenous Peoples, Afro-Descendent Communities and Natural Resources: Human Rights Protection in the Context of Extraction, Exploitation, and Development Activities*, 31 December. OEA Doc. OEA/Ser.L/V/II.Doc. 47/15.

IACtHR. (2007). *Saramaka People v Suriname*, Judgment of November 28, 2007 (Preliminary Objections, Merits, Reparations, and Costs). Inter-Am. Ct. H.R., (Ser. C) No. 172(2007).

ILO. (1988a). International Labour Conference, 75th Session. 1988: ILO Partial Revision of the Indigenous and Tribal Populations Convention, 1957 (no. 107), Reports VI(2), Geneva.

ILO. (1988b). International Labour Conference, 76th Session 1989: ILO Partial Revision of the Indigenous and Tribal Populations Convention, 1957 (No. 107), Report IV(1), Geneva.

ILO. (1989a). International Labour Conference, 76th Session 1989 ILO Partial Revision of the Indigenous and Tribal Populations Convention, 1957 (No. 107), Reports IV (2A), Geneva.

ILO. (1989b). Record of Proceedings, International Labour Conference, 76th Session, 28 June. Provisional Record No. 32, Geneva.

ILO. (2009). Observation (CEACR) – adopted 2009, published 99th ILC session (2010), Indigenous and Tribal Peoples Convention, 1989 (No. 169) – Peru (Ratification: 1994).

ILO. (2013). *Understanding the Indigenous and Tribal Peoples Convention, 1989 (No. 169): Handbook for ILO Tripartite Constituents*. Geneva: ILO.

Inguil, S. (2018). 'Detrás de una consulta no hay un diálogo legítimo, sino la imposición del lado más fuerte'. Entrevista con Juan Carlos Ruíz. Retrieved from https://consultape.com/2018/02/26/detras-de-una-consulta-no-hay-un-dialogo-legitimo-sino-la-imposicion-del-lado-mas-fuerte/.

Inguil, S. & Zevallos, M. (2018). Perú: La Consulta previa en el sector extractivo sigue sin garantizar el respeto a los derechos indígenas. *El Gran Angular*. Retrieved from http://elgranangular.com/consultapreviaperu/reportaje1/.

ISDS. (2017). *Bear Creek Mining Corporation v. Republic of Peru*. ICSID Case No. ARB/14/21.

Leyva, A.V. (2015). La Consulta previa en el lote 192 presenta vicios de nulidad. *Servindi*, 17 September. Retrieved from https://www.servindi.org/actualidad/139591.

Leyva, A. (2018). *Consúltame de verdad. Aproximación a un balance de la consulta previa en el Perú en los sectores minero e hidrocarburífero*. Lima: CooperAccion, Oxfam.

López Tarabochia, M. (2017). Lote 192: ¿Por qué la consulta previa es la causa de las protestas indígenas en la Amazonía peruana? *Mongabay*, 2 October. Retrieved from https://es.mongabay.com/2017/10/peru-lote-192-consulta-previa-causa-protestas-indigenas-amazonia.

MC. (2012). Peruvian Ministry of Culture: Supreme Decree N° 001–2012-MC.

MC. (2015). Peruvian Ministry of Culture: Official Communication No 355–2015-VMI/MC.

MEM. (2015). Peruvian Ministry of Energy: Report No 797–2015-MEM/DGAAE/DNAE/RCO/SED/CIM.

ODECOFROC. (2009). Crónica de un Engaño. Los intentos de enajenación del territorio fronterizo Awajún en la Cordillera del Cóndor a favor de la minería. Lima: IWGIA.

PCT. (2005). Decision No. 0033–2005-PI/TC, 19 September 2006.

PCT. (2007). Decision No. 03343–02007-PA/TC, 20 February 2009.

PCT. (2009a). Decision No. 0022–2009-PI/TC, 9 June 2010.

PCT. (2009b). Decision No. 06316–02008-PA/TC, 11 November 2009.

PCT. (2010). Decision No. 06316–02008-AA Clarification, 24 August 2010.

PCT. (2011). Decision No. 00025–02009-PI, 17 March 2011.

PSC. (2013). Decision No. 2232–2012, 23 May 2013.

PSC. (2018). Decision No. 173–2018, 5 October 2018.

Puno Superior Court. (2017a). Decision No. 01832_2015_0–2101_JM_CI_03, 13 December 2017.

Puno Superior Court. (2017b). Decision No. 00682–02011–66–2101-SP-PE-01, 29 December 2017.

Raynal, D. (2015). Oil Exploitation in the Peruvian Amazon, Violations of Human Rights and Access to Remedy. In C.M. Doyle (Ed.) *Business and Human Rights: Indigenous Peoples' Experiences with Access to Remedy. Case studies from Africa, Asia and Latin America* (pp. 97–141). Chiang Mai, Madrid, and Copenhagen: AIPP, Almáciga, IWGIA.

Ruiz Molleda, J. (2011). *Guía de interpretación de la Ley de Consulta Previa de los pueblos indígenas (Ley N° 29785) Análisis, comentarios y concordancias.* Lima: Instituto de Defensa Legal (IDL).

Ruiz Molleda, J. (2015). Problemas jurídicos en la implementación de la consulta previa en el Perú: o los «pretextos jurídicos» del gobierno para incumplirla. *Derecho & Sociedad*, 42, 179–192.

Schilling-Vacaflor, A. & Flemmer, R. (2013). Why is Prior Consultation Not Yet an Effective Tool for Conflict Resolution? The Case of Peru. GIGA Working Paper, 220. Retrieved from https://www.giga-hamburg.de/en/system/files/publications/wp 220_schilling-flemmer.pdf.

SCL. (2017). Decision No. 32365–32014, 28 March 2017.

Servindi. (2017). Decisión histórica: Poder Judicial ordena consulta del Lote 116. Retrieved from https://www.servindi.org/actualidad-noticias/31/03/2017/decision-his torica-poder-judicial-ordena-consulta-del-lote-116.

UNGA. (2007). Agenda item 68 Report of the Human Rights Council Draft resolution. UN Doc. A/61/L.67.

Unity Pact. (2011). *Pacto de Unidad Principios mínimos no negociables para la aplicación de los derechos de participación, consulta previa y consentimiento previo, libre e informado.* Retrieved from http://servindi.org/pdf/ComunicadoPactodeunidad2.pdf.

UNSR. (2009). United Nations Special Rapporteur on the Situation of Human Rights and Fundamental Freedoms of Indigenous People. UN Doc. A/HRC/12/34/Add.8, 18 August 2009.

UNSR. (2013). Declaración del Relator Especial de las Naciones Unidas sobre los derechos de los pueblos indígenas, James Anaya, al concluir su visita al Perú, 13 December. Retrieved from http://unsr.jamesanaya.org/statements/declaracion-del-rela tor-especial-sobre-los-derechos-de-los-pueblos-indigenas-al-concluir-su-visita-al-peru.

UNSR. (2014). Report of the Special Rapporteur on the Rights of Indigenous Peoples, James Anaya Addendum. UN Doc. A/HRC/24/41/Add.4, 3 July 2014.

UNSR. (2017). Peru Must Halt Oil Talks Until Indigenous Rights and Contamination are Taken into Account. Public statement on behalf of Mr. Baskut Tuncak, Special Rapporteur on toxic wastes & Ms. Victoria Tauli-Corpuz, Special Rapporteur on the rights of indigenous peoples, 13 July. Retrieved from https://www.ohchr.org/EN/ NewsEvents/Pages/DisplayNews.aspx?NewsID=21871&LangID=E.

UNWGB & HR. (2017). Statement at the End of Visit to Peru by the United Nations Working Group on Business and Human Rights, 19 July. Retrieved from https:// www.ohchr.org/EN/NewsEvents/Pages/DisplayNews.aspx?NewsID=21888&La ngID=E.

VMI. (2017). Peruvian Vice Ministry of Culture: Resolution N° 027–2017-VMI-MC.

Wampis Nation. (2015a). Press Release, 29 November. Soledad, Kanus/Santiago River, Perú.

Wampis Nation. (2015b). Estatuto del Gobierno Territorial Autónomo de la Nación Wampis.

Wampis Nation. (2015c). Gobierno Territorial Autónomo de la Nación Wampis Ordenanza N° 001–2015-GTANW_CCNN, 29 November. Soledad, Kanus/Santiago River.

Wright, C. (2014). Indigenous Mobilisation and the Law of Consultation in Peru: A Boomerang Pattern? *The International Indigenous Policy Journal*, 5(4).

Zúñiga, M., Barclay, F., & Campanario, Y. (2017). *5 sinrazones para postergar la decisión de consulta del lote 192*. Lima: Equidad. Retrieved from http://observa toriopetrolero.org/5-sinrazones-para-postergar-la-decision-de-consulta-del-lote-192.

Zúñiga, M. & Okamoto, T. (2018). *Sin derechos, no hay consulta. Pueblos indígenas, petróleo y consulta previa en la Amazonía peruana.* Lima: Oxfam.

Part II

Administrating prior consultation

5 The coupling of prior consultation and environmental impact assessment in Bolivia

Corporate appropriation and knowledge gaps

Almut Schilling-Vacaflor

Introduction

An indispensable condition for exercising the right to prior consultation and to Free, Prior and Informed Consent (FPIC), as a safeguard for protecting all types of rights of Indigenous Peoples, is that the participants are well informed about any decision to be taken. More specifically, it has been emphasised that Indigenous Peoples are entitled to receive complete and non-biased information, and that different forms of knowledge should be brought into an intercultural dialogue (Flemmer & Schilling-Vacaflor, 2016). When discussing how fair and adequate consultation and negotiation processes should be organised, James Anaya, the former United Nations' Special Rapporteur on the Rights of Indigenous Peoples, highlighted the need for information gathering and sharing via environmental and human rights impact assessments (Anaya, 2013).

However, several authors have shown that prior consultation processes carried out in Latin America have been characterised by the existence of a serious information and knowledge gap.[1] They have outlined that the information distributed in consultation processes is often incomplete, too technical and difficult to understand, has a pro-extraction bias, and disregards local perspectives and Indigenous knowledge (Flemmer & Schilling-Vacaflor, 2016; Roa García & Roa Alfonso, 2017).

Based on the analysis of Bolivian cases, this chapter argues that the existing, flawed environmental impact assessment (EIA) system has largely accounted for the information-related deficiencies of prior consultation processes. It shows that the way EIA is used has further exacerbated the power asymmetries between Indigenous Peoples, on the one hand, and extraction corporations and State institutions, on the other. EIA has worked as a proponent-controlled tool for legitimising the proposed extraction projects, being shaped by "corporate science" (Kirsch, 2014), and undermining meaningful prior consultation. As a result, the attempts of Indigenous groups to protect local territories and reduce the projects' socio-ecological impacts through consultation processes have been curtailed.

DOI: 10.4324/9781351042109-8

While this chapter's findings are based on empirical data from Bolivia, their relevance is of a more general nature. Globally, EIA systems are the most often used preventive instrument for environmental protection. Morgan (2012) noted that, to date, 191 of the 193 members of the United Nations have national legislation or have signed some form of international legal instrument that refers to the use of EIAs. In Latin America, all countries have incorporated EIA as part of their environmental management systems and, as in many other countries of the Global South, project-specific, more or less participatory EIAs are often the only available sustainability-oriented tool in place (Morrison-Saunders & Retief, 2012).

Empirically, this chapter is based on data collected during a total of 12 months of fieldwork carried out during different research stays in Bolivia between November 2011 and November 2015. I analysed several draft EIAs that served as the basis for discussing the projects' impacts in prior consultation processes and the respective Indigenous observations, points of critique, and demands for modification. In several cases, I was able to access the final version of the EIA studies. These documents often include responses to Indigenous contributions from the corporations involved or the consultants carrying out the EIA. I also analysed 30 reports from the Ministry of Hydrocarbons and Energy (MHE) on prior consultation processes and on the resulting final agreements. To contextualise these data and to get insights into concrete practices and the perspectives of the diverse actors involved, I conducted over 100 semi-structured interviews with representatives of State ministries, Indigenous organisations and communities (mainly from Bolivia's Guaraní), hydrocarbon corporations and non-governmental organisations (NGOs), and I participated as an observer in two prior consultation processes and in various Guaraní assemblies. For the systematic qualitative analysis of the collected data, I used the ATLAS.ti software.

EIA in environmental governance in the Global South: a critical inquiry

In accordance with international law, prior consultation and consent processes with Indigenous Peoples should be based on complete and reliable information (see articles 6 and 7 of the ILO Convention No.169 Concerning Indigenous and Tribal Peoples in Independent Countries of 1989 (ILO 169), and the United Nations Declaration on the Rights of Indigenous Peoples (UNDRIP)). For its part, ILO 169 stipulates in article 7 that "[g]overnments shall ensure that, whenever appropriate, studies are carried out, in co-operation with the peoples concerned, to assess the social, spiritual, cultural and environmental impact on them of planned development activities. The results of these studies shall be considered as fundamental criteria for the implementation of these activities". Nevertheless, the only studies that have been carried out with the purpose of taking informed decisions on development activities in Latin America have tended to be the EIA.

EIA refers to the evaluation of the effects likely to arise from a major project, considering possible impacts prior to the decision about its approval. EIA is an anticipatory, participatory environmental management tool, of which the EIA report is only one part (Wood, 2003, p. 2). In principle, EIA should lead to the abandonment of unacceptable projects and to the mitigation of negative impacts by implementing feasible alternatives or mitigation measures in projects that are approved. While the EIA primarily focuses on environmental impacts, in the recent past it has increasingly incorporated social and cultural impacts as well (Glucker, Driessen, Kolhoff, & Runhaar, 2013). Vanclay (2003) argues that the role of impact assessment "encompasses empowerment of local people; [and] enhancement of the position of [...] disadvantaged or marginalised members of society" (p. 7). EIA was first developed in the United States in 1969 and gained international visibility during the United Nations Conference on the Environment in Stockholm in 1972 (Wood, 2003, p. 4). Two key events contributed to the international dissemination of the EIA: in 1989, it became a requirement for all World Bank-financed projects, and, in 1992, the UN Conference on Environment and Development, or the Rio de Janeiro Earth Summit, resulted in a series of international laws and policies that encouraged signatories to incorporate the EIA as a national instrument (Sadler, 1996). Indeed, the EIA is a global tool of accountability, having been adopted by multilateral development banks, bilateral donor agencies, and UN agencies.

However, the gap between EIA in principle and in practice has often been huge. Scholars have criticised EIA practices globally, and particularly in Latin America, for different reasons. First, a main point of critique has been that impact assessments have been financed and largely controlled by the proponents. Such proponent-controlled impact assessments shaped by "corporate science" (Kirsch, 2014) tend to minimise expected negative impacts (Li, 2009; Dunlap, 2018; Leifsen, Sánchez Vázquez, & Reyes, 2017). In the same vein, it has been outlined that only in exceptional cases has the EIA led to meaningful changes in planned projects (Devlin & Yap, 2008; Bedi, 2013). While participants in consultations often distrust EIAs, access to independent expert knowledge has usually been very difficult or even impossible (Conde & Walter, 2018).

Second, it has been argued that EIA does not sufficiently account for the complexity of impacts and related uncertainties. Cumulative and synergistic impacts of different activities – that affect the same or overlapping territories – are usually disregarded (Hindery, 2013). Jasanoff (2003) outlines that EIAs focus on the known at the expense of the unknown, producing overconfidence in the accuracy and completeness of the results they find. EIAs tend to downplay what falls outside their field of vision and to overstate whatever falls within it. For instance, EIAs highlight short-term impacts that can be mitigated by the companies, while they avoid discussing long-term irreversible consequences.

Third, a number of authors have criticised that information disclosed in EIAs is made unnecessarily complex, technical and/or abstract, which prevents affected groups from participating effectively in environmental licensing processes (O'Faircheallaigh & Corbett, 2005, p. 33; Gustafsson, 2017). In the same way, local experience and Indigenous knowledge has usually been neglected or subordinated in a conventional impact assessment (Li, 2009; Dunlap, 2018; Roa García & Roa Alfonso, 2017). In general, engineers and planners who are epistemologically opposed to Indigenous understandings of the world are prominent and powerful in environmental planning and impact assessments (O'Faircheallaigh & Corbett, 2005).

Fourth, the State entities responsible for overseeing EIAs have often failed to do so effectively, because of a lack of capacity and/or strong political pressure within the States to facilitate investments and to execute planned extraction projects quickly. For instance, referring to EIAs in Bolivia, Bascopé Sanjines (2010) outlined that the competent authority – the Ministry of the Environment and Water – has tended not to require full and adequate information from the operating companies. States have also often neglected investigations into counter-evidence for providing independent and non-biased information about the expected impacts of the project (Dunlap, 2018; Roa García & Roa Alfonso, 2017).

Fifth, another point of critique has been the lack of stringent follow-up mechanisms to supervise the corporation's compliance with EIA, in both the Global North and the Global South (O'Faircheallaigh, 2007; Schilling-Vacaflor, 2017). Relatedly, many scholars have argued that the real impacts of extraction projects have been much more severe than those identified in the EIA, and that, in many cases, corporations have not complied with the provisions outlined in the EIA (O'Faircheallaigh, 2007; Kirsch, 2014).

In response to the frequent pitfalls of this instrument, diverse Indigenous communities, for instance, in Sweden and Australia, have carried out their own alternative community-based impact assessments (Lawrence & Larsen, 2017; O'Faircheallaigh, 2017; Larsen, 2018). Larsen (2018) examined impact assessment regimes in Sweden, Norway, Canada, Australia, and Aotearoa/New Zealand and argued that Indigenous participation is most meaningful through impact assessment co-management that takes place directly with the State and throughout all impact assessment phases, complemented with strategic community-owned impact assessment.

Prior consultation and EIA in Bolivia

In Bolivia, prior consultation processes started in the hydrocarbon sector in 2007, after the adoption of Supreme Decree No. 29,033 that regulates the right to prior consultation in this sector.[2] Until 2017, over 70 consultation processes had been concluded in the hydrocarbon sector, one (very contested) consultation was carried out about the construction of a highway through the Indigenous Territory and Natural Park Isiboro Securé (TIPNIS), and – as of

2015 – consultation processes have also been organised in the mining sector (Bustamante de Almenara & Cabanillas Linares, 2018). A remarkable volume of literature has been produced in the recent past about the right to prior consultation and FPIC in Bolivia. The existing literature has scrutinised the contested domestic legal interpretation of this right and analysed the implementation of consultation processes on the ground (Bascopé Sanjines, 2010; Haarstad & Campero, 2011; Humphreys Bebbington, 2012; Pellegrini & Arismendi Ribera, 2012; Flemmer & Schilling-Vacaflor, 2016; Haarstad, 2014; Eichler, 2016; Fontana & Grugel, 2016; Tomaselli, 2016; Schilling-Vacaflor, 2017; Tockman, 2018; Falleti & Riofrancos, 2018).

Problematic aspects related to prior consultation processes have been discussed, including the following: (a) the elite capture of prior consultation processes, which has exacerbated local power asymmetries; (b) "divide and rule-tactics" and their negative social repercussions among local communities; (c) information gaps and "information hurdles" due to a lack of reliable and complete knowledge and the neglect of local knowledge; (d) the non-compliance of corporations with the final consultation agreements; (e) the imposed nature of consultation processes; and (f) the fact that short-term economic gains have often been prioritised in consultation processes and related negotiations over compensation payments, while long-term ecological and socio-cultural impacts have usually been relegated. The role of the EIA in the context of prior consultation processes has been mentioned by some authors, but it has not yet been subject to a thorough analysis.

Usually, at the beginning of a prior consultation process in Bolivia, two or three copies of the draft EIA (called "Document for Public Information") are sent to the representative organisations of the affected communities. These documents are voluminous (on average about 600–800 pages each) and written in a highly technical way. Generally, the advisers of Indigenous Peoples and in some cases the Indigenous authorities themselves engage with these documents, in order to criticise specific data, correct information about the local populations, demand additional data, or challenge the impacts identified in these documents and their classification.[3] EIA studies in Bolivia generally contain at least the following sections: 1) executive summary; 2) project description; 3) socioenvironmental baseline; 4) consultation and participation; 5) identification and evaluation of impacts; 6) risk analysis and contingency plan; 7) prevention and mitigation programme; 8) application plan and environmental monitoring; 9) cost–benefit analysis; 10) programme of abandonment and restoration; and 11) applicable legislation.

To establish the socio-environmental baseline, it has been a common practice to invite community members to participate in the collection of data. However, according to my analysis of EIA reports, this section is often not sufficiently linked to the rest of the document, especially the prevention and mitigation program, and the application and monitoring plan. Despite the minor influence of participants in environmental decision-making, the argument that community members participated in drafting the EIA has been used by the MHE consultation team in prior consultation processes in order to debilitate any critique expressed against the draft EIA (Schilling-Vacaflor, 2015).

This chapter argues that many of the aforementioned shortcomings of prior consultation processes were caused or exacerbated by the role that the EIAs played within them. In particular, I criticise that tying EIA practices so neatly to prior consultation processes has led to a corporate appropriation of prior consultation processes. While States are, legally, the main duty-bearers for fulfilling the right to prior consultation and FPIC, corporations have found a powerful way to enter these processes via the back door, by funding and controlling the information that is distributed during consultation processes. When I participated in prior consultation processes in Bolivia, I observed that the consultation team from the Ministry of Hydrocarbons and Energy (MHE) mainly acted as a spokesperson of the EIA consultants when presenting the information about the planned project, summarising the data provided in the draft EIA. Thereby, interested corporations and their consultants substantially shape the discourses that evolve within these processes, as well as the resulting agreements and decisions.

In addition, both the EIA consultants and the consultation team from the MHE are usually trained as environmental or petroleum engineers. Hence, they tend to share a similar worldview and a techno-science approach of assessing hydrocarbon activities' impacts, whereas Indigenous and local experiential forms of knowledge tend to be perceived as irrelevant, unscientific, and inferior.

Pro-extraction information in the EIA

The EIA reports in Bolivia vary considerably with regard to their quality. While some of them are of a very obviously poor quality, containing copy-paste elements from EIA studies from other regions or very superficial and fragmented data, others have been carried out by more rigorous teams of consultants. In many cases, the EIAs provide valuable descriptive data, such as statistical data about rainfall, land property, human development indicators, social services, biodiversity, the quality of soils, etc. In a context of general scarcity of data about rural areas in Bolivia, EIA studies often represent an important and up-to-date source of information that can be useful for many different purposes, such as for developing State- or NGO-funded social or environmental projects.

Nevertheless, when digging deeper into the logic that underlies EIA in Bolivia, especially with regard to the evaluation of impacts and cost-benefit analyses, it turns out that these documents have a clear pro-extraction bias. As is the case with most EIAs worldwide, and to an even greater extent in countries of the Global South, EIAs in Bolivia have been shaped by "corporate science" (Kirsch, 2014). Negative impacts of planned projects have been minimised while positive impacts have been over-emphasised. For instance, the EIA's cost-benefit analyses that I observed all concluded that the overall gains of planned projects outweighed the costs. Strikingly, in some EIAs that were distributed in prior consultation processes, the remediation of environmental damage, such as the planned reforestation of areas that would need to be deforested in order to carry out hydrocarbon activities, was used to argue that the net environmental gains related to the execution of the project would be greater than the environmental losses. To my knowledge, not a single hydrocarbon project has been stopped or substantially modified due to a critical EIA in Bolivia, despite the fact that several of these projects affect territories with severe water scarcity, tropical forests, or places used by Indigenous Peoples living in voluntary isolation.

In almost all prior consultation processes, the quality of the information distributed has been a major point of contestation, and there is a prevailing perception among participants that this information is not trustworthy (Flemmer & Schilling-Vacaflor, 2016). For instance, participants of a prior consultation process in Parapitiguasu explained:

> The biggest problem was that we lacked precise information, because we did not get the complete picture about the significance of a project of such a magnitude [...]. We had several meetings because it was not possible to clarify the real impacts of the project. The corporations just give the information that is convenient for their purpose, but they do not share the other part.
>
> (Schilling-Vacaflor, 2013a)

In those communities that have ample experience of hydrocarbon activities, participants have tended to challenge the EIA for minimising the real impacts. Notably, very often actors who closely observe corporate practices on the ground – such as corporate employees, local socio-environmental monitors[4] or people with a close connection to their territory due to their land use practices (hunting, collecting plants and fruits) – have assumed a proactive role in consultation processes, challenging the biased information presented. Eventually, prior consultation processes have largely revolved around the question of how to classify expected impacts, which is decisive in determining the compensation payments to be transferred to local populations (Schilling-Vacaflor, 2017).

Over and over again, local populations have argued that the irreversible, long-term impacts of hydrocarbon activities were not adequately assessed in the draft EIA (Ministerio de Hidrocarburos y Energía, 2007–2014). However, the participants' arguments have usually been countered harshly by the consultation team and the EIA consultants. As the communities have lacked access to independent experts who could help them verify their claims, they have often approached the State ministries to carry out additional studies. Likewise, in many final consultation agreements that I analysed, the consultation participants required the hydrocarbon corporations to fund additional studies in order to investigate the real impacts of their practices (Ministerio de Hidrocarburos y

Energía, 2007–2014). Unfortunately, the local populations' demands for more reliable environmental information were usually neglected by both the State and the hydrocarbon corporations. In comparison, local populations were more successful when they focused on the projects' social and cultural impacts, which are more difficult for the State, EIA consultants, and interested corporations to refute than are the biophysical ones. For instance, local populations identified impacts such as changes in their customs, the development of a capitalist mentality on the part of their leaders, or the exacerbation of local conflicts, which were included in many EIA studies (Ministerio de Hidrocarburos y Energía, 2007–2014). Instead of discussing whether such impacts could be avoided, the local populations engaged in negotiations with extractive corporations about compensation payments.

EIA knowledge: the only game in town

EIAs influence power asymmetries by enhancing the position and legitimacy of corporate actors, while further relegating the position, values, and forms of knowledge of local and Indigenous actors. It seems paradoxical that community members affected by extractive projects hold a right that should empower them by giving them a voice in decision-making, but that due to the dominant EIA logic, most local actors are silenced, as they do not feel capacitated to actively participate in consultation dialogues. An experienced Guaraní leader who participated in many prior consultation processes summarised the problem as follows: "They [the MHE] come with their engineers, geologists and so on and their technical discourses are overwhelming [...] Within one hour we just bow our heads, because we do not know how to discuss with them" (National Assembly of the Assembly of Guaraní Peoples, 2014).

Eichler (2017) showed that especially Indigenous women, the elderly, and those who are not fluent in Spanish – and who often hold important Indigenous or local knowledge – were formally and informally excluded from these participatory arenas. While this finding would not be surprising for most participatory arenas in Latin America, it is remarkable that such patterns of discrimination are manifest in those instruments that have been particularly designed to foster Indigenous Peoples' participation and that should uphold their rights.

Local communities in Bolivia have searched for ways to counterbalance the conventional impact assessments, and to establish their own assessment of the expected impacts, but only in exceptional cases have these community-based impact assessments been influential. In one prior consultation process about a large gas project in Alto Parapetí, the advisers of the Indigenous communities successfully challenged the data contained in the draft EIA about the quantity of water in a local river. They showed that the water flow of the river was less than outlined in the EIA. As a consequence, in accordance with the environmental legislation, the corporation was prohibited from using water from this river (Schilling- Vacaflor, 2013d).

In other cases, local populations engaged in participatory mapping exercises to create alternative information about their territories, livelihoods, and land use practices, which has been used to challenge the draft EIA studies. Interestingly, in prior consultation processes carried out in Bolivia, it has become a common practice that delegations composed of community members with profound territorial knowledge carry out field inspections to revise the potential impacts of the planned projects *in situ*. During these inspections, they assess and record all expected social and environmental impacts, such as the loss of medicinal plants, trees, and areas that can be used for hunting, or the expulsion of animals from their natural habitat. The delegations often record such impacts meticulously, taking photos, collecting plants and preparing presentations for the consultation team from the MHE.

Nevertheless, when this evidence is presented to the State consultation team, the Ministry staff often seemed to be disinterested (Schilling-Vacaflor, 2013b; 2013e). One of my interview partners – an elderly woman from a Guaraní community – told me that it seems to her that the Ministry staff regarded them as students, who were making childish presentations, based on the assumption that they would clearly be inferior to the more professional EIA staff (Schilling-Vacaflor, 2014b).

Corporate disrespect of achieved agreements

In Bolivia, consultation agreements are signed at the end of the consultation process between the MHE consultation team and the participants. Despite the fact that the State has the duty to uphold achieved agreements, the Bolivian State has left the last word to the hydrocarbon corporations (see below). Hence, in practice, it has often been up to the corporations whether or not they recognise the classification of impacts by consultation participants as long-term and irreversible, and whether they are willing to adopt the participants' requirements to carry out additional studies, or to adopt more stringent mitigation measures.

For instance, in the face of uncertainty regarding the hydrocarbon explorations' impacts on water reserves or plants, many communities requested that the activities should respect a greater distance from these natural sources than established by law. To defend their claim, the communities often referred to the necessity of upholding the principle of precaution. However, corporations have tended to respond that they are only willing to respect the distances established in the country's legislation (Yacimientos Petrolíferos Fiscales Bolivanos, 2011). Likewise, impacts that were classified by consultation participants as long-term and irreversible were often refuted by the corporations with the argument that there was not sufficient empirical evidence for such a classification (Soluciones Ambientales Sol Ambiente, 2014). The carrying out of additional studies, for instance about subterranean water sources, was in several cases denied with the argument that such studies would be too complicated and/or costly.

Furthermore, in my dozens of interviews in different local populations, I kept hearing the remark that corporations would not act according to the EIA and the final consultation agreements, which had become part of the EIA. Most interview partners observed that the real impacts largely superseded the ones outlined in the EIA and reported to them during the prior consultation processes (Schilling-Vacaflor, 2014a).

In at least the two following cases, post-EIA studies were conducted to oversee the project's compliance with the EIA and to evaluate the activities' real impacts. In the first case of a seismic exploration that affected Leco and Mosetén communities in the north of La Paz, the evaluators found that the detonations of explosives were not carried out every 15 meters as announced, but rather at a distance of three meters. Helicopters hovered throughout the day, thereby disrespecting the established temporary flight restrictions (Bascopé Sanjines, 2010, pp. 76–81). In the second case, a seismic exploration that affected the Tacana II in the Bolivian Amazon, the corporation did not respect the minimum distance from the nut trees (*árboles de castañas*), which was established in the final consultation agreement. Due to the agency of the local socio-environmental monitors, 180 detonation holes had to be closed, as they were located closer to the trees than agreed (Acosta, 2017).

In sum, the Bolivian State has delegated the duty to comply with the final consultation agreements to the corporations, and lacks a rigorous monitoring system to oversee corporate compliance with the EIA studies. In several cases, local populations submitted their complaints of noncompliance with EIA and consultation agreements to the responsible State ministry, but such complaints have usually had no effect (Schilling-Vacaflor, 2013c).

Discussion and conclusions

This chapter has argued that many shortcomings of prior consultation processes as implemented in Bolivia have been "imported" from or exacerbated by the existing flawed EIA system. Among these shortcomings are the lack of complete and non-biased information, the non-existence of intercultural dialogue, and non-compliance with achieved agreements. I also show that the coupling of prior consultation and EIA has led to a corporate appropriation of prior consultation processes. Thereby, power asymmetries between different groups of actors, interests, and forms of knowledge – particularly the technoscience approach that has shaped EIA, on the one hand, and Indigenous and local experiential knowledge, on the other – were broadened. This finding is crucial, since a one-level playing field and the recognition of cultural particularities are preconditions for meaningful and fair prior consultation and consent processes.

Previous academic debates have centred around the question of how to improve conventional impact assessment, for instance by taking sociocultural aspects increasingly into account; by establishing more independent forms of impact assessment; by organising more effective participatory processes in

EIA; or by improving the follow-up of EIA agreements (Sadler, 1996; Vanclay, 2003; Morrison-Saunders & Retief, 2012; Glucker et al., 2013). While I agree that such measures are important, based on my analysis I argue that they are in themselves insufficient for enabling the right to prior consultation and consent.

Given that meaningful prior consultation and FPIC is about complete, non-biased information and intercultural dialogue, conventional EIA should not remain "the only game in town". As the understanding of nature and of human-environment relations in EIA studies is not objective, but rather entrenched in techno-science, or an engineer-like worldview, the EIA will be unable to address the deeper conflicts between different frames of problems, interpretations of impacts, and related solutions. Hence, in line with Larsen, Raitio, Stinnerbom, & Wik-Karlsson (2017), I think that such dilemmas call for reflective and critical approaches between parties to jointly explore their underlying view(s) on the problems and, based on that, to seek legitimate decisions.

More concretely, for Indigenous Peoples such as those participating in prior consultation processes in Bolivia, it is vital that their own forms of community-based impact assessment are strengthened. In order to do so, local initiatives like Indigenous socio-environmental monitoring systems, participatory mapping exercises, and extensive field inspections to assess expected project impacts *in situ*, constitute important ways of producing alternative information that counterbalances conventional EIAs. In addition, increased access to independent expert knowledge – for instance, by establishing cooperation with universities – to critically revise and challenge EIA studies would be an important tool for Indigenous Peoples, which has not yet been sufficiently exploited. Overall, initiatives that help to tip the power balance between competing knowledges, values, and worldviews are urgently needed to put conventional EIA in its place, thereby paving the way for more meaningful prior consultation and FPIC processes.

Notes

1 When speaking of the Indigenous Peoples' right in general, I use the term 'right to prior consultation and Free, Prior and Informed Consent (FPIC)', while when analysing concrete consultation practices in Latin America I speak of 'prior consultation processes', because in my opinion these latter practices do not usually qualify as FPIC processes.

2 In a nutshell, the legal framework that regulates the right to prior consultation and FPIC in Bolivia includes the Supreme Decree No. 29,033 of 2007 and subsequent decrees that regulate and limit the right to prior consultation, as well as Bolivia's progressive Constitution of 2009 and the laws that recognise the ILO Convention 169 and the United Nations Declaration on the Rights of Indigenous Peoples as domestic laws (Schilling-Vacaflor, 2017).

3 EIA studies classify expected impacts according to criteria such as duration (short term or long term), reparability (reparable or irreparable), and type of impact (direct or indirect; positive or negative).

4 In order to monitor company compliance with the final consultation agreements and EIA, Indigenous organisations have struggled to enhance an independent system of local, Indigenous socio-environmental monitoring. They pushed for the adoption of Supreme Decree No. 29,103 of 2007, which stipulated that each company has to pay 0.5 per cent of the overall investment for a project into a fund to finance Indigenous monitors. The fund to enable these monitors' independent function has never been established. Currently, the local populations have to fight separately with each project for the inclusion of socio-environmental monitors, and for the companies to pay these monitors directly.

References

Acosta, L. (2017). Estudio de la castaña y otros (ECyO). La Paz: Estudio técnico (unpublished).

Anaya, J. (2013). Extractive Industries and Indigenous Peoples. Report of the Special Rapporteur on the rights of indigenous peoples, James Anaya. Report to the Human Rights Council UN Doc. A/HRC/24/41, 1 July.

Bascopé Sanjinés, I. (2010). *Lecciones aprendidas sobre consulta previa*. La Paz: Centro de Estudios Jurídicos e Investigación Social.

Bedi, H. (2013). Environmental Misassessment, Development and Mining in Orissa, India. *Development and Change*, 44(1), 101–123.

Bustamante de Almenara, M. & Cabanillas Linares, C. (2018). *Buenas prácticas de las defensorías del pueblo de Bolivia, Colombia, Ecuador y Perú en procesos de consulta previa. Incluye experencias de instituciones invitadas de Argentina, Brasil y Guatemala*. Lima: Defensoría del Pueblo.

Conde, M. & Walter, M. (2018). Expertise and Forms of Co-production in Environmental Justice Struggles. Paper presented at the LASA Conference. Barcelona: LASA.

Devlin, J. & Yap, N. (2008). Contentious politics in environmental assessment: blocked projects and winning coalitions. *Impact Assessment and Project Appraisal*, 26(1), 17–27.

Dunlap, A. (2018). "A Bureaucratic Trap": Free, Prior and Informed Consent (FPIC) and Wind Energy Development in Juchitán, Mexico. *Capitalism Nature Socialism*, 29(4), 88–108.

Eichler, J. (2016). The Vernacularisation of Indigenous Peoples' Participatory Rights in the Bolivian Extractive Sector: Including Subgroups in Collective Decision-Making Processes (Doctoral dissertation). University of Essex.

Falleti, T. G. & Riofrancos, T. N. (2018). Endogenous Participation: Strengthening Prior Consultation in Extractive Economies. *World Politics*, 70(1), 86–121.

Flemmer, R. & Schilling-Vacaflor, A. (2016). Unfulfilled Promises of the Consultation Approach: The Limits to Effective Indigenous Participation in Bolivia's and Peru's Extractive Industries. *Third World Quarterly*, 37(1), 172–188.

Fontana, L. & Grugel, J. (2016). The Politics of Indigenous Participation Through "Free Prior Informed Consent": Reflections from the Bolivian Case. *World Development*, 77, 249–261.

Glucker, A. N., Driessen, P. P. J., Kolhoff, A., & Runhaar, H. A. C. (2013). Public Participation in Environmental Impact Assessment: Why, Who and How? *Environmental Impact Assessment Review*, 43, 104–111.

Gustafsson, M. (2017). *Private Politics and Peasant Mobilization in the Peruvian Mining Industry*. London: Palgrave McMillan.

Haarstad, H. & Campero, C. (2011). *La participación en el sector de hidrocarburos en Bolivia: el 'doble discurso' y las limitaciones sobre la gobernanza participativa.* Urbeco Report, 3/11. Bergen: Centro de Ecología Urbana y Universidad de Bergen.

Haarstad, H. (2014). Cross-scalar dynamics of the resource curse: Constraints on local participation in the Bolivian gas sector. *Journal of Development Studies*, 50(7), 977–990.

Hindery, D. (2013). *From Enron to Evo: Pipeline Politics, Global Environmentalism, and Indigenous Rights in Bolivia.* Tucson: University of Arizona Press.

Humphreys Bebbington, D. (2012). Consultation, Compensation and Conflict: Natural Gas Extraction in Weenhayek Territory, Bolivia. *Journal of Latin American Geography*, 11(2), 49–71.

Jasanoff, S. (2003). Technologies of Humility: Citizen Participation in Governing Science. *Minerva*, 41(3), 223–244.

Kirsch, S. (2014). *Mining Capitalism: The Relationship Between Corporations and Their Critics.* Oakland: University of California Press.

Larsen, R., Raitio, K., Stinnerbom, M., & Wik-Karlsson, J. (2017). Sami-State Collaboration in the Governance of Cumulative Effects Assessment: A Critical Action Research Approach. *Environmental Impact Assessment Review*, 64, 67–76.

Larsen, R. (2018). Impact Assessment and Indigenous Self-Determination: A Scalar Framework of Participation Options. *Impact Assessment and Project Appraisal*, 36(3), 208–219.

Lawrence, R. & Larsen, R. (2017). The Politics of Planning: Assessing the Impacts of Mining on Sami Lands. *Third World Quarterly*, 38(5), 1164–1180.

Leifsen, E., Sánchez-Vázquez, L., & Reyes, M. (2017). Claiming prior consultation, monitoring environmental impact: counterwork by the use of formal instruments of participatory governance in Ecuador's emerging mining sector. *Third World Quarterly*, 38(5), 1092–1109.

Li, F. (2009). Documenting Accountability: Environmental Impact Assessment in a Peruvian Mining Project. *PoLAR: Political and Legal Anthropology Review*, 32(2), 218–236.

Ministerio de Hidrocarburos y Energía. (2007–2014). *Informes Finales. Consulta y Participación.* La Paz, unpublished documents (30 ministry reports on consultation cases).

Morgan, R. (2012). Environmental Impact Assessment: The State of the Art. *Impact Assessment and Project Appraisal*, 30(1), 5–14.

Morrison-Saunders, A. & Retief, F. (2012). Walking the Sustainability Assessment Talk-Progressing the Practice of Environmental Impact Assessment (EIA). *Environmental Impact Assessment Review*, 36, 34–41.

National Assembly of the Assembly of Guaraní Peoples. (2014). Camiri, 25 November.

O'Faircheallaigh, C. & Corbett, T. (2005). Indigenous Participation in Environmental Management of Mining Projects: The Role of Negotiated Agreements. *Environmental Politics*, 14(5), 629–647.

O'Faircheallaigh, C. (2007). Environmental Agreements, EIA Follow-Up and Aboriginal Participation in Environmental Management: The Canadian Experience. *Environmental Impact Assessment Review*, 27(4), 319–342.

O'Faircheallaigh, C. (2017). Shaping Projects, Shaping Impacts: Community-Controlled Impact Assessments and Negotiated Agreements. *Third World Quarterly*, 38(5), 1181–1197.

Pellegrini, L. & Arismendi Ribera, M. O. (2012). Consultation, Compensation and Extraction in Bolivia after the 'left turn': the case of oil exploration in the North of La Paz Department. *Journal of Latin American Geography*, 11(2), 103–120.

Roa García, M.C. & Roa Alfonso, N.E. (2017). Tauramena ante el proyecto Odisea 3D. Desafiando la razón pública de la exploración sísmica. In T.R. Avendaño, M.C. Roa García, L.M. Navas Camacho & J. Toloza Chaparro (Eds.). *Como el agua y el aceite. Conflictos socioambientales por la extracción petrolera* (pp. 90–110). Bogotá: CENSAT Agua Vida.

Sadler, B. (1996). Environmental Assessment in a Changing World: Evaluating Practice to Improve Performance. International Study of the Effectiveness of Environmental Assessment, Final Report. Ottawa: Canadian Environmental Assessment Agency.

Schilling-Vacaflor, A. (2013a). Focus group discussion, 25 October. Parapitiguasu.

Schilling-Vacaflor, A. (2013b). Field notes, participation in the prior consultation process in Takovo Mora, November.

Schilling-Vacaflor, A. (2013c). Interview with responsible authority for natural resources and environment of the Assembly of Guaraní Peoples APG, 20 November.

Schilling-Vacaflor, A. (2013d). Interview with an adviser of Indigenous communities from Alto Parapetí, Camiri, 30 November.

Schilling-Vacaflor, A. (2013e). Interview with an adviser from the Assembly of Guaraní Peoples, 7 December.

Schilling-Vacaflor, A. (2014a). Las actividades de gas natural en la capitanía de Parapitiguasu desde la perspectiva de la población local. Report written for local communities. (unpublished).

Schilling-Vacaflor, A. (2014b). Interview with community member from Takovo Mora, 12 November.

Schilling-Vacaflor, A. (2015). Interview with a community leader from a Guaraní community in Takovo Mora, 28 January.

Schilling-Vacaflor, A. (2017). Who Controls the Territory and the Resources? Free, Prior and Informed Consent (FPIC) as a Contested Human Rights Practice in Bolivia. *Third World Quarterly*, 38(5), 1058–1074.

Schilling-Vacaflor, A. & Eichler, J. (2017). The Shady Side of Consultation and Compensation: "Divide-and-Rule" Tactics in Bolivia's Extraction Sector. *Development and Change*, 48(6), 1439–1463.

Soluciones Ambientales Sol Ambiente. (2014). Estudio de Evaluacion de ImpactoAmbiental – Analitico Integral "EEIA-AI" Proyecto: "Prospección Sísmica 3D, Campos Tacobo y Curiche". Santa Cruz, Bolivia.

Tockman, J. (2018). Eliding Consent in Extractivist States: Bolivia, Canada, and the UN Declaration on the Rights of Indigenous Peoples. *The International Journal of Human Rights*, 22(3), 325–349.

Tomaselli, A. (2016). *Indigenous Peoples and their Right to Political Participation*. Baden-Baden: Nomos.

Vanclay, F. (2003). International Principles for Social Impact Assessment. *Impact assessment and Project Appraisal*, 21(1), 5–12.

Wood, C. (2003). *Environmental Impact Assessment: A Comparative Review*. London and New York: Routledge.

Yacimientos Petrolíferos Fiscales Bolivanos. (2011). *Estudio de Evaluación de Impacto Ambiental Analítico Integral (EEIA-AI). Proyecto: "Sismica 3d Itaguazurenda, Area Ovai Boyuibe"*. La Paz, Bolivia.

6 Prior consultation as a scenario for political dispute

A case study among the Sikuani Peoples from Orinoquía, Colombia

Laura Calle Alzate

Introduction

After the profound legal changes regarding the relationship between the State and Indigenous Peoples that were implemented throughout the region as of the nineties, several Latin American countries have constitutionally defined themselves as pluri-ethnic, pluri-cultural, multi-ethnic, multi-cultural and plurinational countries (Giraudo, 2007). These changes broke down all Latin American national projects that considered minority cultural groups as second-class citizens.

In Colombia, the Indigenous movement grew stronger by adopting a stance that sought to achieve not only identity claims but also a profound change in both political and social affairs in the country (Laurent, 2005). The fact that Indigenous movements joined together in a project to build a multi-ethnic nation made them visible at national level and played a leading role in a decade of profound changes, as of the eighties and the beginning of the nineties (Gros, 1991; Laurent, 2005; Caviedes, 2011). On the other hand, the Colombian State's pragmatism was immersed in a crisis of efficiency and legitimacy that could be alleviated by supporting the adhesion of minority cultural groups to the national project, while delegating functions to local authorities in order to decongest the enormous centralised bureaucratic system (Laurent, 2005). Thus, the 1991 Colombian Constitution proposed a more inclusive, multi-cultural and pluri-ethnic country. For these purposes, several participatory mechanisms were established, including prior consultation, which promised to give sectors of the population who had been historically excluded, such as Indigenous Peoples and Afro-Descendant communities, a voice in the country's development (Lopera Mesa & Dover, 2013).

However, not only did the State acknowledge its cultural diversity, but it also reformed the economic, labour, and social regime, submitting itself to the dictates of international financial institutions and imposing a neoliberal State project (Chaves & Zambrano, 2009). As pointed out by Ahumada (1998), "[...] the policies as regards economic openness, privatisation and regional integration, distinctive features of the neoliberal model, became Constitutional laws for the first time in 1991 [...]" (pp. 196–197). On this issue see, for example, articles 226, 227 and 336 of the Colombian Constitution.

DOI: 10.4324/9781351042109-9

In this way, the Colombian State moved toward multi-culturalism and neoliberalism at the same time, a correlation that Hale (2002; 2005) called "neoliberal multiculturalism". Therefore, the State acknowledges cultural rights and deals with the ethnicity of the subjects of these rights while protecting the model for economic development dictated by neoliberalism (Gros, 1997; Hale, 2005). As a result, there is a close relationship between the acknowledgment of differential rights and the consolidation of neoliberalism (Hale, 2002; Comaroff & Comaroff, 2009; Bocarejo, 2011).

Therefore, while the State acknowledged some political and territorial rights of Indigenous Peoples and other ethnic groups, it also promoted a policy of dependence that, from a practical perspective, limits the possibility of building culturally and economically self-governed areas and, consequently, managing the territory as well (Villa Rivera, 2011). This is the result of the distance between the acknowledgment of cultural rights and the refusal to offer up the resources necessary for these rights to be guaranteed and exercised. According to neoliberal multiculturalism, a co-optation or domestication of otherness is carried out by differentiating between the "authorised Indian" (*indio permitido*) as a socio-political category and the "unruly Indian" (*indio insolente*) (Hale, 2004). This differentiation is essential when it comes to shaping the legal subjects who demand a participatory mechanism such as prior consultation (Lopera Mesa & Dover, 2013).

In this chapter, based on the review of secondary sources and data collected from my ethnographic fieldwork, I argue that prior consultation processes involve an asymmetric interaction between actors who are unequal in economic, social and legal terms, as well as from an epistemological perspective (Lopera Mesa, & Dover, 2013). Therefore, we should understand this scenario as a political dispute, where power relationships are debated, legitimised, and redefined at different levels. Thus, I consider it useful to analyse prior consultation processes as "hegemonic" (Roseberry, 1994). That is, as a place for controversy, confrontation, and collusion between dominant and subordinate groups.[1]

The ethnographic data used in this text is part of a long-term study, carried out with the Sikuani people from the Wacoyo reservation of the Puerto Gaitán municipality, in the Colombian region of Orinoquía, during several field trips between the 2004 and 2016. It is important to point out that even though this document is built from the ethnography of prior consultation meetings between the Sikuani peoples and several oil companies present in the region, and from meetings I attended during my fieldwork in 2012–2013 (therefore, it is a site-specific interpretation), I seek to establish certain generalisations and theories related to the participant observation carried out at other times during this period. In this way, the specific characteristics of this case are highlighted to contribute to the understanding of prior consultation in other contexts, examining the social practices and relationships it fosters.

This chapter begins with a short description of the right to prior consultation in Colombia and a general description of oil incursion in the Orinoquía region. Then, taking an ethnographic approach, the consultation process is analysed as a place for dispute. Finally, I offer a view on the limits of prior consultation in Colombia.

Prior consultation in Colombia

Prior consultation is a fundamental right of Colombia's different ethnic groups (*grupos étnicos*) when legislative and administrative measures that may affect their ways of living are adopted. In Colombia, consultation processes have been conducted for over 20 years. Data from the Ethnic Affairs Office report that an average of 1,000 prior consultations are processed per year (Ministerio del Interior de la República de Colombia, 2014; Revista Dinero, 2017). According to Rodríguez (2011), the first prior consultation held with the Sikuani peoples took place in 1994 over a hydrocarbon project carried out by the *Empresa Colombiana de Petróleos* (ECOPETROL).

Although there is no law regulating prior consultation in the country, there is a dilated regulatory framework.[2] For its part, Law No.21 enacted in 1991 ratifies the ILO Convention No.169 Concerning Indigenous and Tribal Peoples in Independent Countries of 1989 (ILO 169), constitutionally establishing the right of Indigenous Peoples to be consulted (see articles 6 and 7 of the convention). The obligation to consult Indigenous Peoples, and other ethnic groups in Colombia, before a decision that may directly affect them was confirmed in article 330 of the Constitution of 1991. At the same time, in its article 332 the Constitution defines the State as the owner of the subsoil and non-renewable natural resources. However, ILO 169, which sets forth prior consultation rights, states in its article 15 that even when subsoil resources are owned by the State, the government should consult with the peoples concerned, "[...] before undertaking or authorising any programme, exploration or exploitation of the resources found in their lands [...]" (ILO, 1989). Additionally, the Constitutional Court considered that prior consultation is appropriate in relation to projects carried out in Indigenous territories, and mandatory for the exploitation of natural resources.

Ideally, and pursuant to international law standards, prior consultation shall be carried out through the appropriate proceedings (see the chapters by Del Castillo and Cantú Rivera in this volume). That is, prior consultation shall not be limited to a mere meeting between the parties involved but the result of a proceeding by means of which the Indigenous Peoples and other ethnic groups may know the projects in depth, expressing themselves and being part of it. For Rodríguez (2008b), a fundamental aspect of the right to consultation is that it should be carried out on a "prior" basis. Despite this, and as reported by researchers, lawyers, and the community organisations themselves, on several occasions, licences for projects are granted first. Indeed, it is when these projects are drawing to a close that the communities are consulted. Likewise, governments and private companies permanently seek mechanisms to avoid consultation and prevent the communities from deciding over the projects that take place in their territories (Houghton, 2008a, 2008b; Rodríguez, 2008b; Villa Rivera, 2011; Lopera Mesa & Dover, 2013).

In this sense, it has been reported that, in most cases, the Colombian State has failed to comply with the prior consultation requirements, as set forth in ILO 169. Thus, fraudulent and illegal negotiations have been conducted, in which false information, offers of money to individuals and communities, and the imposition of projects have prevailed (Houghton, 2008b; Villa Rivera, 2011; Rodríguez, 2011). As duly noted by Villa Rivera (2011), to understand the path taken by Indigenous Peoples towards the acknowledgement of their rights and the systematic, evasive responses of governments regarding the implementation of prior consultation, we need to understand the policies that, as of 2000, turned the extraction of natural resources from the subsoil into one of economic pillars of the national development model.

Oil invasion in Orinoquía's Indigenous territories

In Colombia, hydrocarbon production is historically related to ethnocide, human rights violations, forced eviction, and the extinction of Indigenous Peoples, often with the State's support (Houghton & Villa, 2004; Houghton, 2008a). As of the early twentieth century, foreign oil companies operating in Colombia declared the Orinoquía region a "reservation area" through the acquisition of large extensions of land, obtained by means of licences which were – and continue to be – granted by the State. In the 1950s, the oil exploration activities in the area began. Local and foreign companies have extracted oil from the region since 1976, currently positioning it among the top three producers of oil in the country. From then on, the Sikuani peoples have remained in contact with the companies.

As of 2000, a neo-extractivist development model was consolidated in Colombia, which was characterised by the intensive exploitation of raw material, the creation of new sources of energy, and the construction of infrastructure, such as railroad tracks and ports, aiming to create mobility and take these materials out to international markets (Lopera Mesa & Dover, 2013). Nevertheless, the shift towards this model began in the 1990s along with structural adjustment measures agreed with multilateral organisations such as the International Monetary Fund and the World Bank, favouring market liberalisation and an increase in foreign investment (Lopera Mesa & Dover, 2013). However, it was only after the policies implemented during the first government of Álvaro Uribe Vélez (2002–2006), that the conditions to turn Colombia into an appealing destination for foreign investment, especially in the energy and mining sector, were established (Houghton, 2008a; Lopera Mesa & Dover, 2013). According to Houghton (2008b), the high oil prices and the US dollar devaluation in the world market failed to reverse the tendency to indebtedness and a loss of revenue due to income tax. This is the reason why the number of oil exploration projects increased throughout the country. Likewise, Houghton (2008b) points out that the State's success in granting oil companies a license in a territory is directly proportional to the crisis experienced by its Indigenous Peoples. For example, between 2005 and 2008, 207 reserved areas (*resguardos*) and about 30 communities overlapped with oil blocks throughout the national territory (Houghton, 2008b, pp. 290–294).

During the governments of both Álvaro Uribe Vélez and Juan Manuel Santos, the Orinoquía region was seen as the new pole for the development of Colombia, since it was considered to be the last agricultural border. Therefore, investments were made in agro-industries and the development of infrastructure (Domínguez, 2013; Ñañez Ortiz & Calle Alzate, 2017). On the other hand, and almost simultaneously, the discovery of oil wells intensified, bringing with it an increase in the presence of companies in Indigenous reserved areas (Calle Alzate, 2005; 2014; 2017a). Between 2005 and 2007, there were approximately 68 oil projects in Indigenous territories of the Orinoquía region (Houghton, 2008b). There are 58 Indigenous reserved areas in the high plains alone, which correspond to 12 different communities, including the Wacoyo peoples. Between 2005 and 2007, there were 20 exploration, technical assessment, and oil production projects in the Puerto Gaitán municipality. All of these projects coincided with territories of Indigenous Peoples living in the area.

The development projects do not take into account that this population dates back to pre-Colombian times, or the fact that more than 30 percent of the territory in the region is established as a reserved area. Furthermore, it overlooks the degree to which armed violence, intensive mining, and the agricultural boom are causing a serious impact on Indigenous Peoples. Indeed, early in 2013, permission was granted for oil exploration to a total of 21 companies from Colombia, Canada, Brazil, China, Korea, and India, throughout the department, where 55% of local oil was produced (Massé & Camargo Castro, 2013). The development of heavy crude was concentrated in this region of the country, which was one of the pillars of the strategy to increase oil production and reserves (La Rota-Aguilera & Salcedo, 2016).

Oil companies, consultation, and irregularities

Although Colombian law establishes participatory mechanisms for ethnic groups in general, and for Indigenous Peoples in particular, Indigenous involvement in the making of development policies is limited. Consequently, both prior consultation and Free, Prior and Informed Consent (FPIC) may be interpreted as "illusory promises" (Villa Rivera, 2011). This situation can be seen in the absence of the Indigenous Peoples' point of view in the documents issued by the National Economic and Social Policy Council (CONPES, for its Spanish acronym) and the National Planning Department (DNP, for its Spanish acronym), who defined the Orinoquía region as a territory open to exploration and colonisation, full of opportunities for the private sector and where, unlike the Amazonian region, there are no major environmental restrictions (Calle Alzate, 2017a).

Thus, even though – with the introduction of oil companies in Indigenous territories – prior consultation continues to take place, the processes are full of irregularities. During several assemblies and meetings I attended during my fieldwork, Sikuani Indigenous leaders from Wacoyo reported that the processes failed to comply with the minimum requirements established by the ILO 169 and the national laws. Among these testimonies, it was stated that several consultations were never held or not made in good faith. Some of them did not take place before the project was carried out, others lacked the active involvement of Indigenous and community authorities, and still others were not held in accordance with their customs and traditions.

Furthermore, some of the inhabitants of the reserved areas stated that the community was not sufficiently qualified to carry out the prior consultation processes with the different companies, and reported the buying-off and manipulation of some Indigenous leaders to bring the consultation processes forward so that they would be fast and inexpensive for the company (Calle Alzate, 2014). The irregularity lies in the fact that consultation should be held in the presence of the community authorities and representatives of Indigenous Peoples and, therefore, all members, organisations and, above all, those affected by the projects must be allowed to participate. However, in many cases, the State has accepted private conversations between companies and communities, and even families, as consultation, and in the absence of any witnesses to such negotiations (Rodríguez, 2008a; Houghton, 2008b). In other cases, negotiations have taken place with intimidation by the armed groups present in the area (Massé & Camargo Castro, 2013), which has led to several conflicts between the inhabitants of the reserved areas (Calle Alzate, 2014).

The presence of oil companies has also become an environmental concern within the Indigenous local governments, which have on several occasions reported the adverse effects these companies have on their territories. However, as explained in another text (Calle Alzate, 2017a), the issue that generates most tension among the leaders is related to the distribution of resources by the companies, as part of the compensation or mitigation for being in their territories. Indigenous leaders consider negotiations with oil companies and prior consultation processes as transactions, that is, as a bureaucratic mechanism to legalise projects.

Ideally speaking, prior consultation should take place as a space for dialogue, where Indigenous Peoples may actively express their concerns regarding a project or activity and; "[...] together with the person in charge of developing them, they [may] design measures to prevent, mitigate, correct and compensate the impacts deriving from such works or activities [...]" (Salinas Alvarado, 2011, p. 237). Therefore, in some cases, this dialogue is outlined as a "dialogue of knowledge" or a "two-way learning process" (Rodríguez, 2008b). In practice, the dialogue is replaced by a one-way speech by the company or government authorities. But what happens in these spaces open for dialogue? What agents intervene, and how do they do so? What are legitimate speeches? Are the interlocutors acknowledged?

In the following paragraphs, an ethnographic report will serve to illustrate why consultation processes are far from opening dialogue between equals where there is disagreement, as suggested by the Constitutional Court in one of its judgments (T-129/2011). My analysis aims to show that the consultation process goes beyond the transaction itself, as it is an area in which a discursive and cultural framework is generated. That is to say a common language or way of speaking regarding social relationships delimits the central terms and conditions around which and according to which controversies and struggles occur (Roseberry, 2007 [1994]). In essence, the words, images, symbols, shapes, organisations, institutions, and movements used by the Sikuani peoples to speak, understand, confront, adapt, or resist the presence of hydrocarbon companies are all moulded by the same hegemonic process.

Staging a dispute: Wacoyo's experience of prior consultation

In January 2013, several prior consultation meetings were held between the Tecpetrol Colombia SAS Company and the peoples and authorities of the Wacoyo reserved area. According to statements made by the company's officials, during the meetings, the rules of the game and the methodology to be used throughout the whole process were agreed upon. The three company workers, perfectly dressed in their uniforms, which singled them out from the other people at the meeting, sat to the left of the central table. On the right side of the table were the Governor of the reserved area and the Coordinator of the Assembly, an experienced and knowledgeable leader, who acted as translator and master of ceremonies. The audience was in front of them. Like students in a master class, the Indigenous leaders (*capitanes*) sat in the first rows; professors and traditional doctors could be seen in the middle rows; while at the back, almost outside of the room, were women and some youngsters and children. The Governor greeted the assembly's attendees, describing the way in which people were distributed – perhaps involuntarily – but also the different hierarchies in the different areas. This hierarchy became particularly evident when the participants took turns to speak (Calle Alzate, 2013a).

One of the first matters that reflect the asymmetry between actors is precisely the way in which participants were distributed in the space. The company's officials were separated from the members of the community by a table serving as a stage. Although the company's anthropologist described the activity as a "dialogue of knowledge", the epistemology and speeches that filled the air were not those of the Sikuani. It should be pointed out that during one of his interventions, the company's anthropologist used the expression "you are a world apart" to refer to Sikuani culture and worldview, and referred to himself as a "translator" or an intermediary between the "Indigenous world" and the "non-Indigenous world". This is meaningful, as it invites us to consider consultation as a tool for managing otherness in which there is no will to recognise the interlocutor's "knowledge" or "epistemology", but the intention is rather to assimilate the "other" from the point of view of a dominant discourse. On the other hand, the ritual itself – the detailed rules of behaviour and the tight schedule – seem to be designed so as not to promote meaningful participation in the decision-making process but rather to limit the community's contribution in accordance with a strict and rigid framework.

I agree with Lopera Mesa & Dover (2013), who describe this dialogue as a paradox since the "other" is seen as a legal person with a right to consultation, given that they possess a "difference" that the multicultural State has undertaken to respect. Nonetheless, as other authors point out, "[...] establishing them as legal persons require[s] the 'others' to be different but, at the same time, they should be willing to take on the dominant epistemology [...]" (Lopera Mesa & Dover, 2013, p. 98). Therefore, what makes prior consultation a hegemonic process is not a shared epistemology but a material and cultural framework to live in, talk about, and act upon the social order characterised by dominance (Roseberry, 2007 [1994], p.127). This view depoliticises the relationship between companies and Indigenous authorities, with the aim of legitimising extractive activities. Going back to Hale's (2004) arguments, depoliticisation occurs because neoliberal multiculturalism operates by generating dialogue and negotiation opportunities among participants, as long as the integrity of the productive regime is not compromised.

On the other hand, there is a three-fold discourse on the part of Indigenous elites. First, there is the consultation discourse; second, the discourse about rights; and third, the discourse regarding resources and transactions. The coordinator of a consultation process must be able to handle these three discourses perfectly, as well as to modify the projects submitted by the community, establishing cooperation networks and monopolising the information relevant to the consultation. While having a conversation on this issue with one of the leaders, I noticed the importance that Sikuani people attribute to handling the consultation discourse and how they assume a role more similar to translators than representatives. The leader explained the following issue to me:

> [...] The point is that what is valuable and can be rescued and is a question of legitimacy for the people is handled cautiously, right? That is fine. When it comes to methodology, teaching or being able to coordinate an assembly with the studies carried out by oil companies and the community, only a few of us can understand and apply the methodology that they propose. Therefore, not just anyone will be capable of coordinating the assembly in order to understand and debate the methodology or the project [...].
>
> (Interview with an Indigenous leader, 30 January 2013)

This testimony helps us understand how the coordinator legitimises his own stance on the basis of his capacity to deal with non-Indigenous epistemology, which is something that most of the members of the community cannot do. Therefore, the leader believes that although many people are interested in taking the role of coordinator, they are not qualified for it because they are not familiar with the companies' language, and that is why the leader performs the role of translator. Ultimately, in hegemonic processes, a language that is typical of the dominant sectors also appears in the discourse of subordinate sectors and, at the same time, legitimises their role as interlocutors.

Likewise, it becomes evident that the crucial problem in prior consultation processes is that companies impose their methodologies and engage in a discourse that the majority neither shares nor understands. This facilitates the bilateral negotiation with Indigenous elites who have adopted the official discourse as their own but, at the same time, refute the technical language used by companies. This implies a great difficulty in generating the common language that constitutes a necessary condition for the type of dialogue required in prior consultation processes (Rodríguez, 2008a; 2008b; Lopera Mesa, & Dover, 2013).

In addition, to reach an agreement, one of the parties – in this case, the Sikuani people – must give up its understanding of the concept of "territory", since it is the only possible way to continue the dialogue and move forward with the subsequent phases of the process, considering the fact that weighing impacts is a priority when establishing mitigation and compensation measures, which tend to be understood only in economic terms. This was clear in the second part of the intervention by the leader I interviewed.

> [...] The point is that what can be asked in exchange for an hour of the oil company's impact here is legitimised, understood, and transmitted to the community in general terms. For example, it is already known that an impact can be either positive or negative. However, the environmental management of the territory is yet unknown. This management should be explained clearly to the people. They must be well aware of the potential damage, and environmental management. Therefore, we must conduct many workshops. But, for example, the company forces us to talk about values and payment for the seismic survey impact and, here, in the reserved area, the majority of us are not familiar with this [...].
>
> (Interview with an Indigenous leader, 30 January 2013)

In this case, it is clear how the discourse about impact, which can be interpreted in a negative or positive way, has been internalised and is limited to the environmental aspect. Nevertheless, there is also an imposition by companies to proceed with the mitigation or compensation of these impacts in economic terms, which is why the members of the community must express them in quantities. This generated several debates among the inhabitants of the reserved area, both during the meeting and in later assemblies. In accordance with the ILO 169 standards, the aim is for the community to be fully aware of the project, the procedure to be applied, and the way it is going to affect them. Furthermore, the community should be able to assess the advantages and disadvantages of the project freely, without interference, and to state their point of view on the viability of the said projects. However, there is a prevailing perspective among both the personnel of the companies that are responsible for the projects and the officials in charge of coordinating consultation processes in relation to impacts, which focuses on the immediate effects of the project, or on the environmental damage – pollution and deforestation – leaving aside long-term social repercussions such as changes in cohesion within the community or the alteration of community reciprocity rules (Calle Alzate, 2017a), as well as the effects generated by a project even before it is initiated (Lopera Mesa & Dover, 2013). This may not necessarily be in bad faith, but it suggests that those performing the consultation do not understand that, in accordance with their worldview, Indigenous People perceive territory as something more than just a physical basis for the reproduction of their community. As Escobar (2015) points out,

> [...] territories are vital time-spaces for all the community, both men and women, but it is not just that; it is also the time-space by which they interconnect with the surrounding nature and that is also a part of it [...]. When we talk about a mountain or a lagoon or a river as an ancestor or as a living entity, we are making reference to a social relation, not to a subject-object relation.
>
> (Escobar, 2015, p. 33)

To illustrate this, during the meeting, the description provided by the company's officials explained how the path followed by the different commissions in the territory aimed to identify important sites where specific impact assessments must be applied. This way, during the second part of the meeting, the company's engineer and the attendees discussed a series of maps created in a participatory way with the commissions who had traced paths around the "sites of interest", which could be affected by any of the six pipelines that would cut across the reserved area, should the project be carried out. Afterwards, a representative of each commission described the paths they had been walking through and the sites indicated in the Sikuani language, and there was a debate on the sum necessary to compensate possible material damage to these sites. In this regard, I would like to point out three aspects that reveal how the dominant epistemology is imposed.

The first aspect has to do with the demarcations of the paths. The State and the oil industry have aimed to minimise the application of the right to consultations only for those cases that foresee an exploitation project to be carried out in a place where an Indigenous population is physically present, including reserved areas, paying almost exclusive attention to the material impact.[3] This is related to the debate about whether it is mandatory to carry out a prior consultation, not only in those cases where there is full physical or legal evidence of the direct impact that exploration or exploitation projects have on Indigenous Peoples, but also in those cases where they affect territories for which no title deed has been granted or which are the result of complex cultural processes (Houghton, 2008b). In this case, the paths and the identification of the "sites of interest" were only within the limits of the reservation, leaving aside all those surrounding territories that the Sikuani people consider as part of their "ancestral lands" and from which they have been evicted over time (Calle Alzate, 2017b).

The second point has to do with the predominance of the perspective of tangible impacts. The sites of interest identified by the Sikuani people that were involved in the commissions were burial grounds and cemeteries, plots or farming areas, water springs, gallery forests, flooding areas, and sacred places, among others. For these sites to be considered as part of the impact area, they would have to match some of the six lines drawn by the seismic survey or be close to them. Those places identified by the Sikuani people as "sites of interest" were disregarded by the company's officials because they were not affected directly, although later in the meeting an anthropologist representing the company went further by pointing out that the distance required to determine whether there had been an impact or not would not be agreed upon with the community, which in itself represents a contradiction.

The critical debate among the attendees took place when the question was raised about the assessment of impacts on the identified sites of interest. This leads me to the third point, which is the prevalence of mitigation and economic compensation. The debate focused on the importance for leaders and the community to learn to negotiate with oil companies, since the Sikuani people consider that the compensation that these companies give the communities for the damage caused to their land can help them to fulfil unsatisfied needs. This is connected with the above-mentioned testimony provided by the leader, in which a reference was made to the importance of knowing how to "evaluate impacts". Some of the leaders argued that

> [...] if the seismic survey was carried out in the reserved area, they would ask the company to build a healthcare centre, to install electricity in every home, to give young people a job and to pay two million pesos to each family [...].
> (Calle Alzate, 2013b)

This is problematic as Indigenous leaders see companies as an alternative source of funding from the State in order to pay for and carry out projects in their communities (Calle Alzate, 2014). Therefore, the problem is not that oil companies are located in the territory but how impacts are assessed in economic terms, who will benefit from this, and how resources are distributed. For this reason, communities stop being a direct threat to the extractive regime promoted by multicultural neoliberalism since, as such, the discussion is not about fostering significant participation, but rather about depoliticising extractive activities, defusing tensions and allowing members of the community to be part of State projects for resource extraction. Communities do not pose a direct threat because they end up adhering to the dominant epistemology and siding with the definition of "authorised Indian".

Conclusions

The 1991 Constitution established that prior consultation is a suitable mechanism for the protection of the integrity of Indigenous Peoples in contexts of natural resource exploitation on Indigenous lands. However, as in the case of the Wacoyo Reservation, prior consultation in Colombia has become an opportunity for some of the population to mobilise resources around a productive project or infrastructural works, as well as issue contracts for the members of the affected communities. The consultation, in itself, is problematic because of the clear gap between what "should be" the procedure used to carry it out, and the target to whom it is directed.

The topics of debate and the rules for participation in prior consultation processes are often limited by a discourse, an epistemology, or the notion of territory held by the State and the oil industry officials who carry out the meetings. These discourses are part of a common language that selects the key terms on which the debate about this prior consultation will revolve. Because of this, the language used by the Sikuani people to negotiate with hydrocarbon companies is shaped by the dominant epistemology, that is to say, that of the companies and the State.

The position of the latter two actors is still assimilationist, since they are incapable of recognising that Indigenous communities are worthy of consideration and respect. For their part – as has been shown – these communities limit themselves to negotiating specific impacts, but their epistemological view is not permitted to contribute to the debate of what development means and how it can be effectively materialised. Nevertheless, the companies and the State disguise the process as a "dialogue of knowledge".

Here, the State finds itself in a complex position since it is committed to the neoliberal project and its corresponding neo-extractivist agenda, while at the same time it must guarantee that this project is carried out without infringing the rights of the Indigenous Peoples. Hence, as pointed out by Lopera Mesa & Dover (2013), one of the key questions is related to the position that the Colombian State should, and effectively does, occupy in the processes of prior consultation, since it is understood that the decision to carry out the project is ultimately in its hands. From this perspective, the problem is reduced to identifying, together with the Indigenous Peoples, which measures must be taken in order to carry out the project. Therefore, the acclaimed "dialogue of knowledge" simply does not exist, given the exclusion of Indigenous epistemologies and their notions of territory as "[...] a space that is both biophysical and epistemic – where life becomes world" [...] (Escobar, 2015, p. 35). Instead, the extractive activities are depoliticised, thus defusing tensions and allowing members of the community to be part of State projects for resource extraction (Perreault, 2015). Consequently, as pointed out by Serje (2003), the foundations are laid for the launch of a "dialogue of the deaf", since the agreements reached at the end of the meetings are understood in a different way by the parties.

Thus, consultation is a normalised, formalised, and choreographed practice but also a space for political debate. In this sense, consultation works as a means for everyday governance and ends up by establishing boundaries between the acceptable and unacceptable means of social mobilisation, defining the role that Indigenous Peoples must play (Hale, 2004; 2005). In this choreography, the design and execution of the projects is not subject to discussion with the communities since, as has been illustrated, some impact mitigation and compensation measures are negotiated, as well as minor social works, in exchange for which the communities must commit themselves not to obstruct the project's execution (Rodríguez, 2011; Lopera Mesa & Dover, 2013). This is why they are required to stick firmly to their role of "authorised Indian". This reinforces the main consequence of neoliberal multiculturalism, which is to protect the integrity of the productive regime, particularly those sectors which participate in the globalised economy. In order for consultation to be a real dialogue of knowledge, the State and the companies should revise their concept of "development" and territory to take proper account of the epistemological point of view of Indigenous Peoples.

Notes

1 They could also be seen as "multi-dimensional and dynamic fields of force". Roseberry uses the "field of force" metaphor coined by Thompson to refer to the dialectic relationship between the popular and the dominant culture (Thompson, 1978, pp. 133–165).
2 See the chapter by Herrera in this volume.
3 This is proven by the adoption of Decree No.1,320 of 1998, according to which the consultation processes should be carried out only in the Indigenous reserved areas (*resguardos*) or their titled lands (*territorios titulados*).

References

Ahumada, C. (1998). *El modelo neoliberal y su impacto en la sociedad colombiana.* Bogotá: El Áncora Editores.

Bocarejo, D. (2011). Dos paradojas del multiculturalismo colombiano: la espacialización de la diferencia indígena y su aislamiento político. *Revista Colombiana de Antropología*, 47(2), 98–121.

Calle Alzate, L. (2005). *Autoridad y Poder entre los Sikuani del Resguardo Wacoyo.* (Trabajo de Grado para optar al título de Antropóloga). Universidad Nacional de Colombia, Facultad de Ciancias Humanas, Departamento de Antropología, Bogotá.

Calle Alzate, L. (2013a). Fieldnotes, Prior Consultation carried out by Tecpetrol in the Wacoyo Reserve, Puerto Gaitán, Meta, Colombia.

Calle Alzate, L. (2013b). Interview with an Indigenous leader, Puerto Gaitan, Meta, Colombia.

Calle Alzate, L. (2014). El espejismo de la autonomía indígena: mirada a la situación de una comunidad en la Orinoquía Colombiana. *Anuario de acción humanitaria y derechos humanos*, 12, 71–96.

Calle Alzate, L. (2017). Empresas, recursos económicos y gobiernos indígenas: una aproximación al estudio de las redes clientelares en un resguardo indígena en la Altillanura colombiana. *Puerto Gaitan, Meta Colombia*, 84, 171–199.

104 Laura Calle Alzate

Caviedes, M. (2011). *Oro a cambio de Espejos: Discurso hegemónico y contra-hegemónico en el movimiento indígena en Colombia, 1982–1996*. Tesis de Doctorado en Historia. Universidad Nacional de Colombia, Bogotá.

Comaroff, J. & Comaroff, J. (2009). *Ethnicity Inc*. Chicago: University of Chicago Press.

Chaves, M. & Zambrano, M. (2009). Desafíos a la nación multicultural. Una mirada comparativa sobre la reindianización y el mestizaje en Colombia. In C. Martínez Novo (Ed.), *Repensando los Movimientos Indígenas* (pp. 2015–2245). Quito: Flacso Ecuador.

Domínguez, J.C. (2013). Plan de 9,6 billones de pesos para la altillanura. Documento Conpes para la zona reúne seis estrategias y será presentado hoy antes de ser llevado a debate y aprobación por el consejo de ministros. *Portafolio*, 15 November. Retrieved from http://www.portafolio.co/economia/finanzas/plan-billones-pesos-a ltillanura-89022/.

Escobar, A. (2015). Territorios de diferencia: la ontología política de los "derechos al territorio". *Cuadrenos de Antropología Social*, 41, 25–38.

Giraudo, L. (2007). Entre rupturas y retornos: la nueva cuestión indígena en América Latina. In L. Giraudo (Ed.), *Ciudadanía y derechos indígenas en América Latina: poblaciones, estados y orden internacional* (pp. 7–57). Madrid: Centro de Estudios Políticos y Constitucionales.

Gros, C. (1991). *Colombia Indígena. Identidad cultural y cambio social*. Bogotá: Fondo Editorial CEREC.

Hale, C. (2002). Does Multiculturalism Menace? Governance, Cultural Rights and the Politics of Identity in Guatemala. *Journal of Latin American Studies*, 34, 485–524.

Hale, C. (2004). Rethinking Indigenous Politics in the Era of the 'Indio Permitido'. *NCLA Report on the Americas*, 38(2), 16–21.

Hale, C. (2005). Neoliberal Multiculturalism: The Remaking of Cultural Rights and Racial Dominance in Central America. *Political and Legal Anthropology Review*, 28(1), 10–28.

Houghton, J. (2008a). Desterritorialización y pueblos indígenas. In J. Houghton (Ed.), *La Tierra contra la muerte. Conflictos territoriales de los pueblos indígenas en Colombia* (pp. 15–55). Bogotá: Centro de Cooperación al Indígena CECOIN.

Houghton, J. (2008b). Estrategia petrolera en los territorios indígenas. In J. Houghton (Ed.), *La Tierra contra la muerte. Conflictos territoriales de los pueblos indígenas en Colombia* (pp. 279–311). Bogotá: Centro de Cooperación al Indígena CECOIN.

Houghton, J. & Villa, W. (2004). *Violencia Política contra los Pueblos Indígenas en Colombia*. Bogotá: Centro de Cooperación al Indígena CECOIN.

La Rota-Aguilera, M.J. & Salcedo, L. (2016). Parte I. La Altillanura: Desafíos y Posibilidades del Ordenamiento Territorial en la Nueva Frontera Agrícola Colombiana. In C. Duarte (Ed.), *Desencuentros Territoriales. Tomo II. Caracterización de los conflictos en las regiones de la Altillanura, Putumayo y Montees de María* (pp. 23–158). Bogotá: Instituto Colombiano de Antropología.

Laurent, V. (2005). *Comunidades indígenas, espacios políticos y movilización electoral en Colombia, 1990–1990. Motivaciones, campos de acción e impactos*. Bogotá: Instituto Colombiano de Antropología e Historia ICANH, Instituto Francés de Estudios Andinos IFEA.

Lopera Mesa, G. & Dover, R. (2013). Consulta Previa, ciudadanías diferenciadas y conflicto socioambiental. *Boletín de Antropología Universidad de Antioquia*, 45(28), 76–103.

Massé, F. & Camargo Castro, J. (2013). *Actores armados ilegales y sector petrolero del Meta: Informe Monográfico*. Bogotá: Observatorio Internacional DDR.

Ministerio del Interior de la República de Colombia. (2014). En Colombia se hacen más de 2000 consultas previas al año, 20 February. Retrieved from https://www.mininterior.gov.co/sala-de-prensa/noticias/en-colombia-se-hacen-mas-de-2000-con sultas-previas-al-ano/.

Ñañez Ortiz, J. & Calle Alzate, L. (2017). Conflicto armado e implementación del extractivismo como modelo de desarrollo: un estudio de caso en la altillanura colombiana. In M. Delgado & A. Lara, *Congreso El Extractivismo en América Latina: Dimensiones Económicas, Sociales, Políticas y Culturales* (pp. 311–325). Sevilla: Universidad de Sevilla.

Perreault, T. (2015). Performing Participation: Mining, Power, and the Limits of Public Consultation in Bolivia. *The Journal of Latin American and Caribbean Anthropology*, 20(3), 433–451.

Revista Dinero. (2017). Consultas previas al paredón: Gobierno busca fijar reglas claras. *Dinero*, 6 August. Retrieved from https://www.dinero.com/edicion-impresa/pa is/articulo/consultas-previas-se-regularan-con-ley-estatutarias/246291/.

Rodríguez, G. A. (2008a). La autonomía y los conflictos ambientales en territorios indígenas. In J. Houghton (Ed.), *La Tierra contra la muerte. Conflictos territoriales de los pueblos indígenas en Colombia* (pp. 57–78). Bogotá: Centro de Cooperación al Indígena CECOIN.

Rodríguez, G. A. (2008b). Los obstáculos y las potencialidades de la consulta previa en el reconocimiento de los derechos de los pueblos indígenas. *Etnias y Política*, 6, 52–57.

Rodríguez, G. A. (2011). Proyectos y Conflictos en relación con la consulta previa. *Opinión Jurídica*, 57–72.

Rosberry, W. (1994). Hegemony and the Language of Contention. In G. Joseph, & N. Daniel (Eds.), *Everyday Forms of State Formation: Revolution and the Negotiation of Rule in Modern México*. Durham: Duke University Press.

Roseberry, W. (2007). Hegemonía y el lenguaje de la controversia. In M. Lagos, & P. Calla (Eds.), *Antropología del Estado. Dominación y prácticas contestatarias en América Latina* (pp. 119–137). La Paz: INDH/PNUD.

Salinas Alvarado, C. (2011). La consulta previa como requisito obligatorio dentro de trámites administrativos cuyo contenido pueda afectar en forma directa a comunidades indígenas y tribales en Colombia. *Revista Derecho del Estado*, 27, 235–259.

Serje, M. (2003). ONGs, indios y petróleo: El caso U'wa a través de los mapas del territorio en disputa *Bulletin de l'Institut français d'études andines*, 32(1). 101–131.

Thompson, E.P. (1978) Eighteenth-Century English Society: Class Struggle Without Class? *Social History*, 3(2).

Villa Rivera, W. (2011). *El movimiento social indígena colombiano: entre autonomía y dependencia*. In A. Betancur (Ed.), *Movimientos indígenas en América Latina: Resistencia y nuevos modelos de integración*. Copenhagen: IWGIA.

7 Prior consultation as a door opener

Frontier negotiations, grassroots contestation, and new recognition politics in Peru

Riccarda Flemmer

Introduction

Peru's national legislation on prior consultation from 2011 and 2012 is celebrated internationally as a forerunner for such policies outside Peru, but it is often forgotten that the tragic events of the Bagua conflict in 2009 were necessary for the right to prior consultation to finally be incorporated into the national agenda. In the Bagua conflict or the "Baguazo", 34 people died when police forces cleared a road blockade of Indigenous protesters in the Northern Amazon who had mobilised against the policies of the government of Alan García (2006–2011) to open the rainforest for further private investments in the context of a Free Trade Agreement between Peru and the USA (Hughes, 2010). Amazonian organisations wanted to stop the government's legal attempts, drawing on their experiences that the Peruvian rainforest is one of the areas most affected by Latin America's expanding resource frontiers (Gavaldà Palacín, 2013; Orta-Martínez & Finer, 2010). The right to prior consultation enshrined in the ILO Convention No.169 Concerning Indigenous and Tribal Peoples in Independent Countries of 1989 (ILO 169) was a key reference in the protesters' fight to stop the government's initiatives. In the aftermath of the conflict, the Peruvian legislation on consultation was negotiated and initially celebrated as a victory for Indigenous mobilisations, but shortly afterwards national Indigenous organisations harshly criticised its shortcomings. Nevertheless, 24 consultations were concluded under the government of President Ollanta Humala (2011–2016), the majority of them in the hydrocarbon sector.

This chapter aims to show that despite the evident deficiencies of State-led consultation processes, the adoption of the national legislation has provoked a push towards a new generation of recognition politics in Peru. The "politics of recognition" is understood in the way that Nancy Fraser framed her call for a critical theory of recognition, which requires both recognition and redistribution (Fraser, 1995; 2005). Bringing together both sides of the story – narrow State-led consultations about extractive projects and the promotion of national intercultural policies – this chapter argues that prior consultation is an ambiguous instrument that operates as a "door opener" in two contrasting

DOI: 10.4324/9781351042109-10

ways. On the one hand, there is a critique that consultation processes are used by the hydrocarbon and mining industries to legitimise new projects and expand extractive frontiers. On the other hand, prior consultation at the level of national policies means that Indigenous Peoples' rights have moved onto the national agenda and opened the "back door" for the institutionalisation of intercultural policies. The "back door" refers to the fact that reforms were initially driven by the violence that occurred in the Bagua conflict, but were carried further by democratic contestation – controversies, informal politics, and strategic litigation articulated by grassroots, progressive State entities, and activists – surrounding the administrative implementation of the relatively limited mechanisms of prior consultation.

The chapter is based on fieldwork material on the implementation of prior consultation collected between 2013 and 2015. This material deals primarily with processes regarding new hydrocarbon concessions in the Peruvian Amazon and observations on the national implementation process during the Humala government (2011–2016). The data comprises protocols of participant observation as well as transcripts of 68 semi-structured interviews with State and Indigenous actors as well as with "third parties" – advisers, translators, and the German Agency for International Development Cooperation (Deutsche Gesellschaft für Internationale Zusammenarbeit, GIZ) – involved in the implementation of prior consultation. This fieldwork material is embedded into an analysis of legal texts and primary documents, including public statements from national and international NGOs and State bodies, as well as secondary literature. The chapter is written from a perspective inspired by critical literature on contested norms in the fields of international relations (Wiener, 2014; 2018; Zimmermann, 2016), legal anthropology (Benda-Beckmann, 2001; Merry, 2006), and postcolonial socio-legal studies (Eckert et al., 2012; Santos & Rodríguez-Garavito, 2005). The central idea is that international norms, including rights stipulated in international law, such as prior consultation, are inherently contested and actors have to (re) negotiate their meaning in each context. In other words, contestations over norms are controversies about principles and rules, and are critically analysed by studying actors' different underlying understandings and the way in which those understandings compete.

The chapter begins with a brief overview of Peru's legislative consultation framework of 2011 and 2012, as well as of the first processes implemented under the Humala administration between 2011 and 2016. The second section critically examines the deficient prior consultations about extractive projects in the hydrocarbon and mining sectors. The third section highlights the positive changes towards the recognition of Indigenous Peoples' rights in Peru at the levels of national discourse, public policy-making, and the State administration. Finally, the last section reflects on the potential and limitations prior consultation may have for Peruvian recognition politics in the long term.

Peru's legislative framework: from mobilisations to law

The Peruvian prior consultation legislation comprises the 2011 Consultation Law (Law of the Right to Prior Consultation of Indigenous and Tribal Peoples, recognised in Convention 169 of the International Labour Organisation (ILO), Law No. 29785[1]); and its 2012 Regulatory Norm (Regulating Decree of Law 29785, Law of the Right to Prior Consultation of Indigenous and Tribal Peoples, recognised in the Convention 169 of the International Labour Organisation (ILO), No. 001–2012-MC[2]). Both pieces of legislation were highly disputed during the process of their adoption, which took place in two steps. First, Indigenous and government representatives negotiated a draft law after the Bagua conflict. The draft was submitted to the Congress but rejected by President García in 2010. Only after a change in government in 2011, was a restricted version of the Consultation Law adopted. Second, the Humala administration consulted Indigenous representatives about the Regulatory Norm. However, the process was soon abandoned by important Indigenous organisations due to the government's limited willingness to renegotiate the shortcomings of the Consultation Law. The Regulatory Norm was adopted without an Indigenous consensus in 2012 (Schilling-Vacaflor & Flemmer, 2015).

The Peruvian legislation is highly contested by Indigenous organisations for four main reasons. The first critique focuses on the limited definition of "whom to consult". In line with ILO 169, only Indigenous Peoples "directly affected" by a measure have to be consulted. This means that "indirectly affected" communities at the margins of an oil concession – who may also suffer from potential contamination and other negative impacts – will not be consulted. In contrast to ILO 169, the consultation legislation establishes that only people "directly descended" from precolonial populations are to be considered "Indigenous Peoples", and grants the State the authority to decide on Indigenous status via the establishment of an official database on Peru's Indigenous Peoples, which is not a binding but a referential instrument. Further, the consultation legislation stipulates that the Indigenous Peoples' representative organisations that participate in the consultations have to be from the geographic area of the planned measure. Accordingly, higher-level Indigenous organisations with more experienced leaders remain outside the processes.

The second critique concerns the question of "how to consult". The Peruvian government ignored demands for an obligatory phase of joint planning, and the legislation establishes that the consulting State entity prepares the consultation schedule and then presents it to the peoples to be consulted.

The third critique is that the Peruvian consultation legislation establishes that all measures previously adopted without consultation remain in force. This is also contested from a legal perspective, because the Peruvian government ratified ILO 169 in 1994, and it has been formally obliged to implement consultations since 1995.

The fourth and most important critique concerns the question of the "final decision" and the communities' right to say no to a project. In this regard, the consultation legislation repeats ILO 169 and stipulates that reaching a consensus is mandatory only when relocation or the disposal of hazardous waste is planned. For all other situations, the legal framework stipulates that only agreements are binding, which means that in the case of disagreement – if communities oppose a measure – the State can still implement the projects but the communities do not have a legal possibility to veto them (Schilling-Vacaflor & Flemmer, 2015). It is important to keep in mind that the consultation legislation was the only result of the much broader topics discussed at the roundtables of the post-Bagua dialogue process, which also dealt with initial demands for a joint development plan for the Amazon and constitutional changes to re-establish the inalienability of Indigenous territories.

Under the Humala administration, 24 consultations involving 23 different Indigenous Peoples were concluded. The entity responsible for carrying out and financing the consultation process is the one that plans the project or legal initiative, i.e. the sub-administration of the Ministry of Energy and Mines for the hydrocarbon and mining sectors. The Vice-Ministry of Intercultural Affairs (VMI) was created to provide "technical support" for the implementation process. The Ministry of Energy and Mines has implemented the highest number of processes, 11 of which concerned hydrocarbon concessions in the Amazon and four of which dealt with mining projects in the Andes. These were followed by four consultations about natural protected areas and one infrastructure project. Further, prior consultations in Peru have also concerned national laws that directly affect Indigenous Peoples: the four processes conducted during the Humala government concerned intercultural healthcare (2013–2014), the regulatory framework of the new Forestry and Wildlife Law (2014–2015), the regulatory norm of the law on Indigenous languages (2015–2016), and the plan for intercultural bilingual education (2015–2016) (Ministerio de Cultura, 2016).

Grassroots contestation and narrow consultations in the extractive sector

The implementation of consultations about hydrocarbon and mining projects was highly contested. In this regard, the Ministry of Energy and Mines, via Ministerial Resolution (*Resolución Ministerial*) No. 350–2012-MEM/DM, defined that prior consultation would be conducted exclusively at one single point in the licensing process. The key difference between the hydrocarbon and the mining sectors is that consultations over hydrocarbon projects are conducted regarding the concessions, while in mining consultation is only undertaken upon the commencement of the activity itself. This means that, in the hydrocarbon sector, consultations take place at an early stage, that is when the State gives a company permission to exploit and extract oil or gas in a certain area, or when the contract for a running concession is changed. For new concessions, there is no definition of the sites of production facilities or whether gas or oil will be extracted. In contrast, in the case of mining, consultation is undertaken later, i.e., during the process to issue the licence, and deals with the resolution that permits the start of activities. At this point, the projects have already been defined (Merino Acuña, 2018).

Hydrocarbon consultations, with one exception, concerned newly designated oil and gas concessions ("blocks") comprising rainforest areas of between 340 and 680 hectares, overlapping with the territories of various native communities (*Comunidades Nativas*). Conflicts arose from the beginning, and of the 26 processes initially announced (El Comercio, 2013) only eleven were concluded. Nine of them went "as planned", one was partly opposed (Block 191), and one is still the subject of a contentious renegotiation for a new consultation (Block 192/Ex-1AB). Additionally, other consultation processes began but were cancelled due to local opposition (Block 157). The two later cases were strongly contested by the grassroots and have implications for how Indigenous organisations can make use of prior consultation in Peru to sustain resistance. The consultation over planned oil concession 157 (PC157) in the southern rainforest was cancelled after two of the three communities consulted refused to participate in the consultation process supported by their representative organisation, the Native Federation of the Madre de Dios River and its Tributaries (*Federación Nativa del Río Madre de Dios y Afluentes*, FENAMAD), and legal advisers from the NGO International Law and Society Institute (*Instituto Internacional de Derecho y Sociedad*, IIDS). From the beginning of the consultation process, FENAMAD had emphasised that the organisation rejected any further hydrocarbon projects in the Madre de Dios region and opposed consultation processes because they did not include the right to veto projects (Flemmer, 2014b). However, the Peruvian State announced a total of three consultations in the region, and most communities agreed to be consulted. Only in the case of PC157, the community assembly of Tres Islas decided not to accept hydrocarbon activities in their territory and therefore refused to participate in the consultation (Viceministerio de Interculturalidad, 2015), and one of the other two communities joined Tres Islas in their opposition. In the aftermath, the area planned for the hydrocarbon concession was reduced, and the process of CP157 was cancelled. For Peru, this was the first case in which communities were able to successfully use the State's formal obligation to conduct prior consultations in order to avoid a new hydrocarbon project.

The conflict over Block 192 in the northern Amazon region of Loreto is still ongoing (see also the chapter by Doyle in this volume). This consultation was the only process that concerned a block already in operation and in an area with a long history of conflict (Schilling-Vacaflor, Flemmer, & Hujber, 2018). Local Indigenous federations, supported by NGOs, refused to participate in the consultation process before severe contamination caused by over 40 years of exploitation activities was addressed and adequate compensation paid. A first consultation or consultation attempt took place between 2014 and 2015, but it remained highly controversial and a majority of local organisations refused to recognise the process. The consultation ended with just one of the groups being consulted (O'Diana Rocca & Vega Díaz, 2016, p. 11). Since then, local organisations have organised an ongoing fight for true consultation that has involved UN entities and opened negotiations with the Peruvian Congress and the Executive (Lévano, 2017).

A glance at the biggest extractive sector in Peru – the mining industry – shows that the implementation process was very polarised and the consultation processes themselves even more limited than in the hydrocarbon sector. Before the first mining consultations finally began in September 2015, the Ministry of Energy and Mines, together with the National Society for Mining, Petrol and Energy (*Sociedad Nacional de Minería, Petróleo y Energía*, SNMPE), had refused to implement consultations over mining projects for several years. Their main argument was that Andean peasant communities should not be considered Indigenous (Gálvez & Sosa Villagarcia, 2013). The mining industry took an openly hostile position against prior consultation, stating that with this law, the government "shot itself in the foot" (Redacción Gestión, 2013a). Meanwhile, the SNMPE was furious that "now anyone who puts on a feather has the right to be consulted" (Paucar Albino, 2014). At the same time, new mining projects located in the Andes were to be authorised without consultations, with companies making communities sign that they did not consider themselves Indigenous (CEPES, 2014; Defensoría del Pueblo del Perú, 2014). Additionally, in April 2013, President Humala stated that he did not consider "agrarian communities", i.e., the peasant communities from the Andes, to be Indigenous Peoples (La Mula, 2013).

Only at the end of 2015 did mining consultations begin, and four processes had been concluded at the end of the Humala government. These processes were criticised for being rushed, i.e., taking between 34 to 43 days (Defensoría del Pueblo del Perú, 2015; Irigoyen, 2016). The Peruvian Ombudsman also noted that the company personnel interfered in communities' internal decision-making processes and that the agreements would only repeat the Ministry of Energy and Mines' previously established obligation to supervise the projects (Defensoría del Pueblo del Perú, 2016). National debates were heated and even during the implementation, the SNMPE maintained its negative rhetoric against prior consultation, and, in April 2016, the influential International Council on Mining and Metals (ICMM) intervened after Oxfam and other civil society organisations informed it of the SNMPE's statements (SERVINDI, 2016). However, affected communities did not mobilise against the processes. The main reason for this was that consultations were only conducted about projects where the company was already working in the area and communities had already signed "previous agreements" with them (Ocampo & Urrutia, 2016).

Through the back door: intercultural reforms at the national level

At the national level, Peru has long lagged "behind" in recognition politics, especially compared to its neighbouring countries Bolivia, Colombia, and Ecuador (Yashar, 2005). While the highly visible Bagua conflict in 2009 led to a political push for major changes regarding Indigenous Peoples' rights, the relatively minor mechanism of prior consultation has served as a kind of catalyst for intercultural policies. Discussions about the legal and administrative implementation of prior consultation have not ignited new political conflict or street protests, but have initiated contestation that has led to changes in national discourse, the State administration, and public policy-making. This way, the implementation of prior consultations has opened a "back door" for intercultural policies in Peru.

In the national discourse, the implementation of prior consultation has placed discussions about Indigenous Peoples and their rights in the public debate. This has given Peruvians the chance to rethink their colonial and Indigenous heritage, which was previously imagined to be the invention of the nation based on a heroic Incan past (Greene, 2006). On the one hand, these discussions brought to the fore the polarised ideas regarding "savage Amazonian tribes" and the peasant communities in the Andes, as promoted by the García government after Bagua (El Comercio, 2009; Hughes, 2010). On the other hand, it led to a more reflective and critical discussion about contemporary Indigenous Peoples in Peru, as well as to more substantial advances such as the establishment of ethnic self-identification as a category for the first time in the 2017 census (Redacción Gestión, 2016). With regard to indigeneity and rights, the polarised discourse initiated under President García and other conservative political voices, which questioned whether prior consultation should apply at all because Peru is a country of mestizos (Tribunal Constitucional, 2010), shifted to encompass a rather broad recognition of prior consultation which is now unquestioned, even by the sectors which initially opposed implementation. However, State entities are still very clear that consultation does not grant Indigenous Peoples the right to oppose a planned measure. From the Indigenous side, organisations still criticise the consultation legislation for not considering FPIC, but at the same time they invoke it and use it as a legal reference for their demands for more participation and influence in State decision-making. This is the case, for example, regarding the claim to also consult environmental impact studies as well as public policies implemented by subnational governments and municipalities (Inguil, 2018).

In the Peruvian State administration, new entities were created to implement and administer prior consultation. The most important institution is the Vice Ministry of Intercultural Affairs (VMI), which, as noted above, was created in July 2010 as part of the Ministry of Culture in order to provide technical assistance for the implementation of prior consultation, and represents the national entity for Indigenous "concerns" of the Executive. The VMI, together with the Human Rights Ombudsman (*Defensoría del Pueblo*) and the Ministry of Environment (MINAM, created in 2008), were the three State entities which promoted prior consultation (Redacción Gestión, 2013b; Paredes, 2015). The VMI started out with few resources and employed personnel educated in the fields of anthropology, sociology, and political science – disciplines which are more than underrepresented in the Peruvian State administration – as well as lawyers who had recently finished university in Lima (Flemmer, 2014a). At the beginning of the implementation process, the progressive position of VMI representatives on the promotion of Indigenous Peoples' rights was constrained by the entity's limited political space, and relations with the potential allies – national Indigenous organisations – were tense, because the government created the VMI without the participation of Indigenous representatives (SERVINDI, 2011).

From its creation in 2010 until the end of the Humala administration, the VMI – with the technical and financial support of international development organisations, especially the GIZ – gained political standing and provided binding opinions about the potential impacts of State measures on Indigenous Peoples, developed a clearer institutional profile, and began to interact with Indigenous organisations. In this regard, prior consultation has proved to be a topic that attracts funding and technical support. Two initiatives created for the implementation of prior consultation have a wider reach: the official database on Indigenous Peoples, and the register of Indigenous interpreters and translators. The database on Indigenous Peoples was first published by the VMI in 2013. It recognised 48 Amazonian Indigenous Peoples and – despite opposition from the mining sector – four Indigenous Peoples in the coastal region and the Andes (Amancio & Romo, 2015). The database is the first national, State-administered compilation of information on Indigenous Peoples and communal lands in Peru, and has a fundamental role in the production and distribution of knowledge on Indigenous Peoples, providing information about the needs of the communities (Sanchez, 2015). The second initiative is the VMI's official register of translators and interpreters (Directive No. 006–2012/MC), which was established in October 2012. Created from scratch in June 2016, the register lists 171 interpreters and/or translators of 29 Indigenous languages who have been trained by the VMI to work in prior consultations as well as other parts of the State administration and the judiciary (Ministerio de Cultura, 2017). Also very important is the improvement in the access of Indigenous Peoples to the judicial system, for example through the translation of legal texts and interpreting services in court (Pedro, Ciudad, & Howard, 2018).

Adopting a long-term perspective, the VMI has developed a "National Policy for the Mainstreaming of Interculturalism" via the Supreme Decree No. 003–2015-MC, and has promoted its implementation as a top priority for the administration in the 2016–2021 administration (Ministerio de Cultura, 2016). The National Policy for the Mainstreaming of Interculturalism consists of four pillars: (1) strengthening the State's "intercultural capacities"; (2) recognising cultural diversity; (3) fighting racial discrimination; and (4) granting specialised attention to Indigenous Peoples and the Afro-Peruvian population to systematically incorporate an intercultural focus into public policies and State administration (Ministerio de Cultura, 2016).

In terms of public policy-making, Indigenous organisations have gained influence in legal and political decision-making processes, as consultations have been conducted about intercultural policies in healthcare and education, as well as about forestry sector regulations, with national and regional Indigenous organisations. The latter set of consultations – those about forestry regulations – resulted in legislation supported by the State and Indigenous Peoples (Cauper & Hidalgo, 2016). This means that Indigenous Peoples have gained influence in the design of laws, and their access to public policies has improved. At the local level, communities have articulated demands for land titles and access to State services during consultation processes, which has resulted in the improvement of social services and new land titling processes (Ministerio de Cultura, 2016). At the level of organisations, the Working Group of Indigenous Policies (*Grupo de Trabajo de Políticas Indígenas*) was created in November 2014 as a permanent space for coordination and dialogue between Indigenous Peoples and the Executive branch.

Conclusion: the combined effects of legal change, policy reform, and grassroots contestation

From the perspective of contested norms, controversies about prior consultation have developed at different sites, but the negotiations about their meaning in law, concrete practices surrounding extractive projects, and in national policies, are interconnected. At Peru's extraction frontiers in the Amazon and the Andes, prior consultation processes have been narrow, participatory events resulting in vague agreements regarding the planned projects, and have mainly repeated the already established legal standards for companies' codes of conduct. The extraction model is not called into question and prior consultation, as established in Peruvian law, does not allow the communities to say no to a project. However, communities that have refused to enter into State consultations have been able to use the State's formal obligation to conduct prior consultations in order to avoid new hydrocarbon projects. Simultaneously, at the level of national politics, controversies about the implementation of "prior consultation" generated an entry point for a new State approach towards Indigenous Peoples. This has manifested itself in a more differentiated public discourse; concrete institutional reforms such as the creation of the VMI and the official register of Indigenous translators; and improvements in the design of public policies and Indigenous Peoples' access to State services. These progressive intercultural reforms at the national level are oriented toward symbolically paying respect to Indigenous Peoples, providing them with access to public State services, and, to a lesser extent, giving them influence in the State's decision-making processes. Hopefully, these may be the first steps towards more inclusive, redistributive recognition politics in Peru.

Time will tell whether prior consultation served as a kind of "Trojan horse" for recognition politics in the Peruvian State administration. On the one hand, under the García government prior consultation was a sufficiently inoffensive mechanism for the neoliberal government to turn it into law. On the other hand, prior consultation has served as a catalyst for more substantial debate and reform towards intercultural policies. Additionally, prior consultation as a formal obligation of the State to consult extractive projects has proven to be useful for Indigenous Peoples' demands. However, these processes cannot stand alone but rather need to be embedded in broader and permanent participatory frameworks that have a real influence on State decision-making processes. If we look at recent developments in Peru, the opportunities seem to be diminishing. The Humala administration, as well as the government under Pedro Pablo Kuczynski, have adopted various legal packages to promote private investment by lowering social and environmental standards (Orihuela & Paredes, 2017). However, mobilisations continue to be effective, and protesters have made the Peruvian congress abolish some of these laws (SERVINDI, 2017).

In conclusion, the long-term impacts of prior consultation remain uncertain. Grassroots contestation, legal change, and policy reform have interacted and have opened up new opportunities for Indigenous Peoples' politics in Peru. However, it remains to be seen whether they can lead to a real transformation of the State's colonial heritage and the Peruvian economy's extractive imperative. A new era of recognition politics in Peru will have to turn rhetorical, legal, and administrative changes into a profound restructuring of the State, and establish broad influence for Indigenous organisations in State decision-making at all levels of public policy. Processes of intercultural translation and intercultural mediation are key issues in this regard. For Indigenous Peoples, prior consultation has become an additional instrument in their repertoires of resistance. Nevertheless, the priority of resource extraction is unchanged. Stopping projects requires a combination of legal strategies with extra-legal contention.

Notes

1 In Spanish, *Ley del Derecho a la Consulta Previa a los Pueblos Indígenas u Originarios, reconocido en el Convenio 169 de la Organización Internacional del Trabajo* (OIT), *Ley* No. 29785.
2 In Spanish, *Reglamento de la Ley N° 29785, Ley del Derecho a la Consulta Previa a los Pueblos Indígenas u Originarios reconocido en el Convenio 169 de la Organización Internacional del Trabajo* (OIT), No. 001–2012-MC.

References

Amancio, N. & Romo, V. (2015). Los secretos mineros detrás de la lista de comunidades indígenas del Perú. *Ojo Público*, 22 July. Retrieved from http://ojo-publico.com/77/los-secretos-detras-de-la-lista-de-comunidades-indigenas-del-peru.

Benda-Beckmann, F. (2001). Legal Pluralism and Social Justice in Economic and Political Development. *IDS Bulletin*, 32(1), 46–56.

CEPES. (2014). Identidades indígenas en tiempos de la consulta previa. *La Revista Agraria*, 15(163). Retrieved from http://www.cepes.org.pe/portal/node/224.

Defensoría del Pueblo del Perú. (2014). Oficio N° 249–2014-DP/AMASPPI: Oficio dirigido al Ministerio de Energía y Minas en relación a la falta de implementación del derecho a la consulta previa de los pueblos indígenas en el subsector minería.

Defensoría del Pueblo del Perú. (2015). Oficio-026–2015-DP: Oficio dirigido al Ministro de Energía y Minas, Eleodoro Mayorga, en relación al D.S. N° 001–2015-MEM.

Defensoría del Pueblo del Perú. (2016). Oficio N° 504–2016-DP: Oficio al Ministro de Energía y Minas sobre los procesos de consulta previa en el sector minería.

El Comercio. (2009). El video del Gobierno sobre los sucesos en Bagua polariza más el conflicto. *El Comercio*. Retrieved from http://archivo.elcomercio.pe/politica/gobierno/video-gobierno-sobre-sucesos-bagua-polariza-mas-conflicto-noticia-297990.

El Comercio. (2013). Perú-Petro iniciará en julio 26 consultas previas en la Amazonía. *El Comercio*. Retrieved from http://archivo.elcomercio.pe/economia/peru/peru-p etro-iniciara-julio-26-consultas-previas-amazonia-noticia-1589101.

Flemmer, R. (2014a). Interview D31, 22 April. Lima.

Flemmer, R. (2014b). Interview D36, 21 October. Puerto Maldonado.

Fraser, N. (1995). From Redistribution to Recognition? Dilemmas of Justice in a 'Post-Socialist' Age. *New Left Review*, I/212, 68–93.

Fraser, N. (2005). Reframing Justice in a Globalized World. *New Left Review*, 36, 1–19.

Gálvez, Á. & Sosa Villagarcia, P. (2013). "El problema del indio": una mirada a la implementación de la consulta previa desde la lógica del Estado y sus funcionarios. *Revista Argumentos*, 7(5), 7–15.

Gavaldà Palacín, M. (2013). *Gas Amazónico. Los pueblos indígenas frente al avance de las fronteras extractivas en Perú*. Barcelona: Icaria.

Greene, S. (2006). Getting over the Andes: The Geo-Eco-Politics of Indigenous Movements in Peru's Twenty-First Century Inca Empire. *Journal of Latin American Studies*, 38(2), 327–354.

Hughes, N. (2010). Indigenous Protest in Peru: The "Orchard Dog" Bites Back. *Social Movement Studies*, 9(1), 85–90.

Inguil, S. (2018). 'Detrás de una consulta no hay un diálogo legítimo, sino la imposición del lado más fuerte'. Entrevista con Juan Carlos Ruíz. Retrieved from https://consultape.com/2018/02/26/detras-de-una-consulta-no-hay-un-dialogo-legitim o-sino-la-imposicion-del-lado-mas-fuerte/.

Irigoyen, M. (2016). Consulta previa en minería: algunos rasgos de la mirada empresarial. In K. Vargas (Ed.), *La implementación del derecho a la consulta previa en Perú. Aporte para el análisis y la garantía de los derechos de los pueblos indígenas* (pp. 191–204). Lima: GIZ.

La Mula. (2013). El Presidente Ollanta Humala habla sobre la Consulta Previa, 16 November. Retrieved from https://youtube/pf2WeHWlYwM.

Lévano, M. (2017). Derechos en riesgo: La lucha por la verdadera consulta a la comunidad continúa en Perú. *Oxfam*. Retrieved from https://peru.oxfam.org/blog/derechos-en-riesgo-la-lucha-por-la-verdadera-consulta-la-comunidad-contin%C3%BAa-en-per%C3%BA.

Merino Acuña, R. (2018). Re-politicizing Participation or Reframing Environmental Governance? Beyond Indigenous' Prior Consultation and Citizen Participation. *World Development*, 11, 75–83.

Merry, S. (2006). *Human Rights and Gender Violence: Translating International Law into Local Justice*. Chicago: University of Chicago Press.

Ministerio de Cultura. (2016). Presentación ante la Comisión de Pueblos Andinos, Amazónicos y Afroperuanos, Ambiente y Ecología del Congreso de la República. Retrieved from http://www.congreso.gob.pe/Docs/comisiones2016/PueblosAndinosEcologia/files/ministerio_de_cultura.pdf.

Ministerio de Cultura. (2017). Interpretes y Traductores capacitados 2013–2015. Retrieved from http://www.infocultura.cultura.pe/infocultura/.

Ocampo, D. & Urrutia, I. (2016). La implementación de la consulta en el sector minero: una mirada a los primeros procesos. In K. Vargas (Ed.), *La implementación del derecho a la consulta previa en Perú. Aporte para el análisis y la garantía de los derechos de los pueblos indígenas* (pp. 169–190). Lima: GIZ.

O'Diana Rocca, R. & Vega Díaz, I. (2016). *¿Cómo va la aplicación de la Consulta Previa en el Perú? Avances y retos*. Lima: CAAAP.

Orihuela, C. & Paredes, M. (2017). Fragmented Layering: Building a Green State for Mining in Peru. In E. Dargent, J. Orihuela, M. Paredes & M. Ulfe (Eds). *Resource Booms and Institutional Pathways: The Case of the Extractive Industry in Peru* (pp. 97–118). London: Palgrave Macmillan.

Orta-Martínez, M. & Finer, M. (2010). Oil Frontiers and Indigenous Resistance in the Peruvian Amazon. *Ecological Economics*, 70(2), 207–218.

Paredes, M. (2015). Transnational Networks Acting from Below: Indigenous Prior Consultation and the Peruvian Paradox. Unpublished conference paper.

Paucar Albino, J. (2014). ¿Qué dicen los opositores de la consulta previa a las comunidades campesinas? *La Mula*. Retrieved from https://redaccion.lamula.pe/2014/06/16/que-dicen-los-opositores-de-la-consulta-previa-a-las-comunidades-campesinas/jorgepaucar/.

Pedro, R., Luis, A., & Howard, R. (2018). The Role of Indigenous Interpreters in the Peruvian Intercultural, Bilingual Justice System. In E. Monzó-Nebot & J. Jiménez Salcedo (Eds.) *Translating and Interpreting Justice in a Postmonolingual Age* (pp. 91–110). Malaga: Vernon Press.

Redacción Gestión. (2013a). Iván Lanegra renunció al viceministerio de Interculturalidad. *La Gestión*, 4 May. Retrieved from https://gestion.pe/peru/politica/ivan-lanegra-renuncio-viceministerio-interculturalidad-37568.

Redacción Gestión. (2013b). Daniel Saba: 'La Consulta Previa es una idea romántica: no perderíamos nada sin ella'. *La Gestión*, 28 May. Retrieved from https://gestion.pe/economia/daniel-saba-consulta-previa-idea-romantica-perderiamos-39369.

Redacción Gestión. (2016). INEI incluirá por primera vez dos preguntas de autoidentificación étnica en Censo 2017. *La Gestión*, 11 November. Retrieved from https://gestion.pe/peru/politica/inei-incluira-primera-vez-dos-preguntas-autoidentificacio n-etnica-censo-2017-121543.

Sanchez, D. (2015). ¿Sabemos cuánta población indígena hay en el Perú? *Revista Ideele*, 11 February. Retrieved from http://revistaideele.com/ideele/content/¿sabem os-cuánta-población-indígena-hay-en-el-perú.

Schilling-Vacaflor, A. & Flemmer, R. (2015). Conflict Transformation through Prior Consultation? Lessons from Peru. *Journal of Latin American Studies*, 47(4), 811–839.

Schilling-Vacaflor, A., Flemmer, R., & Hujber, A. (2018). Contesting the Hydrocarbon Frontiers: State Depoliticizing Practices and Local Responses in Peru. *World Development*, 108, 74–85.

Servindi. (2011). Perú: Organizaciones indígenas se pronuncian sobre institucionalidad pública para PPII. *Servindi*, 7 October. Retrieved from https://www.servindi.org/actualidad/52672.

Servindi. (2016). Declaraciones anti-consulta de empresarios preocupa a ente mundial. *Servindi*, 6 April. Retrieved from https://www.servindi.org/actualidad-noticias/06/04/2016/declaraciones-anti-consulta-de-empresarios-preocupa-ente-mundial.

Servindi. (2017). Pueblos originarios exigen archivar "Nueva ley del Despojo". *Servindi*, 12 December. Retrieved from https://www.servindi.org/actualidad-noticias/19/10/2017/pacto-de-unidad-exige-archivamiento-de-nueva-ley-del-despojo.

Tribunal Constitucional. (2010). Sentencia del Tribunal Constitucional. EXP. N° 05427–02009- PC/TC Recurso de agravio constitucional interpuesto por la Asociación Interétnica de Desarrollo de la Selva (AIDESEP), Lima. Retrieved from http://www.tc.gob.pe/jurisprudencia/2010/05427-2009-AC.pdf.

Vargas, K. (2016). *La implementación del derecho a la consulta previa en el Perú. Aportes para el análisis y la garantía de los derechos colectivos de los pueblos indígena*. Lima: GIZ.

Viceministerio de Interculturalidad. (2015). Consulta Previa para el Lote 191. Retrieved from http://consultaprevia.cultura.gob.pe/proceso/consulta-previa-para-el-lote-191/.

Wiener, A. (2014). *Theory of Contestation*. Heidelberg: Springer.

Wiener, A. (2018). *Contestation and Constitution of Norms in Global International Relations*. New York: Cambridge University Press.

Yashar, D. (2005). *Contesting Citizenship in Latin America: The Rise of Indigenous Movements and the Postliberal Challenge*. Cambridge: Cambridge University Press.

Zimmermann, L. (2016). Same Same or Different? Norm Diffusion Between Resistance, Compliance, and Localization in Post-conflict States. *International Studies Perspectives*, 17, 98–115.

8 Processes and failures of prior consultations with Indigenous Peoples in Chile

Alexandra Tomaselli

Introduction

In 2008, the first mandate of President Michelle Bachelet and the (much delayed) ratification by Chile of the ILO Convention No.169 Concerning Indigenous and Tribal Peoples in Independent Countries of 1989 (ILO 169) relaunched the debate on Indigenous rights throughout the country. In particular, it focused on their right to consultation. In the intervening years, Chile began to hold a number of consultations with Indigenous Peoples, some of which were carried out even before the treaty entered into force. However, it is important to note that some of these consultations repeatedly addressed the same issues over the course of eight years, involving *de facto* overlapping consultation processes, nullifying and undermining the credibility of the previous processes. This is the case of the consultations on the constitutional recognition of Indigenous Peoples and the creation of a Ministry of Indigenous Peoples, a National Council of Indigenous Peoples, and (other) Councils of Indigenous Peoples. The nine Indigenous Peoples of Chile were consulted five times on some or all of these issues: twice in 2009, once in 2011, again in late 2014–early 2015, and once again in 2017.

Furthermore, the legislation over the regulation of the right to be consulted has changed considerably over the years, creating further confusion and overlaps. Eventually, it was codified by Supreme Decree No. 66 of 15 November 2013 (Ministerio de Desarrollo Social, 2013), which entered into force on 4 March 2014 (hereinafter, Decree 66). This decree decentralises the possibility to carry out consultations with Indigenous Peoples to a number of public authorities, and it also stipulates that even those consultations that do not obtain the consent of the concerned peoples should be considered as having fulfilled their objective (see further below). This casts a shadow over the possibilities for fair implementation of this internationally-recognised and fundamental right of Indigenous Peoples.

Against this background, this chapter analyses the process and results of selected cases of consultations with Indigenous Peoples in Chile, and explores the reasons for their failures. First, it offers a short overview of Indigenous Peoples in Chile and the legislative framework vis-á-vis the (weak) protection

DOI: 10.4324/9781351042109-11

of their rights. Second, it focuses on how Chile has regulated their right to consultation. Third, it addresses the five processes of consultations over those potential institutional reforms that would have directly affected the entire discourse and protection system of Indigenous Peoples in Chile – and thus called here "institutional" consultations. Finally, it provides some concluding remarks on the reasons for the failures of these consultation processes.

A short overview of Indigenous Peoples in Chile and their rights

The 2012 census in Chile registered that 11.07% of the Chilean population (i.e., 1,842,607 people) self-identify as Indigenous. According to this official data, the most numerous Indigenous People in Chile are the Mapuche (almost 82%), followed, in order, by the Aymara, Diaguita, Quechua, Colla, Rapa Nui, LikanAntai (also called Atacameños), Yagán o Yámana, and Kawésqar (Instituto Nacional de Estadísticas Chile, 2012, p. 172). These are the nine Indigenous Peoples recognised by the so-called *Ley Indígena* (Indigenous Law) (Ministerio de Planificación y Cooperación, 1993, article 1, para. 2), although the Diaguita were included only in 2006 by modifying article 2 (para. 1) of the Indigenous Law (Ministerio de Planificación y Cooperación, 2006).

Approximately 31% of the population of Indigenous Peoples in Chile are affected by multidimensional poverty, which includes housing, income, health, and education (Observatorio Ciudadano de Chile, 2017). They have long suffered, and continue to suffer, from land dispossession (Aylwin, 1995; Bengoa 2000; Carruthers & Rodriguez, 2009; Rosti, 2008; Toledo Llanca-queo, 2006a). Furthermore, they are often prosecuted in military tribunals under the Antiterrorist Law (Ministerio del Interior, 1984). This law was adopted under Pinochet's dictatorship and involves severe penalties and the use of arbitrary means of evidence, such as declarations by unidentified witnesses (Ministerio del Interior, 1984, articles 15–18), who are known as the *sin rostro* (faceless) (Tomaselli, 2016).

The Indigenous Law is the main domestic legal source of recognition of Indigenous rights in Chile. Despite several attempts after the restoration of democracy (Toledo Llancaqueo 2006b; Tomaselli, 2016), Indigenous Peoples still lack constitutional recognition, which was the object of three of the recent consultation processes (see below).

In short, the Indigenous Law recognises the existence of the aforementioned Indigenous Peoples, although it defines them as ethnicities (*etnías*); establishes a definition of an Indigenous community and how this can be created; recognises land rights and titles in accordance with nineteenth-century documents; con-stitutes a fund for land and water (re)distribution; recognises a number of cultural rights, including intercultural and bilingual education; states some participatory rights; and creates the National Corporation of Indigenous Development (CONADI), which is the public agency in charge, *inter alia*, of the promotion of Indigenous policy at State level and the implementation of the Indigenous Law (Ministerio de Planificación y Cooperación, 1993, articles 1, 9–10, 12–19, 20–22,

28–31, 32–33, 34–35 and 38 and ff.).[1] However, this law does not recognise Indigenous Peoples as such, and remains essentially unimplemented (Instituto de Estudios Indígenas, 2003; Instituto Nacional de Derechos Humanos, 2013; Vergara, Gundermann & Foerster, 2006), particularly with regard to bilingual education (Webb & Radcliff, 2013).

Another relevant law is the so-called *Ley Lafkenche* that refers to the Mapuche peoples of the coast (Ministerio del Trabajo y Previsión Social, 2008; Ministerio de Planificación y Cooperación, 2008). This (long-awaited) law establishes coastal marine space(s) for those Indigenous Peoples that have preserved a customary use of them (art. 3). Nevertheless, it is also poorly implemented (Aylwin & Silva, 2014; Kaempfe & Ready, 2011).

Chile is also bound to those international obligations descending from the human rights treaties it has signed and ratified, including the core nine treaties (United Nations Human Rights Office of the High Commissioner, n.d.), in accordance with article 5 of its Constitution (Meza-Lopehandía, 2010). Among these treaties, ILO 169 was ratified by Chile in late 2008, and it entered into force on 15 September 2009 (Ministerio de Relaciones Exteriores, 2008). Chile also voted in favour of the UN Declaration on the Rights of Indigenous Peoples (UNDRIP) (United Nations Press Release, 2007).

Since the analysis of the complex Chilean context with regard to Indigenous Peoples goes beyond the scope of this paper – and due to space constraints – it suffices to highlight here the two most relevant developments in the Chilean Indigenous agenda. The first is the apology to Indigenous Peoples by (former) President Bachelet in June 2017 for the past atrocities and errors committed by the Chilean State, an act that was long overdue and which had a political resonance, but nevertheless resulted in little concrete action (Milesi, 2017). The second is the recent, worrying declaration by newly re-elected President Piñera of his intent to withdraw Chile from ILO 169 (Bertin, 2018), which – as of December 2018 – had not been followed up by any official action to this effect.

The Chilean legislative framework on the right to prior consultation of Indigenous Peoples

Beyond ILO 169 and UNDRIP, the right to prior consultation of Indigenous Peoples, but not their Free, Prior and Informed Consent (FPIC), is regulated in Chile by the abovementioned Decree 66 (Ministerio de Desarrollo Social, 2013).[2] This executive order takes inspiration from some of the international standards but ultimately fails to meet some key requirements contained in both ILO 169 and the UNDRIP.[3] It establishes a detailed procedure, which includes five specific steps: (1) a common elaboration of the consultation plan; (2) the dissemination of information about the consultation process; (3) Indigenous Peoples' own internal deliberation process; (4) a "dialogue" between Indigenous Peoples and State representatives; and (5) classification and communication of the results, which concludes the consultation process

(Ministerio de Desarrollo Social, 2013, article 16). Nevertheless, it considers "all the possible efforts" made by the authorities involved to reach an agreement with Indigenous Peoples or obtain their consent – rather than their actual agreement or consent – as fulfilling the duty of the consultation (Ministerio de Desarrollo Social, 2013, article 3); it enumerates the public bodies which have a duty to consult Indigenous Peoples, but it leaves out other relevant State actors that adopt measures likely to directly affect Indigenous Peoples, such as the Armed Forces, public companies, and other local administrative bodies (e.g., the municipal governments) (Ministerio de Desarrollo Social, 2013, article 4); and it requires consultation only for those administrative and legislative measures likely to cause a "significant and specific impact" on Indigenous Peoples, without providing further criteria (Ministerio de Desarrollo Social, 2013, article 7).

In addition to Decree 66, Decree 40 of 2013 regulates the Environmental Impact Assessment System that is required for all investment projects in Chile (Ministerio del Medio Ambiente, 2013). Therefore, it governs all the consultation processes with Indigenous Peoples in the case of extractive or infrastructural projects that are to be built within or close to their territories (e.g., a new mine or a hydroelectric plant). An examination of this executive order reveals that – while it enshrines some international standards, such as the duty to consult in good faith and observe appropriate socio-cultural mechanisms (see Ministerio del Medio Ambiente, 2013, article 85) – other articles contain some worrying stipulations. This is the case of article 83, in relation to art. 7 of ILO 169, which limits the duty to consult Indigenous Peoples only to those projects which have a "high impact" on them, i.e., when relocation is required or their lifestyle and customs are likely to suffer a "significant alteration" (Ministerio del Medio Ambiente, 2013, article 83). This means that, for all those projects that are considered to have an unspecified "low impact" on Indigenous Peoples, a few "informative meetings" would suffice (Ministerio del Medio Ambiente, 2013, article 86). This leaves room for arbitrary interpretation, e.g., when it comes to defining what a "significant alteration" may be or imply.

The institutional consultations in Chile (2009–2017)

Since early 2009, a number of consultations with Indigenous Peoples have been carried out by the regional offices of CONADI. This means that some of them were realised prior to the entry into force of ILO 169.

For the purposes of this chapter, five processes of consultation are analysed. These consultations essentially addressed very similar, and, in some cases, exactly the same issues. These were: (1) the participation of Indigenous Peoples in the Chilean Congress, including the establishment of a Secretariat of Indigenous Affairs, a Ministerial Council of Indigenous Affairs, a regional Unit for Indigenous Affairs in each Governor's office, and a Council of Indigenous Peoples (2009); (2) the constitutional recognition of Indigenous Peoples and their rights (2009); (3) (again) the constitutional reform regarding Indigenous Peoples and their rights, the creation of a Council of Indigenous

Peoples and of an Indigenous Development Agency, the procedure for consultation, and the Environmental Impact Assessment System (2011); (4) the establishment of a Ministry of Indigenous Affairs, and (again) an Indigenous Development Agency, a (National) Council of Indigenous Peoples and other nine Councils, one for each Indigenous People that is recognised in Chile (2014–2015); and (5) (again) the constitutional reform (2017). Out of these consultation processes, four were carried out during the first and second mandates of (former) President Bachelet (2006–2010, and 2014–2018, respectively), and one under the first term of the current President Sebastián Piñera (2010–2014; his new term runs from 2018 to 2022).

The first consultation under scrutiny here referred to a number of proposals to favour Indigenous participation within the Congress, which included the establishment of a Secretariat of Indigenous Affairs, a Ministerial Council of Indigenous Affairs, a regional Unit for Indigenous Affairs in each Governor's office, and a Council of Indigenous Peoples (CONADI, 2009a). The process of this consultation consisted of sending letters to those Indigenous organisations that were registered by CONADI in accordance with the Indigenous Law (Ministerio de Planificación y Cooperación, 1993, article 12), inviting them to be part of "participatory dialogues" (CONADI, 2009a, p. 12). In early 2009, 4,599 letters were sent (CONADI, 2009a, p. 10), and these dialogues were held between 12 and 27 March 2009, which brought together 789 Indigenous representatives (CONADI, 2009a, p. 12). The offices of CONADI received only 410 replies to the letters they dispatched, i.e., less than 10% of those that were sent. As for the dialogues, although (almost) 800 representatives is a high number *per se*, it barely represented 0.04% of the Chilean population that self-identified as Indigenous in the 2012 census (Instituto Nacional de Estadísticas Chile, 2012). Although they officially disseminated information about the consultation in both national and local media, on the CONADI website, and at the offices of CONADI by putting up a number of posters (CONADI, 2009a, p. 6) – which may be a sign of the potential good faith and efforts on the part of CONADI – the timeframe for this fundamental and complex consultation was extremely short. Moreover, it occurred during the southern hemisphere's summer months, during which many Indigenous individuals travel away from their community for seasonal work. Finally, CONADI and the ministerial staff in charge of this consultation process did not take into consideration that many Indigenous communities are isolated or located in remote rural places, and Indigenous individuals may not travel frequently to the urban areas where the CONADI offices are situated.

Subsequent actions included submitting two draft laws, one regarding the Ministry of Indigenous Affairs and the Agency for Indigenous Development, and one the Council of Indigenous Peoples. Both have remained pending since 20 January 2010, i.e., after the presidential election. In July 2010, the newly established government of President Piñera created a Ministerial Council for Indigenous Affairs. However, all it did was recreate a body that had already been founded by President Bachelet (Tomaselli, 2012). Hence, Indigenous Peoples were called on to express their consent on the creation of a body that already existed.

The second consultation analysed here relates to the first time Indigenous Peoples were consulted about the constitutional reform that would have mentioned their existence and included their rights. The process of this consultation foresaw the following phases: dispatching the instructions and the guidelines for the consultation via letters or emails to Indigenous communities, organisations, and associations, and through the media between 13 April and 1 June; a one-day training workshop for civil servants working closely with Indigenous Peoples (scheduled on 21 April); an unspecified number of workshops on the consultation process to be held at provincial or local level between 24 and 30 April; and potential talks to be organised with the local CONADI offices between 22 April and 5 June (CONADI, 2009b, p. 2). CONADI and the ministerial staff in charge of this consultation process never published the official results, but they mentioned them in a document prepared by CONADI and the government for the Committee of Experts on the Application of Conventions and Recommendations of the ILO after the first year following the ratification of ILO 169. In particular, they declared that they had received 428 replies, which were submitted to the Chilean Congress (CONADI, 2010, p. 38). Hence, also, in this case, the responses from Indigenous Peoples were extremely sparse, which may be a sign of the lack of appropriateness of the methodology and timeframe of this consultation. Indeed, they scheduled the workshops during only one week, and allowed only one and a half months for sending the letters and receiving potential replies.

The third consultation addressed here occurred under the first mandate of President Piñera, the so-called "Broad Consultation" (*Gran Consulta*). It addressed a number of unresolved questions, including the constitutional reform concerning Indigenous Peoples and their rights; the creation of an Indigenous Development Agency (which should have replaced the CONADI), and a Council of Indigenous Peoples; the overall procedure of consultations with Indigenous Peoples; and the Environmental Impact Assessment System (Gobierno de Chile, 2011). This consultation ran online from April to September 2011. Indigenous individuals could vote via the internet or at the offices of CONADI. Apparently, CONADI and the ministerial staff in charge of this consultation process organised a few workshops between June and July 2011 (Gobierno de Chile, 2011), but no minutes or participants' lists have ever been made available, and whether they were actually carried out or not remains uncertain. Due to the intensification of the debate on the legislation over the right to consultation during the same months, the Government decided to suspend this Broad Consultation in September (Marimán Quemenado, 2012). As mentioned above, many Indigenous families live in remote places, and if they cannot access the internet from home, must travel to an urban centre in order to do so. Hence, there were implications for the potential of these peoples not only to be adequately informed about the process but also, and most importantly, to participate in it. In sum, this consultation regarded those fundamental issues that had already been addressed by the 2009 consultation, and, likewise, it did not come to any conclusion. The constitutional reform was the object of another consultation in 2017 (see below).

The fourth consultation process under analysis here was launched by the (re-elected) government of Bachelet (again) on the establishment of the Ministry of Indigenous Affairs, a National Council of Indigenous Peoples and another nine Councils of Indigenous Peoples in June 2014 (Tomaselli, 2016, p. 460). This was the third consultation on the creation of these bodies, which, as was the case with the previous proposals, would reform CONADI. This consultation ran from September 2014 to January 2015, and was organised according to Decree 66. This implied the participation of those Indigenous communities enrolled in CONADI's register (Ministerio de Desarrollo Social, 2013, article 5; Ministerio de Planificación y Cooperación, 1993, article 12), and the use of the aforementioned five-step procedure. As a further guarantee of the transparency and good faith of this process, the Ministry signed agreements with international bodies and the Universidad Diego Portales, a well-respected university in the areas of law and human rights (Ministerio de Desarrollo Social, 2014a; Ministerio de Desarrollo Social, 2014b).

This consultation officially came to an end on 30 January 2015. The Ministry of Social Development proclaimed it a great success (Ministerio de Desarrollo Social, 2015a) due to the broad participation of Indigenous representatives, since the phases of the consultation were carried out in 122 places (Ojeda, 2015). This is also reported in the Final Report of the consultation, which registered the participation of 6,833 Indigenous representatives (Ministerio de Desarrollo Social, 2015b, p. 23). Nonetheless, Indigenous and other civil society organisations alleged a number of irregularities. They denounced a lack of transparency in the overall process; the manipulation of Indigenous individuals during the phases of the consultation; the lack of legitimacy of some of the Indigenous representatives involved, and – at the same time – the denial of the opportunity to designate proper representatives on the part of other communities. Aymara, Mapuche, and Quechua marched against the consultation in Santiago, while other Mapuche, and Kawésqar did the same in the South. Other Indigenous organisations filed writs of *Amparo* against local and national authorities alleging, *inter alia*, the omission of documents, false declarations, and a lack of proper information during the consultation process (Ojeda, 2015).

The results of this consultation were published in the first half of 2015 (Ministerio de Desarrollo Social, 2015b). They reported the choices of the Indigenous representatives who were consulted, which are summarised as follows:

- a favourable opinion on the establishment of the Ministry of Indigenous Affairs as well as the Council(s) of Indigenous Peoples, although the former shall not only supervise (*velar*) but duly protect (*resguardar*) the fair application of Indigenous rights;
- the request to create a stable body in charge of consultation processes within this Ministry of Indigenous Affairs;
- an agreement to keep a Register of Traditional Authorities by the Ministry of Indigenous Affairs, with the exception of the Rapa Nui people;
- conformity on the reform of CONADI and the transfer of its funds to the Ministry of Indigenous Affairs;
- the demand to increase the numbers of representatives in the Council(s) of Indigenous Peoples (Ministerio de Desarrollo Social, 2015b).

Two draft laws were thus submitted to create the National Council of Indigenous Peoples, and the other nine councils (Draft Law/*Boletín* No. 10526–06), and the Ministry of Indigenous Affairs (Draft Law/*Boletín* No. 10525–06) in January 2016. Nevertheless, as this chapter went to press (December 2018), the former was still under the scrutiny of the Chilean Senate, while the latter was withdrawn in May 2016, i.e., four months after its submission. This was supposedly due to a new legislative strategy (Cooperativa.cl, 2016), which has not yet been implemented, and which is unlikely to be promoted during the second mandate of President Piñera. Moreover, the Council of Atacameño Peoples denounced, *inter alia*, that the latest draft law on the Council(s) of Indigenous Peoples limits the deliberations of these bodies to recommendations and observations only, rather than resolutions with at least some binding effects (Mapuexpress, 2018).

The fifth and last process of consultation under analysis here dealt with (again) the constitutional recognition of Indigenous Peoples and a number of their rights. This process included a preparatory phase called the Indigenous Constituent Process (*Proceso Constituyente Indígena*), within the framework of the debates on the general constitutional reform. It consisted of pre-organised and self-arranged meetings to collect ideas and proposals concerning recognition at constitutional level as well as online contributions (Ministerio de Desarrollo Social, 2016, article 3, para. 5; Observatorio del Proceso Constituyente en Chile-Fundación RED, n. d.; Gobierno de Chile, 2017). This process took place between August 2016 and January 2017. The official data reported the participation of approximately 17,000 Indigenous representatives (Gobierno de Chile, 2017).

The proposals collected were reworked into a text that was put forward for consultation by Resolution No. 726 of 22 July 2017. This consultation followed the five-step procedure in accordance with Decree 66 (Ministerio de Desarrollo Social, 2013, article 16), and was held between August and November 2017. The overall process was organised with the collaboration of experts from the United Nations and the University of Chile, and representatives of the National Institute of Human Rights were invited as observers (Gobierno de Chile, 2017).

The official data report that meetings were held in 123 places in order to plan the consultation during August 2017 (steps 1 and 2). Apparently, Indigenous Peoples organised their own deliberation processes (step 3) in more than 300 places during September, and more than 10,000 Indigenous individuals took part in them (Gobierno de Chile, 2017).

The "dialogue" (step 4) was realised in a twofold way: first, in regional meetings from 30 September onwards; second, in a National Meeting convened between 16 and 21 October (Gobierno de Chile, 2017), which involved about 145 Indigenous delegates (Gobierno de Chile, 2017). This intense meeting resulted in a provisional agreement on a number of new features to be inserted into the Chilean Constitution, but it also reported partial agreement or dissent on other parts. Moreover, this agreement was signed without any Quechua, Yagán, or Kawésqar representatives (Gobierno de Chile, 2017).[4]

Consent was apparently reached on the constitutional recognition of the following aspects: the pre-existence of Indigenous Peoples, including their ancestral/pre-Colombian presence, the conservation of their culture, and the land as the fundamental source of their existence and culture; their right to preserve, strengthen, and develop the own history, identity, culture, languages, institutions, and traditions (including their ancestral authorities); the State duty to preserve the cultural and linguistic diversity of the Indigenous Peoples; the right of Indigenous Peoples to culture and (official use of their) languages, including their material and immaterial patrimony in accordance with their worldview, and the right to education in their own language; and a general affirmation of the right to equality of all Indigenous individuals and Peoples, as well as the prohibition of discrimination on the basis of (Indigenous) origins or identity (Gobierno de Chile, 2017).

The parties partially agreed on other points, as follows: the duty to interpret the (new) constitution in light of those international treaties that safeguard Indigenous rights and which have been ratified by Chile; the reservation for Indigenous representatives of up to 10% of the total seats in Congress; a right to health and to the best possible healthcare as well as to traditional Indigenous medicine and health practices; the constitutional consecration of their rights to consultation and self-determination, including their right to autonomy and to their political, legal, socio-cultural, and economic authorities (Gobierno de Chile, 2017).

Finally, there was clear dissent over the constitutional guarantee for the (later) establishment of Indigenous Territories (by ordinary law), and over the plurinational character of the Chilean State. The closing document of this intense meeting was signed by only 38 Indigenous representatives out of approximately 145 delegates (Gobierno de Chile, 2017).

On 3 November, the Ministry of Social Development summoned a negotiating table (*Mesa de diálogo*), which was turbulently left by 27 Indigenous delegates due to their persisting dissent on several of the abovementioned points. Some other 31 Indigenous representatives signed the closing document of 3 November, which included the aforementioned points of agreement, and also reframed and included the interpretation of the Constitution in light of Chile's international obligations vis-à-vis Indigenous rights, and their right to self-determination, to be exercised within the new Constitution. Partial agreement was reiterated on the measures regarding the reserved Congress seats, as well as Indigenous rights to health and consultation. Indigenous land rights and the plurinational character of Chile were referred to as having reached a partial agreement (Gobierno de Chile, 2017). Nevertheless, all the topics that were classified as partially agreed remained excluded, *de facto*, from the text that was prepared for the constitutional reform.

Despite the efforts of both parties (Indigenous Peoples on the one hand, and the State bodies on the other hand), this consultation suffers from the abovementioned weaknesses that are intrinsic to all those consultation processes that have been organised within the framework of Decree 66. In particular, this consultation could easily fall under the scope of "all the possible efforts" (Ministerio de Desarrollo Social, 2013, article 3) and thus be archived, at least with regard to all those abovementioned aspects that remained pending or partially agreed upon.

Moreover, Indigenous leaders denounced a lack of mutual trust, as was evidenced by the schism between those Indigenous delegates who exited the 3 November negotiating table and those who remained. This was apparently due to threats made by some government representatives to terminate the consultation without reaching an agreement. Others complained about the technical difficulties of the documents under discussion during both the regional "dialogue" phase meetings and the National Meeting (Observatorio Ciudadano de Chile, 2018, p. 217). Indeed, the last phases of the consultation took place in less than two months. This suggests that the authorities appointed were under pressure to come to a conclusion before the Presidential and general elections of December 2017.

Concluding remarks

This chapter has discussed five consultation processes that addressed the constitutional recognition of Indigenous Peoples and their rights, and the creation of a number of public bodies related to Indigenous issues in Chile. They have been framed as "institutional" since they addressed a number of potential institutional reforms at national level that would have had a direct impact on the entire discourse and protection system of Indigenous Peoples in this country.

In the course of nine years (2009–2017), Indigenous Peoples were called to express their consent three times on different constitutional reforms that would have recognised them as Peoples with collective rights and led to the creation of a Council (and, later, more Councils) of Indigenous Peoples; twice on the transformation of CONADI into another Indigenous Development Agency; twice on the creation of a ministerial body that would run their affairs (first, a Ministerial Council of Indigenous Affairs that already existed, then a Ministry of Indigenous Affairs, the draft law of which was withdrawn a few months after its submission); once on the creation of a Secretariat of

Indigenous Affairs and of a regional Unit for Indigenous Affairs in each Governor's office; and once on the procedure for consultation itself, and the Environmental Impact Assessment System, which were both regulated by Decrees, the text of which was not subject to a consultation process.

The repetition of these consultations *per se* undermines the credibility of the Chilean government(s) in pursuing a fair implementation of the right to (prior) consultation of Indigenous Peoples. Moreover, all the attempts made to regulate the consultation process continue to present lacunae in comparison to the required international standards.

There seem to be three reasons for the failure of prior consultation in Chile: the overall lack of good faith on the part of the State representatives; the many irregularities, which Indigenous representatives and organisations complained about; and, the unwillingness or impossibility of pursuing any concrete action after the end of the consultation process (this remains to be seen with regard to the 2017 consultation, but it is unlikely that it will be taken forward). This continuing discontent and the apparent reluctance of the Chilean State to carefully observe its duty regarding the right to prior consultation of Indigenous Peoples have become even more concerning after the recent, above-mentioned declaration of (newly re-elected) President Piñera to withdraw from ILO 169, which is the lynchpin of safeguards for Indigenous Peoples and their rights, in a country that still denies them constitutional recognition. It sadly seems that the Chilean government(s) and its bodies continue to envision Indigenous Peoples' right to consultation as a tick in the box: once it is "done" – in whatever form they do it – their duty is fulfilled, irrespective of the international obligation to pursue Indigenous Peoples' consent (or, at least, agreement). This limited understanding destabilises not only the whole protection system of Indigenous rights within this country but also the possibility of continuing the dialogue between the State and a sector of society that fairly claims its (long-awaited) recognition and acknowledgment. It also points to bad practice in terms of realising Indigenous rights, thereby undermining Chile's credibility in pursuing its (long overdue) human rights agenda.

Notes

1 Moreover, CONADI possesses legal status, holds regional offices, counts on an annual budget, and is steered by a National Council composed of 17 members, 8 of whom are elected by Indigenous communities (Ministerio de Planificación y Cooperación, 1993, articles 38 and ff.).
2 On the previous Decree No.124 of September 2009 and the following debate on the legislation on Indigenous Peoples see Tomaselli (2016). For an overview of how Chilean national courts have (extensively) recognised misapplications of art.6 of the ILO 169, see Contesse (2012) and Tomaselli (2016).
3 On the international requirements of the right to prior consultation, see the chapters by Cantú Rivera and Del Castillo in this volume, and, *inter alia*, Doyle (2015), and Instituto Interamericano de Derechos Humanos (2016).
4 The official document by the Gobierno de Chile (2017) includes the original closing signed acts of both the 16–21 October and the 3 November meetings.

References

Aylwin, J. (2009). Chile. In International Work Group for Indigenous Affairs (Ed.), *The Indigenous World 2009* (pp. 218–230). Copenhagen: IWGIA.

Aylwin, J. & Silva, H. (2014). Chile. In International Work Group for Indigenous Affairs (Ed.), *The Indigenous World 2014* (pp. 203–213). Copenhagen: IWGIA.

Bengoa, J. (2000). *Historia del Pueblo Mapuche Siglo XIX y XX*. Santiago de Chile: LOM Ediciones.

Bertin, X. (2018). Eventual retiro de Chile del Convenio 169 inquieta a pueblos originarios. *La Tercera*. Retrieved from http://www2.latercera.com/noticia/eventual-retiro-chile-del-convenio-169-inquieta-pueblos-originarios.

Carruthers, D. & Rodriguez, P. (2009). Mapuche Protest, Environmental Conflict and Social Movement Linkage in Chile. *Third World Quarterly*, 30(4), 743–760.

CONADI. (2009a). Informe final del proceso de consulta sobre participación política de los pueblos indígenas en la Cámara de Diputados, en los Consejos Regionales y la creación del Consejo De Pueblos Indígenas. Santiago de Chile.

CONADI. (2009b). Minuta para la Consulta sobre Reconocimiento Constitucional de los Pueblos Indígenas. Santiago de Chile.

CONADI. (2010). Memoria Presentada por el Gobierno de Chile en conformidad con las disposiciones del artículo 22 de la Constitución de la Organización Internacional del Trabajo correspondiente al periodo 15 de septiembre de 2009 al 15 de septiembre de 2010 acerca de las medidas adoptadas para dar efectividad a las disposiciones del Convenio sobre Pueblos Indígenas y Tribales, 1989 (NÚM. 169), ratificado por Chile el 15 de septiembre de 2008. Santiago de Chile.

Contesse, J. (2012). El Derecho de Consulta Previa en el Convenio 169 de la OIT. Notas para su Implementaciónen Chile. In J. Contesse (Ed.), *El Convenio 169 de la OIT y el Derecho Chileno. Mecanismos y Obstáculos para su implementación* (pp. 189–240). Santiago de Chile: Ediciones Universidad Diego Portales.

Cooperativa.cl. (2016). Gobierno retiró proyecto que crea el Ministerio de Pueblos Indígenas. Retrieved from http://www.cooperativa.cl/noticias/pais/gobierno/gobierno-retiro-proyecto-que-crea-el-ministerio-de-pueblos-indigenas/2016-05-16/062958.html.

Doyle, C. M. (2015). *Indigenous Peoples, Title to Territory, Rights, and Resources. The transformative role of free, prior and informed consent*. London: Routledge.

Gobierno de Chile. (2011). Minuta Explicativa sobre la Consulta de parte del Reglamento del Sistema de Evaluación de Impacto Ambiental y Guías de Procedimiento de Participación Ciudadana y de apoyo para la Evaluación de Efectos Significativos sobre Pueblos Originarios en el Sistema de Evaluación de Impacto Ambiental (SEIA). Santiago de Chile.

Gobierno de Chile. (2012). Propuesta de gobierno para una nueva normativa de consulta y participación indígena de conformidad a los artículos 6° y 7° del Convenio 169 de la organización Internacional del Trabajo. Retrieved from http://unsr.jamesanaya.org/docs/data/2012-08-08-propuesta-de-normativa-de-consulta-gobierno-chile.pdf.

Gobierno de Chile. (2017). Informe Final. Sistematización Proceso de Consulta Constituyente Indígena. Retrieved from http://www.constituyenteindigena.cl/wp-content/uploads/2017/11/Iibro-15-1.pdf.

Instituto de Estudios Indígenas. (2003). *Los derechos de los pueblos indígenas en Chile. Informe del Programa de Derechos Indígenas*. Santiago de Chile: LOM Ediciones.

Instituto Interamericano de Derechos Humanos. (2016). *El derecho a la consulta previa libre e informada*. San José: Instituto Interamericano de Derechos Humanos.

Instituto Nacional de Derechos Humanos. (2013). *Situación de los Derechos Humanos en Chile. Informe 2013.* Santiago de Chile: Instituto Nacional de Derechos Humanos.

Instituto Nacional de Estadísticas Chile. (2003). Síntesis de Resultados. Retrieved from http://www.iab.cl/wp-content/files_mf/resumencenso_2012.pdf.

Instituto Nacional de Estadísticas Chile. (2012). Censo 2012. Resultados XVIII Censo de Población. Retrieved from https://www.cooperativa.cl/noticias/site/artic/20130425/a socfile/20130425190105/resultados_censo_2012_poblacion_vivienda_tomosiyii.pdf.

Kaempfe, I. & Ready, G. (2011). Comentarios a la Ley N° 20.249, que crea los ECMPO y su Relación con el Convenio N° 169 de la OIT. *Ideas para Chile*, 1(9), 1–33.

Mapuexpress. (2018). Consejo de Pueblos Atacameños señala que proyecto de Ley sobre Consejo de Pueblos vulnera Derechos. Retrieved from http://www.mapuexp ress.org/?p=22439.

Marimán Quemenado, P. (2011). Chile. In International Work Group for Indigenous Affairs (Ed.), *The Indigenous World 2011* (pp. 211–219). Copenhagen: IWGIA.

Meza-Lopehandía, M. (2010). VI. El Convenio N° 169 de la OIT en el sistema normativo chileno. In Matías G. Meza-Lopehandía (Ed.), *Las implicancias de la ratificación del Convenio N° 169 de la OIT en Chile*, 2nd Edition (pp. 103–160). Temuco: Programa de Derechos de los Pueblos Indígenas del Observatorio Ciudadano.

Milesi, O. (2017). Chilean President's Apology to the Mapuche People Considered 'Insufficient', *Inter Press Service News Agency*, 29 June. Retrieved from http://www.ip snews.net/2017/06/chilean-presidents-apology-mapuche-people-considered-insufficient.

Ministerio de Desarrollo Social. (2013). Aprueba Reglamento que Regula el Procedimiento de Consulta Indígena en virtud del artículo 6 N° 1 Letra A) y N° 2 del Convenio N° 169 de la Organización Internacional del Trabajo y deroga normativa que indica. Decree (Decreto) No. 66 of 15 November 2013, published on 14 March 2014.

Ministerio de Desarrollo Social. (2014a). Ministerio de Desarrollo Social suscribe convenio de colaboración con Universidad Diego Portales para observación de Consulta Indígena. Retrieved from http://www.consultaindigenamds.gob.cl/articu los/articulo040914_2.html.

Ministerio de Desarrollo Social. (2014b). Organismos Internacionales respaldan proceso de Consulta a los Pueblos Indígenas. Retrieved from http://www.consultaindi genamds.gob.cl/articulos/articulo130814_2.html.

Ministerio de Desarrollo Social. (2015a). Los nueve pueblos aprueban la creación del Ministerio y los Consejos de Pueblos Indígenas. Retrieved from http://www.consulta indigenamds.gob.cl/articulo.php?id=13849.

Ministerio de Desarrollo Social. (2015b). Sistematización Proceso de Consulta Previa Indígena, Ministerio de Desarrollo Social. Informe Final. Retrieved from http://www. ministeriodesarrollosocial.gob.cl/pdf/upload/Informe%20Nacional%20CONSULTA% 20PREVIA%20INDIGENA.pdf.

Ministerio de Desarrollo Social. (2016). Dispone la Realización del Proceso Participativo Constituyente Indígena que Indica e Inicia Procedimiento Administrativo. Resolution (Resolución Exenta) No. 329 of 14 May 2016.

Ministerio de Planificacion y Cooperacion. (1993). Establece Normas sobre Proteccion, Fomento y Desarrollo de los Indigenas, y Crea la Corporacion Nacional de Desarrollo Indigena [Ley Indígena]. Law No. 19,253 of 5 October 1993.

Ministerio de Planificacion y Cooperacion. (2006). Reconoce la Existencia y Atributos de la Etnia Diaguita y la Calidad de Indígena Diaguita. Law No.20,117 of 8 August 2006.

Ministerio de Planificacion y Cooperacion. (2008). Aprueba Reglamento de la Ley 20,249 que Crea el Espacio Costero Marino de los Pueblos Originarios. Supreme Decree (Decreto Supremo) No.134 of 29 August 2008.

Ministerio de Relaciones Exteriores. (2008). Promulga el Convenio N° 169 Sobre Pueblos Indígenas y Tribales en Países Independientes de la Organización Internacional del Trabajo. Decree (Decreto) No. 236 of 14 October 2008.

Ministerio del Interior. (1984). Determina Conductas Terroristas y Fija su Penalidad. Law No. 18,314 of 16 May 1984.

Ministerio del Medio Ambiente. (2013). Aprueba Reglamento del Sistema de Evaluación de Impacto Ambiental. Decree (Decreto) No.40 of 12 August 2013.

Ministerio del Trabajo y Previsión Social. (2008). Crea el Espacio Costero Marino de los Pueblos Originarios. Law No.20,490 of 31 January 2008.

Observatorio Ciudadano de Chile. (2017). Chile. In International Work Group for Indigenous Affairs (Ed.), *The Indigenous World 2017* (pp. 259–272). Copenhagen: IWGIA.

Observatorio Ciudadano de Chile. (2018). Chile. In International Work Group for Indigenous Affairs (IWGIA), *The Indigenous World 2018* (pp. 212–222). Copenhagen: IWGIA.

Observatorio del Proceso Constituyente en Chile – Fundación RED. (n.d.). Participación Indígena en el Proceso Constituyente Chileno. Retrieved from http://red constituyente.cl/wp-content/uploads/2017/04/PPII-1.pdf.

Ojeda, A. (2015). Pueblos originarios acusan vicios en Consulta Indígena. *Diario Universidad de Chile*, 2 February. Retrieved from http://radio.uchile.cl/2015/02/02/p ueblos-originarios-acusan-vicios-en-consulta-indigena.

Rosti, M. (2008). Reparations for Indigenous Peoples in Two Selected Latin American Countries. In F. Lenzerini (Ed.), *Reparations for Indigenous Peoples: International and Comparative Perspectives* (pp. 345–362). Oxford: Oxford University Press.

Toledo Llancaqueo, V. (2006a). *Pueblo Mapuche. Derechos Colectivos y Territorios: Desafíos para la Sustentabilidad Democratica*. Santiago de Chile: LOM Ediciones.

Toledo Llancaqueo, V. (2006b). *Trayectoria de una Negación. La transición chilena y el compromiso de reconocimiento de los Derechos de los Pueblos Indígenas 1989–2006*. Santiago de Chile: Centro de Políticas Públicas y Derechos Indígenas.

Tomaselli, A. (2012). Natural Resources' Claims, Land Conflicts and Self-empowerment of Indigenous Movements in the Cono Sur. The case of the Mapuche People in Chile, *International Journal of Minority and Group Rights*, 19(2), 153–174.

Tomaselli, A. (2016). *Indigenous Peoples and their Right to Political Participation. International Law Standards and their application in Latin America*. Baden-Baden: Nomos.

United Nations Human Rights Office of the High Commissioner. (n.d.). View the Ratification Status by Country or by Treaty – Chile. Retrieved from http://tbinter net.ohchr.org/_layouts/TreatyBodyExternal/Treaty.aspx?CountryID=35&Lang=EN.

United Nations Press Release. (2007). General Assembly adopts Declaration on Rights of Indigenous Peoples; "Major Step Forward" towards Human Rights for All, Says President. Retrieved from https://www.un.org/press/en/2007/ga10612.doc.htm.

Vergara, J., Gundermann, H., & Foerster, R. (2006). Legalidad y legitimidad: ley indígena, Estado chileno y pueblos originarios (1989–2004). *Estudios Sociológicos*, 24(71), 331–361.

Webb, A. & Radcliff, S. (2013). Mapuche Demands during Educational Reform, the Penguin Revolution and the Chilean Winter of Discontent. *Studies in Ethnicity and Nationalism*, 13(3), 319–341.

9 Institutional scope and limitations of the right to consultation and the free, prior and informed consent of Indigenous Peoples in Mexico

Anavel Monterrubio Redonda

Introduction

In the context of economic liberalisation, large corporations and transnational companies are supported by a State whose ability to make decisions has become weakened and whose legal framework transfers wealth from the poor to the rich, obtaining benefits from the market economy but without paying the social, cultural, and environmental costs derived from this economic activity (Cárdenas, 2016). Such is the case of the extractive and development activities carried out in territories that have historically been occupied by Indigenous Peoples, which are often rich in natural resources. State governments authorise the free exploitation of these lands, serving interests other than those of Indigenous Peoples, who have no part in the distribution of benefits arising from the said exploitation and, moreover, are often expelled from their ancestral lands, which constitute the basis of their survival.

In fact, as a result of these activities, many Indigenous people end up living in conditions of extreme poverty (FAO, 2016; CDI, 2011a; IACtHR, 2009). These situations lead to the violation of many Indigenous individual and collective rights, including their "right to food, water, health, life, honour, dignity, freedom of thought and religion, freedom of association, family rights and freedom of residence" (IACtHR, 2009, para. 2), notwithstanding their "collective right to survive as an organised group with the control of their habitat as the basic requirement for the reproduction of their culture, their own development and for carrying out their life plans" (IACtHR 2009, para. 3). These rights are necessary to end inequality, overcome differences (once they have been acknowledged), promote inclusion during the democratic process (Pérez, 2014), and to ensure human development.

In the case of Mexico, in an effort to materialise Indigenous Peoples and communities' rights as of 2001 (with the reform of article 2 of the Constitution), the government has established legal measures, institutions, and policies to guarantee their participation, particularly through the right to consultation. Taking this into account, consultation and Free, Prior and Informed Consent (FPIC) are regulated by international law and, in accordance with article 133 of the Constitution, these norms are binding for the Mexican government.

DOI: 10.4324/9781351042109-12

With the aim of "establishing the methodological and technical procedures for the peoples and the communities to be consulted" (CDI, 2013, p. 6), the Consultation Council of the National Commission for the Development of Indigenous Peoples (CDI) devised a system for Indigenous consultation that operates via the Protocol for the implementation of consultations to Indigenous Peoples and communities in conformity with ILO Convention No.169 Concerning Indigenous and Tribal Peoples in Independent Countries of 1989 (ILO 169), which has served as the basis for the CDI to carry out over 30 consultations with Indigenous Peoples. However, the representatives of these communities continue to express their concern due to the lack of regulations and institutions capable of safeguarding this right, arguing that the efforts to reach its full effectiveness have been insufficient. In view of this scenario, the following question arises: what has been the scope of Indigenous consultation processes in Mexico?

The aim of this chapter is thus to analyse the Indigenous consultation processes promoted by the CDI as the official body in charge of these activities,[1] bearing in mind the standards established by international law. To this effect, the first section explains the relationship between direct democracy and Indigenous consultation, which is understood as a key tool for these peoples to exercise their rights. The second section analyses the consultation processes carried out by the CDI, while the third deals with recommendations made by the National Human Rights Commission (*Comisión Nacional de Derechos Humanos* – CNDH). The fourth section offers some final considerations.

Direct democracy and Indigenous consultation

The Mexican State is ruled by a democratic, representative system, which nevertheless is perceived to be weak. This situation has led the federal legislative branch to develop mechanisms and instruments aimed at improving the relationship between the representatives and those represented.[2] In this context, in 2012, reforms were made to the Constitution to incorporate citizen initiatives and popular consultation as mechanisms for direct democracy to complement standard representation, which entails increasing the opportunities for citizens to intervene in certain public matters. These two citizen participation instruments should, ideally, work as alternatives to the traditional decision-making mechanisms provided by electoral and other institutions that are biased towards parties. Indeed, they seek to promote public participation in the decision-making processes, which may or may not have a binding effect, depending on the characteristics of its institutions (Alonso, 2017; Escobar, 2015).

With popular consultation, citizens express their opinion or their proposals on a certain topic of public interest related to the exercise of governmental power, with the aim of avoiding a monopoly of representation by the legislative branch in particular (Alonso, 2017). Thus, the Mexican Constitution establishes the vote in popular consultations on issues of national significance as a citizen's right, sets out several formalities that must be met, and determines in which cases the outcomes shall be binding and in which they shall not (Cámara de Diputados, 2016, article 35).[3] This precept is regulated through the Federal Law of Popular Consultation, which – while working at national level – is not directed at Indigenous consultation *per se* and must not be mistaken for it. This law defines popular consultation as "[...] a participatory mechanism through which citizens exercise their right, by casting a vote in order to express their opinion regarding one or more issues of national public interest" (Cámara de Diputados, 2014, article 4).

Indigenous consultation is, on the one hand, a collective human right that is interrelated with other human rights, such as the right to self-determination, sustainable development, property and cultural biodiversity, and which can be infringed with actions or omissions by the State (CNDH, 2016b). As well as being a tool for approving or rejecting projects, its efficient implementation guarantees the acknowledgement of communities as legal persons and not as an object of compensatory public policies. The right to consultation, according to article 6 of the ILO 169, implies the establishment of a genuine dialogue between the parties, which is characterised by communication and under-standing, mutual respect and good faith, and a true intention of reaching common agreement or obtaining consent on the measures proposed.

Thus, the right to consultation and the FPIC of Indigenous Peoples is of a procedural nature, as it refers to a technical-methodological procedure for the establishment of a dialogue and validation of a decision-making process. It is also a substantive right, considering that its goal is to protect these peoples' rights in the face of the State's actions which may infringe them, including the monitoring of companies' activities that may affect them (Monterrubio, 2014).

However, although this right is constitutionally acknowledged in Mexico and has been addressed in policies formulated by several sectors of the public admin-istration, in the view of Indigenous Peoples and both national and international bodies, there is a very large implementation gap between what is formally recog-nised and both legal and institutional practice, as will be discussed below.

The right to Indigenous consultation within the Mexican State

In Mexico, the regulatory framework on the right to consultation can be found within both domestic and international law. From the perspective of international law, the consultation of Indigenous Peoples – seen as an obligation and legal responsibility – is established in ILO 169 and the United Nations Declaration on the Rights of Indigenous Peoples (UNDRIP) of 2007, as well as some case law-based decisions by the Inter-American Court on Human Rights (IACtHR). With respect to this right, the international legal framework establishes that there are two types of projects requiring the use of consultations and FPIC: those specifically aimed at Indigenous Peoples, and those which, although they are not specifically directed at them, affect the said peoples (UNPFII, 2005).

Within the national legal framework, the Mexican Constitution establishes that with the goal of decreasing the necessities and difficulties affecting Indigenous Peoples and communities, the government has the obligation of "consulting them in the elaboration of a National Development Plan and a State and Municipal plan and, wherever necessary, of incorporating the recommendations and proposals made" (Cámara de Diputados, 2016, article 2, para. b, section 9). Likewise, it recognises the right to consultation as an inherent instrument of the rights to autonomy and self-determination (Cámara de Diputados, 2016, article 133).

Nevertheless, although it is officially recognised in the Constitution, there is currently no secondary law that regulates the exercise of and claims to this right. Nevertheless, it has been incorporated into different laws, including the Planning Law (Cámara de Diputados, 2018) and the Law of the National Commission for the Development of Indigenous Peoples (Cámara de Diputados, 2003).

The Planning Law sets out as one of its goals, "[...] promoting and guaranteeing the democratic participation of several social groups and that of Indigenous Peoples and communities through their representatives and authorities, in the elaboration of a plan and the programmes to which this law refers" (Cámara de Diputados, 2018, article 1, para. 5). Likewise, it determines that "the indigenous communities shall be consulted and may participate, from their definition, in federal programmes that directly affect the development of their peoples and communities" (Cámara de Diputados, 2018, article 20, para. 3).

The Law of the National Commission for the Development of Indigenous Peoples mentions, as one of its principles, ruling its actions by "consulting the Indigenous Peoples and communities every time the Federal Executive promotes legal reforms and administrative acts, development programmes or projects that have a significant impact on their living conditions and environment" (Cámara de Diputados, 2003, article 3, para. 6). In relation to this, the Internal Regulations of the Consultation Council of the CDI define consultations as the "interlocution and dialogue between the Commission and the Indigenous Peoples and communities by means of their instances of representation and participation for the formulation, execution and assessment of the development plans and programmes" (CDI, 2011c, article 2, para. 14), and the System of Consultation as the "process that promotes, enriches, registers, returns and monitors the information generated by the Consultation" (RICC, 2014, para. 14), in accordance with which, "the State must take public policy measures" (CDI, 2013a, p. 25).

Thus, according to the federal legislation, there are three matters that are considered relevant for the consultation of Indigenous Peoples: the elaboration of a National Development Plan; the federal programmes directly affecting the development of these peoples; and the legal reforms and administrative acts that have an impact on their living conditions and environment. It is important to note that the goals and scope of consultation are established at the level of local legislation on Indigenous rights.[4]

Additionally, the CNDH – with reference to international law – points out that Indigenous participation imposes a double obligation on the government: firstly, to make Indigenous Peoples participants and attend to their opinions in conformity with their customs and traditions; and secondly, to lay the foundation for the Indigenous communities to participate in any "administrative, legal, or any other type of procedure, which may have an impact on their interests or right, in an effective, informed, and free manner" (CNDH, 2016, para. 78). The CNDH also points out that the administrative measures to be consulted include plans and programmes as well as the elaboration, approval, and monitoring of public policies originating from the public administration that may have an impact on the rights of those being consulted, such as infrastructure works and large-scale or investment projects (CNDH, 2015a).

Nevertheless, although the right to consultation and FPIC have been acknowledged in Mexico's Constitution, this right has been adopted gradually; indeed, its regulation is not fully in line with international standards and its implementation by federal entities is both unclear and incomplete (CNDH, 2016). Therefore, according to the CNDH, the Mexican government has not been able to guarantee the exercise of this right, specifically in the face of projects such as the allocation of lands for the construction of hydroelectric plants, mining developments, re-settling Indigenous populations, utilising water resources, and tourism, among others (CNDH, 2018). This is also due to the lack of a secondary legislation, which means that the authorities in charge of guaranteeing the right do so on the basis of their own interpretation of both the Constitution and international law.

Thus, it is clear that consultations in Mexico do not live up to the standards established by international bodies. In this sense, the practices carried out by the CDI and the resolutions of the Supreme Court of Justice of the Nation (SCJN, 2014) through protocols and orders as a result of complaints from the communities before the CNDH, particularly regarding the impact of the development projects, notoriously conflict.

The right to consultation in the Indigenous consultation system of the CDI

According to information offered by the CDI (2018), from 2005 to date, it has carried out more than 30 consultations with Indigenous Peoples: seventeen consultations related to legislative harmonisation; six over development plans and programmes; six on the creation of public policy; four on cultural production and protection; and two regarding the conservation of natural resources. In order to analyse the scope of these processes, reports on the following consultations were analysed and are summarised below[5]:

1 Consultation with Indigenous Peoples on their means and aspirations regarding development (CDI, 2004);

2   The migrant Indigenous population (CDI, 2006);
3   Alcoholism and Indigenous Peoples (CDI, 2008);
4   Consultation of the Indigenous Peoples of the coastal region of the Gulf of California regarding marine ecological ordering (CDI, 2009);
5   The sacred places of the Wixárika peoples (CDI, 2010);
6   Mechanisms to protect traditional knowledge (CDI, 2011a);
7   Consultation on the bill of a General Law of Consultation with Indigenous Peoples and Communities (CDI, 2011b);
8   Consultation to identify Afro-Descendant communities (CDI, 2012a);
9   The preservation of the sacred places and ceremonial centres of the Yoreme peoples of Sinaloa (CDI, 2012b);
10   The rights of Indigenous women (CDI, 2012c);
11   Consultation on reforming the General Education Law (CDI, 2013a);
12   The development priorities of Indigenous and Afro-Descendant communities (CDI, 2013b).

Issues dealt with in consultations

Topics highlighted as goals of the different consultations include the preservation, recovery and protection of sacred places and ceremonial centres; the creation of programmes for ecological regulation and programmes to manage the territorial planning of living and working areas; the identification of development priorities to be incorporated in the National Plan of Development and in the Special Programme for Indigenous Peoples; the identification of factors to protect rights related to the protection of traditional knowledge, cultural expressions, and natural and genetic resources as part of Indigenous heritage; the characterisation of the needs of Indigenous Peoples and communities in reforming the General Law on Education and in drafting the bill of the General Law on Consultation with Indigenous Peoples and communities; and, finally, the identification of problems related to Indigenous women of different communities with regard to their development and the fulfilment and exercise of their rights.

Issues avoided in consultations

For their part, Indigenous Peoples and their communities refer to various problems that were not necessarily addressed in the consultation exercises, including the following: pressure and threats towards their sacred places and ceremonial centres due to tourism projects; environmental pollution due to mining activities; the absence of legal regulations regarding article 2 of the Constitution on the right to consultation; the transformation, disappearance, and commercialisation of traditional knowledge[6]; the licensing or privatisation of Indigenous territories and natural resources for exploitation, exploration or for the construction of large-scale projects for tourism and infrastructure, without any information regarding the negative impact on society and environment; the lack of information about ecological regulations and changes in the use of soil; a lack of transparency in contracts for licences to use sacred places for tourism purposes and authorisations for businesses related to tourism, such as bars, restaurants, hotels, boats, etc.; the commercialisation of cultural resources by third parties; the commercialisation of medicinal plants for recreational purposes without receiving benefits but rather discrimination and abuse (as is the case with *peyote*); Indigenous women's disadvantages regarding development, access and control of economic and productive resources, structural violence, reproductive and political rights and decision-making; the indiscriminate exploitation of marine resources in the Gulf of California; pollution caused by drainage systems; the reduction in seafood owing to the excessive use of agrochemicals; the industrial chemical waste from thermal power plants, fishing boats and shrimping; the dumping of diesel from boats, gas and oil from marine engines, and of agrochemicals from industries such as Pemex and the Federal Electricity Commission; and, finally, illegal fishing, logging of mangroves, and the destruction of wetlands and estuaries.

Participants in consultations

Finally, participants in the analysed consultations included, on the part of Indigenous Peoples, traditional authorities including the council of elders or governor, inspector, judge-mayor, sheriff, captain, coordinators of the community culture, rural authorities, the commissary òf communal property and land, community delegates, and representatives of oversight councils. Likewise, a whole series of federal government authorities have participated in these processes, including the Ministry for Ecology and Environmental Management (SEGAM), the Ministry of Foreign Relations (SRE), Ministry for the Environment and Natural Resources (SEMARNAT), Ministry of Agriculture, Livestock, Rural Development, Fisheries and Food (SAGARPA), Ministry of Social Development (SEDESOL), Ministry of Health (SS), the Consultation Council and delegations of the CDI in the entities involved in the process, the National Commission for the Knowledge and Use of Biodiversity (CONABIO), the Mexican Institute of Industrial Property (IMPI), the National Council for Culture and Arts (CONACULTA), the National Institute of Copyright (INDAUTOR), the National Commission of Protected Natural Areas (CONANP), the Indigenous Peoples Coordination Centre (CEAPI), the National Indigenous Languages Institute (INALI) and House of Senators and House of Deputies.

It is important to note that several consultations were conducted with the support of international organisations such as the United Nations Development Programme and the World Bank.

Method used in consultations

In general, the consultations were carried out in four stages: (a) designing the process, (b) training people to act as coordinators of the working meetings through forums and workshops, (c) conducting forums and workshops, and (d) analysing the information and presentation of the results. The participation mechanisms used included face-to-face meetings (regional discussions, micro-regional, State and interstate forums), interviews, and workshops with the communities and traditional authorities in community assemblies. In some cases, online interviews were carried out, and presentations and lectures were held.

Observations on the characteristics of these consultation processes in Mexico

The consultations carried out refer to administrative and legislative measures that only superficially address the economic, social, and environmental impacts on the areas where Indigenous Peoples live, particularly in those cases related to infrastructure works, large-scale projects or investments that entail exploration for and exploitation of natural resources. Furthermore, the topics addressed were those that were legally and institutionally required by the federal government rather than those based on the priorities of the Indigenous Peoples. In this sense, there is a clear gap between the problems mentioned by Indigenous participants and the proposals that were finally included. Likewise, the scope of the consultations was limited, as they were based on national legislation rather than on the special content set forth by international law. In the same vein, the mechanisms used for consultation were general instruments of participative democracy and the processes were limited to a request for opinions, proposals, and points of view, without any binding agreements, as is clear in the following statement:

> Although the non-binding characteristic of the [Indigenous] consultation is clear, with the different determinations, actions and policies by the State, from a moral perspective, consultation should be a mandatory reference for policies and actions of different State institutions in their relation with the Wixarika peoples and an instrument for the management of the Indigenous Consultation Council of the CDI.
>
> (CDI, 2010, p. 128)

On the important issue of FPIC, in the consultations analysed here, the processes failed to grant consent and only dealt with the planning phase of a programme. Indeed, consultations were established as a goal and purpose in themselves. For its part, the methodology used refers to the integration of participative diagnoses to identify problems and ways to resolve needs rather than a deliberative process allowing the incorporation of these groups' opinions when adopting State decisions. The language used was completely institutional and the use of an online survey made the process less accessible, reflecting a lack of culturally-adequate procedures.

While the demand for consultations is acknowledged in specific development projects (such as tourism projects and the construction of roads near sacred places), the legal scope of the obligation to consult is limited only to general issues regarding development plans and programmes rather than specific projects that truly affect Indigenous Peoples' lives and territory. For instance, while the State has an obligation to conduct a consultation and FPIC process to adopt measures related to the effective access and enjoyment of ancestral territory, the results involve corrective actions so that projects already in place affect Indigenous Peoples to a lesser extent. Nevertheless, they were not involved in defining these projects or the benefits they offer.

With regards to actors, the government and its institutions are responsible for summoning the consultation, in accordance with international law. However, in practice, the administrative authorities have been reluctant to act when projects have had an impact on Indigenous lands and sacred places. It was the Indigenous Peoples involved who had to sensitise the authorities, rather than the other way around. Likewise, although it was mentioned that workshops were held in Indigenous communities, many of them were conducted in areas with institutional coverage (by the CDI) rather than in areas of great importance for Indigenous Peoples (despite attempts to do so), and thus Indigenous participation was limited. Furthermore, it can be seen from the reports that the huge display of human and material resources used for the consultations was mainly because of the interest on the part of the federal executive branch and some governmental institutions to strengthen their image or dismantle possible conflicts, rather than because they acknowledged the perspective of Indigenous Peoples. All of this infringes the principle of good faith.

A final difficulty is that, in some consultations regarding legislative measures, such as the drafting of the bill of the General Law on Consultation with Indigenous Peoples and Communities, the process was aimed at answering incorrect questions (e.g., who is subject to the right? what is prior consent?).

Nevertheless, in some cases the results of these processes are interesting because – although they cannot be considered consultation processes in accordance with international standards – the discussions that took place reflected the range of contexts and conflicts over the precise meanings of basic issues such as traditional knowledge, territory, culture, and diversity, etc. This was clearly the case in the consultation regarding the ways of development and aspirations, and that regarding the mechanisms to protect the traditional knowledge, cultural expressions, natural, biological and genetic resources of Indigenous Peoples (CDI, 2011a).

Indigenous consultation from the perspective of the National Human Rights Commission

According to the Mexican government, the results produced by these consultation processes constitute progress towards guaranteeing Indigenous rights. However, the practices carried out by the CDI and other authorities significantly fail to live up to some of the resolutions and recommendations made by the CNDH. To date, several complaints have been filed by Indigenous Peoples and communities who, at the same time, have also resorted to the courts to file legal actions regarding the projects and licences that affect their territories. Some of the recommendations that deal with the violation of the right to consultation and the FPIC of Indigenous Peoples (CNDH, 2012; 2015a; 2015b; 2016a; 2016b; 2018) refer to several common issues in consultations carried out by the CDI and other federal authorities.

For instance, mining licences have been granted in conservation and sacred areas, which is detrimental to Indigenous communities (as in the case of the Wixárika Indigenous Peoples). In several cases (e.g., Independence Aqueduct in Sonora, the cultivation of genetically modified soya beans to the detriment of several Mayan Indigenous communities, the construction of the Toluca-Naucalpan motorway, the Morelos Integral Project, etc.) there has also been a lack of independent environmental impact studies, wherein measures are established but which, in general, fail to mitigate the damage caused. Another issue has involved the limitation of the competences of Indigenous municipalities owing to modifications in the Constitution and local secondary legislation (as is the case in Cherán).

In terms of procedural elements, there have been a series of consultation processes with inadequate procedures that fail to clearly establish dates and the subject of the consultation, with no prior information or answers to participants' questions. These processes – including the consultation over the wind farm project in Juchitán de Zaragoza and the La Parota hydroelectric dam project – limit consultation to the acceptance or rejection of a project that has already been established in terms set forth by the Government. Likewise, inadequate expropriations and administrative proceedings have affected, sometimes in an irreparable way, Indigenous territories (as is the case of La Parota hydroelectric dam).

Another common aspect to consultations is that the communities have to resort to the writ of *Amparo* (or protection proceeding) as a legal action against acts of authority designed to protect development projects. For instance, some of these proceedings were heard by the Supreme Court of Justice (SCJN), which resolved that consultations were not carried out pursuant to the criteria set forth by international regulations, and reminded the SCJN that it has itself already defined a specific protocol.[7]

Hence, the recommendations made by the CNDH (2012; 2015a; 2015b; 2016a; 2016b; 2018) confirm that, despite these protocols, consultation and FPIC of Indigenous Peoples remain unfulfilled, and the opinions of the CDI are not taken into account. Ambiguity is the order of the day, which means that government authorities – together with the economic actors involved – can justify the absence of consultations. Sometimes, private companies are given the responsibility of defining which community has representative bodies to carry out the process of consultation, and therefore whether they should be consulted or not.

In this context, Indigenous Peoples' right to consultation and their FPIC continue to be violated, mainly in relation to projects related to economic, infrastructure, and natural resource development, areas which as of yet have not been subject to consultations by the CDI. This situation also has negative implications for the effective enjoyment of the economic, social, and cultural rights of those affected peoples.

Conclusions

Based on the analysis carried out in this chapter, it can be concluded that the legal and institutional situation of Indigenous consultation and FPIC in Mexico is still highly unsatisfactory. Indeed, this right is constantly violated for several reasons, including the following.

There is an absence of a national legal framework for consultation and an institutional structure to make it effective. The lack of institutional will to implement the existing protocol makes the demand for this collective human right complex, given that it ends up being resolved in the courts and by entities that defend human rights, in proceedings that are of a considerable duration. In other words, by the time the courts render judgement on a suit, the negative effects that motivated the complaints and claims, are, in many cases, irreversible and the damage caused is irreparable.

The lack of a real conventionality control within the national legal system means that Indigenous consultations are considered to be just another mechanism of direct democracy, with limited scope regarding the essential contents of this right (namely, that consultations are prior, free, informed, in good faith and culturally adequate, taking into account language, identity and oral tradition, their conditions, demands, decision-making mechanisms, and means of argumentation).

The consultations carried out by the Mexican State are reduced to participatory practices and do not constitute an exercise in real democracy, as they fail to achieve an authentic inter-cultural dialogue. In this context, the authorities' discretion is imposed, thereby violating the right to political participation of those consulted.

The historical autonomy of the Indigenous Peoples and communities was invalidated because of the conditions of the constitutional reform of article 2 in 2001 (Cámara de Diputados, 2016). Given that they are not formally acknowledged as legal subjects (with the corresponding rights and obligations), their participation in decision-making and the distribution of benefits is subject to the government's decisions and to particular economic interests. All of this aggravates their situation of exclusion and inequality.

Rather than being citizen consultations related to deliberation over specific issues that affect Indigenous Peoples' lives and development, the consultations conducted by the CDI seek participative diagnoses to identify problems and possible ways to resolve needs that nevertheless, are not real inputs for the definition of public policies but, rather, lead to political use and manipulation of the results.

The aims and characteristics of consultations are a reflection of the level of commitment on the part of the government with regard to mechanisms of participative democracy. Clearly, the Mexican State's degree of commitment in this sense is negligible.

In conclusion, consultation and FPIC are far from being a reality in Mexico, in spite of the State having endorsed international treaties on human rights and made constitutional reforms. There is no real instrument for the public protection of collective rights able to maintain Indigenous Peoples' cultures and ways of life because, in part, we still fail to acknowledge that the defence of our land, territory, and natural resources by these communities is also the defence of our identity.

Notes

1 It must be pointed out that other consultation processes carried out by other authorities have taken place in which the CDI has only been a supporting institution. These authorities were, e.g., the Ministry for Energy (SENER); the National Water Commission (CONAGUA), and the National Electoral Institute (INE).
2 For Alonso (2017, p. 169), the weakening of representative democracy derives from two aspects inherent to its own nature: a) citizen participation in decision-making processes is scant or inexistent without any spaces where citizens can exercise their rights, favouring a certain distance between those who wish to exert this power as the basis for making decisions and those who consider themselves to be beneficiaries of the exercise of said power; and, b) the existence of a participative culture that is limited, deeming it deficient or insufficient for citizens to be able to influence the decision-making process.
3 Public consultation processes can be carried out in two cases: 1) if summoned by the Congress upon the request of the Executive branch at the federal level, by the equivalent to 33% of the members of any of the Chambers of the Congress or by 2% of the nominal electoral register, and, 2) when directly promoted by 40% of the nominal electoral register.
4 For a full list of these laws, please see Monterrubio (2014, p. 8).
5 The revision of the reports corresponds to a limited list of consultations in Mexico, since it refers only to those consultations made by the CDI and those available on the website of the agency.
6 Including the knowledge of mother tongue and ritual-related language, traditional medicine, typical clothing of each group, music, chants, dance, astral knowledge, traditional techniques, sacred places, trekking, traditional food, organisation and traditional methods of election and communication, and species of animals and plants that are directly related to rituals and traditional ceremonies.
7 This is a protocol for those in charge of enforcing the law in cases involving rights of people, communities and Indigenous Peoples (SCJN, 2014).

References

Alonso, A. (2017). Mecanismos de participación ciudadana en el ámbito del Poder Legislativo Federal. *Revista Pluralidad y Consenso*, 6(29), 166–191.

Cámara de Diputados. (2003). Ley de la Comisión Nacional para el Desarrollo de los Pueblos Indígenas. 21 May 2003. Ley Abrogada. 4 December 2018. Retrieved from http://www.diputados.gob.mx/LeyesBiblio/abro/lcndpi/LCNDPI_abro.pdf.

Cámara de Diputados. (2014). Ley Federal de Consulta Popular. Nueva Ley publicada en el Diario Oficial de la Federación, 14 March 2014. Retrieved from http:// www.diputados.gob.mx/LeyesBiblio/pdf/LFCPo.pdf.

Cámara de Diputados. (2016 [1917]). Constitución Política de los Estados Unidos Mexicanos. Constitución publicada en el Diario Oficial de la Federación, 5 February 1917. Texto Vigente. Última reforma publicada DOF, 27 January 2016. Retrieved from http://www.diputados.gob.mx/LeyesBiblio/htm/1.htm.

Cámara de Diputados. (2018 [1983]). Ley de Planeación. Nueva Ley publicada en el Diario Oficial de la Federación, 5 January 1983. Texto Vigente. Última reforma publicada DOF, 16 February 2018. Retrieved from https://www.snieg.mx/con tenidos/espanol/normatividad/marcojuridico/Leydeplaneacion.pdf.

Cárdenas, J. (2016). *El modelo jurídico del neoliberalismo.* Ciudad de México: UNAM.

CNDH. (2012). Recomendación 56/2012 sobre la violación de los derechos humanos colectivos a la consulta, uso y disfrute de los territorios indígenas, identidad cultural, medio ambiente sano, agua potable y saneamiento y protección de la salud del pueblo wixárika en wirikuta. Retrieved from http://www.cndh.org.mx/sites/all/ doc/Recomendaciones/2012/Rec_2012_056.pdf.

CNDH. (2015a). Recomendación 23/2015 sobre el caso de vulneración al derecho a una consulta libre, previa e informada, en perjuicio de diversas comunidades indígenas. Retrieved from http://www.cndh.org.mx/sites/all/doc/Recomendaciones/2015/Rec_2015_ 023.pdf.

CNDH. (2015b). Recomendación 3/2018 sobre el caso de violaciones a los derechos a la consulta previa, libre, informada, de buena fe y culturalmente adecuada para pueblos y comunidades indígenas y a la información, en relación con el Proyecto Integral Morelos. Retrieved from http://www.cndh.org.mx/sites/all/doc/Recomenda ciones/2018/Rec_2018_003.pdf.

CNDH. (2016a). Recomendación 56/2016 sobre el caso de vulneración al derecho a la propiedad colectiva en relación con la obligación de garantizar el derecho a la consulta previa de las comunidades indígenas afectadas con la construcción de la "Autopista Toluca-Naucalpan". Retrieved from http://www.cndh.org.mx/sites/all/ doc/Recomendaciones/2016/Rec_2016_056.pdf.

CNDH. (2016b). Recomendación General No. 27/2016 sobre el derecho a la consulta previa de los pueblos y comunidades indígenas de la República Mexicana. Retrieved from http:// www.cndh.org.mx/sites/all/doc/Recomendaciones/generales/RecGral_027.pdf.

CNDH. (2018). Comunicado de Prensa DGC/244/18. Retrieved from http://www. cndh.org.mx/sites/all/doc/Comunicados/2018/Com_2018_244.pdf.

CDI. (2004). Consulta a los pueblos indígenas sobre sus formas y aspiraciones de desarrollo. Retrieved from https://www.gob.mx/cms/uploads/attachment/file/37012/ consulta_pueblos_indigenas_formas_aspiraciones_desarrollo.pdf.

CDI. (2006). Memoria de la consulta sobre migración de la población indígena. Retrieved from https://www.gob.mx/cms/uploads/attachment/file/37020/consulta_m igracion_poblacion_indigena.pdf.

CDI. (2009). Consulta a los pueblos indígenas de la zona costera del Golfo de California referente al ordenamiento ecológico marino. Retrieved from http://www.cdi. gob.mx/dmdocuments/consulta_zona_costera_golfo_california.pdf.

CDI. (2010). Informe final de la Consulta sobre Lugares Sagrados del Pueblo Wixárika. Retrieved from https://www.gob.mx/cms/uploads/attachment/file/37010/inform e_consulta_lugares_sagrados_wixarika_cdi.pdf.

CDI. (2011a). Consulta sobre mecanismos para la protección de los conocimientos tradicionales, expresiones culturales, recursos naturales, biológicos y genéticos de los pueblos indígenas. Retrieved from https://www.gob.mx/cms/uploads/attachment/file/37014/cdi_consulta_proteccion_conocimientos_tradicionales.pdf.

CDI. (2011b). Informe final de la Consulta sobre el Anteproyecto de Ley General de Consulta a Pueblos y Comunidades Indígenas. Retrieved from https://www.gob.mx/cms/uploads/attachment/file/37019/informe_final_de_la_consulta_sobre_el_anteproyecto.pdf.

CDI. (2011c). Reglamento Interior del Consejo Consultivo de la CDI. Retrieved from http://www.cdi.gob.mx/consultivo/reglamento_interior_del_consejo_consultivo_doc.pdf.

CDI. (2012a). Informe final de la Consulta para la identificación de comunidades afrodescendientes. Retrieved from https://www.gob.mx/cms/uploads/attachment/file/37016/cdi_informe_identificacion_comunidades_afrodescendientes.pdf.

CDI. (2012b). Informe final de la Consulta sobre la conservación de los sitios sagrados y centros ceremoniales del pueblo Yoreme de Sinaloa. Retrieved from https://www.gob.mx/cms/uploads/attachment/file/37017/cdi_conservacion_sitios_sagrados_pueblo_yoreme_sinaloa.pdf.

CDI. (2012c). Informe de la Consulta Nacional sobre la situación que guardan los Derechos de las Mujeres Indígenas en sus Pueblos y Comunidades. Retrieved from https://www.gob.mx/cms/uploads/attachment/file/37015/cdi_consulta_nacional_situacion_derechos_mujeres_indigenas.pdf.

CDI. (2013a). Consulta para la Reforma a la Ley General de Educación 2011–2012. Informe Final. Retrieved from https://www.gob.mx/cms/uploads/attachment/file/37018/cdi_ley_fed_edu_2011_2012.pdf.

CDI. (2013b). Informe final de la Consulta sobre las Prioridades de Desarrollo de las Comunidades Indígenas y Afrodescendientes. Retrieved from https://www.gob.mx/cms/uploads/attachment/file/37013/cdi_consulta_prioridades_indigenas.pdf.

CDI. (2018). Consulta a los pueblos indígenas, Portal de Transparencia. Retrieved from http://www.cdi.gob.mx/transparencia/gobmxcdi/participacion/documentos/consulta_pueblos_indigenas.pdf.

Escobar, L. (2015). La consulta popular en México. *Revista de la Facultad de Derecho de México*, 64(262), 185–201.

FAO. (2016). *Consentimiento libre, previo e informado. Un derecho de los Pueblos Indígenas y una buena práctica para las comunidades locales. Manual dirigido a los profesionales en el terreno*. Ciudad de Guatemala: FAO. Retrieved from http://www.fao.org/3/a-i6190s.pdf.

IACtHR. (2009). *Derechos de los Pueblos Indígenas y Tribales sobre sus Tierras Ancestrales y Recursos Naturales: Normas y Jurisprudencia del Sistema Interamericano de Derechos Humanos*. San José: OEA.

Monterrubio, A. (2014). Derechos de los pueblos indígenas en México en materia de consulta, participación y diálogo. Avances y desafíos desde el ámbito legislativo. *Documento de trabajo, Cámara de Diputados*, 167. Retrieved from http://www5.diputados.gob.mx/index.php/camara/Centros-de-Estudio/CESOP/Estudios-e-Investigaciones/Documentos-de-Trabajo/Num.-167.-Derechos-de-los-pueblos-indigenas-en-Mexico-en-materia-de-consulta-participacion-y-dialogo.-Avances-y-desafios-desde-el-ambito-legislativo.

Pérez, O. (2014). *Por senda de justicia. Inclusión, redistribución y reconocimiento*. Ciudad de México: UNAM.

SCJN. (2014). *Protocolo de actuación para quienes imparten justicia en casos que involucren derechos de Personas, Comunidades y Pueblos Indígenas.* Ciudad de México: SCJN.

UNPFII. (2005). Informe del Seminario internacional sobre metodologías relativas al consentimiento libre, previo e informado y los pueblos indígenas, 17 a 19 de enero de 2005. UN Doc., E/C.19/2005/3, 17 February 2005.

Part III

Institutionalising prior consultation

10 The construction of a general mechanism of consultation with Indigenous Peoples in Costa Rica

William Vega

Introduction

The aim of this chapter is to analyse the historical context and the framework of regulations and case law which have given rise to the construction of a General Mechanism of Consultation Mechanism for Indigenous Peoples (CMGDPI) in Costa Rica. To this end, first of all the chapter briefly revises the discourse of national identity and the exclusion of any reference to Indigenous culture. Next, it considers the current status of acknowledgement of the right to consultation in Costa Rica, as well as regulations related to international human rights law. Next, the process to construct the CMGDPI is described in detail, and some final conclusions are offered.

Discourse of national identity and exclusion of Indigenous culture in Costa Rica

Costa Rica has sought to create a discourse of national identity – and position itself internationally – based on such principles as the promotion of peace, neutrality, sustainable development, the preservation of nature, and respect for and promotion of human rights. With regard to human rights, specifically, this discourse has not necessarily been materialised in public policies related to guarantees and protection, and these aspects remain in a purely discursive sphere. This discourse has historical roots, as will be explained below.

As is the case for other nations in the Americas, the creation of a discourse of national identity in Costa Rica dates back to the nineteenth century. However, Acuña Ortega (2002) and Soto Quirós (2008) state that the case of Costa Rica was an exception in this sense within the Central American region, given that the discourse – based on the liberal tradition – had a civilising purpose and invented several characteristics of the population, which it presented as peaceful, democratic, rural, homogenous, hardworking, predominantly white, of Spanish origin, poor, Catholic, and prudent. It also emphasised the reduced number of Indigenous Peoples (although data from that time estimate it at between 10 and 20%), and the low numbers of mixed

DOI: 10.4324/9781351042109-14

citizens, presenting these features as facts and as favourable in comparison with other States in the region. By the 1870s, the liberal way of thinking was consolidated, and these characteristics were socialised by means of educational reforms that promoted the idea of a white race, rendering the Indigenous population invisible by locating them in the past and placing them outside the national project, as if they were unrelated and – in any case – on the verge of disappearance (Soto Quirós, 2008).

The qualities mentioned above served as the basis for Costa Rica's historical singularity. Thus, the myth of a pure, white, national race was established, in the context of the promotion of a nationalist rhetoric by liberal intellectuals who supported the articulation of "a real cultural nation and a national identity" (Soto Quirós, 2008, p. 52).

These ideas successfully made it through to the twentieth century (Soto Quirós, 2008). The absence of Indigenous Peoples in Costa Rica's national discourse explains the lack of laws to protect them during the first half of the century; indeed, an invisible population does not need to be protected. Similarly, this exclusion from the official discourse was one of the main challenges in promoting laws to protect them during the second half of the century. Nevertheless, motivated by rights-based claims by Indigenous Peoples, and multicultural constitutional reforms in Latin America, the collective rights of Indigenous Peoples were gradually recognised, particularly at the end of the twentieth century and beginning of the twenty-first (Botero, 2008).

Recognising the right to prior consultation in Costa Rica

The right to prior consultation of Indigenous Peoples in Costa Rica has been acknowledged since 1992, when ILO Convention No.169 Concerning Indigenous and Tribal Peoples in Independent Countries of 1989 (ILO 169) was ratified by Costa Rica. In 2007, the executive signed the United Nations Declaration on the Rights of Indigenous Peoples (UNDRIP) – an instrument that would reinforce the recognition of this right. In the same year, the Inter-American Court of Human Rights (IACtHR) rendered judgement on the *Saramaka* v. *Suriname* case, introducing consultation with Indigenous Peoples as a guarantee protected by the American Convention on Human Rights (ACHR) (IACtHR, 2007). This was the first judgement rendered by the Inter-American system regarding consultation with Indigenous Peoples. These norms are the basis for the protection of the right to consultation in International Human Rights Law (IHRL) in Costa Rica.

As will be detailed further, the regulations mentioned are integrated into domestic laws, according to introductory constitutional regulations[1] for IHRL[2] (Góngora, 2014, p. 304) and validated by the interpretation of the Constitutional Court of the Supreme Court of Justice of Costa Rica (CCSCJ) [3] regarding the constitutional hierarchy of international treaties and the conventionality control of the ACHR (CCSCJ, 2013; 2014).

Despite the current regulatory framework, there are no constitutional or legal regulations that have developed the right to consultation of Indigenous Peoples, although their claims have increased in the last two decades. However, at the constitutional level, there are two regulations on the protection of rights of Indigenous Peoples in general: article 76, which determines the duty of the State to safeguard the "maintenance and fostering of national Indigenous languages", and the reform of article 1 of 2015, which determines that Costa Rica is a "democratic, free, independent, multi-ethnic and pluri-cultural" Republic (Legislative Assembly, 2015). Nevertheless, the extension of this reform has not yet been interpreted by the Constitutional Court.

For this analysis, it is important to contextualise the hydroelectric project known as *El Diquís* (HP Diquís). This was a massive, public initiative by the Costa Rican Institute for Electricity (ICE) to create energy in the southern region of the country, affecting eight of the 24 Indigenous territories. Although the project was ushered in during the 2006–2010 administration (through the issuance of a Public Interest Order), it was during the 2014–2018 administration that the consultation mechanism needed to continue with the project was created. Currently, the 2018–2022 administration has repeatedly stated that it has no intention of continuing with the project.[4]

Owing to the progress made by the HP Diquís at the end of the last decade, the UN Special Rapporteur on the Rights of Indigenous Peoples (SRRIP) visited the peoples affected by the project. Afterwards, the SRRIP issued a series of recommendations to the State of Costa Rica (Human Rights Council, 2011). Some ideas of the UN Special Rapporteur were reconsidered by the CCSCJ that year (in particular, the standard that any consultation methodology should be previously agreed with the peoples by the State) when issuing Decision No 12,795-2011 that determined the government's duty to carry out a consultation process on HP Diquís within six months, as well as to agree on the consultation process with the affected peoples (CCSCJ, 2011b).

Although there are no domestic regulations dealing with the right to prior consultation, this situation did not hinder its recognition in the case law of the CCSCJ as of 1992, when a mandatory consultation of the legislative branch regarding the constitutionality of ILO 169 was resolved (CCSCJ, 1992). After its ratification, the Constitutional Court held a regular position regarding its binding nature and the content it deals with.[5] Several cases are worth highlighting, including: the judgements on the administration of the aqueducts of the Indigenous Territories of Cabagra and Térraba, of 2016 and 2013, respectively; the decisions on the HP El Diquís of 2011 and 2016; the judgement on the Indigenous Education Subsystem of 2016; the judgement on an appointment of an Indigenous representative to the FILAC (Fund for the Development of Indigenous Peoples) of 2014; and the vote on the requirements for the application of housing loans in 2017 (CCSCJ, 2011a; 2011b; 2016; 2017).

More recently, the CCSCJ resolved a writ of *Amparo* filed by the Indigenous territory of Alto Chirripó, where the constitutionality of some of the aspects of the CMGDPI process were challenged, and several elements of the constitutional standards on the subject were specified (CCSCJ, 2017). It is also worth noting that, in the cases admitted so far, the Court has issued an order for the government to carry out a consultation process[6] and to agree on the consultation process with the affected Peoples.[7]

Together with the standards regarding consultation, there are other domestic regulations that are relevant for this analysis; for instance, the case of the Indigenous Law of 1977 which constitutes one of the most important regulations regarding protection of the rights of Indigenous Peoples (Legislative Assembly, 1977). This law systematised prior improvements regarding land and territories[8] and regulated issues such as identity, the right to self-government, procedures for land recovery and the right to natural resources, among others. The law was regulated through Executive Order No. 8487, establishing that the representation of Indigenous Peoples would be made through the already existing Associations of Integral Development (ADI) (Executive Branch, 1978), although the Indigenous Law stated that Indigenous territories must be governed by Indigenous People according to traditional community structures, or by relevant national laws (Legislative Assembly, 1978). The establishment of ADIs as the mechanism to represent the 24 Indigenous territories in Costa Rica has been subject to criticism, as it is an imposition and alien to the traditional structures of the Indigenous Peoples (Chacón, 2005), as well as having low levels of membership.

However, it is also noteworthy that, since 1992, several institutions have carried out consultation processes. Thus, the Legislative Branch fostered at least two consultation processes on the Bill for the Autonomous Development of Indigenous Peoples, although it was dismissed in the legislative process (Mesa Nacional Indígena de Costa Rica, 2007). Furthermore, some processes carried out by the Executive Branch have been registered, such as: the consultation process carried out – since 2014 – by the Costa Rican Institute of Aqueducts and Sewage Systems in the Indigenous territory of Térraba to determine how to administrate the Aqueduct[9]; the consultation process carried out between 2012 and 2013 by the Ministry of Culture and Youth on National Policies of Cultural Rights, and the bill of the General Law of Cultural Rights; as well as the consultation process carried out by the Ministry of Public Education between 2007 and 2009 regarding the Reform of the Order on the Indigenous Subsystem of Education (Consulta Indígena Costa Rica, n.d.). However, all consultation processes were designed for each specific case, as at the time there was no general consultation mechanism.

These norms and processes encouraged the Ministry of the Presidency[10] to promote the creation of the CMGDPI between 2015 and the beginning of 2018. This mechanism was created with the participation of 22 of the 24 Indigenous territories in Costa Rica, who participated in a consultation process and gave their consent to the project.

It should be noted that, since the approval of ILO 169, Indigenous Peoples have continuously insisted on the need to seek ways to exercise their rights. In this context, conflicts over consultations have been taken to the Constitutional Court and this has become an effective tool for attracting the attention of the public authorities. However, as well as the judicial route, it was also essential to engage in national dialogues to look for ways to implement prior consultation through administrative measures.

Thus, the role of the Indigenous movement in developing and constructing the CMGDPI is clear: first of all, through a series of claims over the past 25 years, and, more recently, through its involvement in the process of establishing and implementing the mechanism.

Introductory regulations and case law on International Human Rights Law

IHRL has been included in the Costa Rican legal system, in line with constitutional norms and case law. Thus, article 7 of the Constitution determines that international agreements duly approved by the Legislative Assembly are superior to the law. By contrast, article 48 regulates the *amparo* and *habeas corpus* proceedings of protection included in international documents.

With regard to the hierarchy of international treaties on human rights, the Constitutional Court has resolved since the 1147–1990 vote that IHRL law in our legal system, as opposed to other instruments of International Law, is superior not only to law pursuant to article 7 of the constitution but its provisions, so long as they provide wider coverage or protection of a specific right, and should prevail over these, with the aforementioned taking into account what is set forth in article 48 of the Constitution (CCSCJ, 1990; 2007a; 2007b).

In terms of the control of conventionality and the constitutional body of law or protection, there has been considerable jurisprudence. In the 18643–2014 vote, the Constitutional Court emphasised that it has developed a criterion of conventionality as part of the block for the protection of fundamental rights that seeks to award human rights full efficacy in the way in which they have been regulated in the Inter-American System, mainly when found within a more protective system than that which the internal regulations can provide (CCSCJ, 2014). Furthermore, the Constitutional Court has also emphasised that all State bodies, including the Executive and the Legislative, must guarantee control of conventionality with the aim of promoting respect and guaranteeing the protection and effective exercise of fundamental rights (CCSCJ, 2013). Consequently, the regulations mentioned by the CCSCJ and its own case law offer fertile ground for the assertion and acknowledgement of the human rights contained within international treaties.

With respect to its receptiveness of ILO 169, the Constitutional Court has held a similar stance since 1992 (CCSCJ, 1992). In the case of the Indigenous territory of Alto Chirripó in relation to the CMGDPI process,[11] the CCSCJ mentioned the criteria established in 1992 when the constitutionality of the ratification of Convention No. 169 was perceptively pronounced. According to the Constitutional Court, far from including disagreements with the Constitution, ILO 169 reflects the most prized values of our democratic nation, developing the human rights of the Indigenous Peoples of Costa Rica, and it may be a starting point for a revision of the secondary legislation so as to adapt it to these needs (CCSCJ, 2017).

On the other hand, the CCSCJ considered that "article 6 of the Convention demands of the State the obligation to guarantee Indigenous Peoples their right to organise and participate actively in the decisions that affect them. In fact, whenever legislative or administrative measures that may directly affect them are foreseen, it is established that the State must consult them by means of appropriate procedures and, specifically, by means of their representative institutions" (CCSCJ, 2017, p. 20). Among other relevant issues, the Constitutional Court established that Indigenous Peoples – and not the public authorities – are responsible for specifying how their collective rights will be affected, as well as the benefits or harm that a certain project will cause them (CCSJC, 2017).

In relation to the implementation of article 6 of the Convention, Decision No. 14,522 has established that

> Undoubtedly, *adopting a general consultation process for indigenous peoples* to enforce what has been set forth in article 6 of ILO Convention 169 on Indigenous and Tribal Peoples, which establishes the cases in which they must be consulted and the means to be employed, *is a regulatory measure that concerns and affects the interests of the indigenous communities*, since once said mechanism has been established, it shall be used in any future consultation [...] In view of what has been previously mentioned, *adopting this mechanism* established by the Presidency of the Republic by means of ruling No. 042-MP, **must also be in line with what has been established in article 6 of Convention No. 169** in the above-mentioned terms [...] [emphasis added].
>
> (CCSCJ, 2017, p. 27)

Thus, the State's obligation to carry out a consultation on public policies – whether they are administrative or legislative in nature – which aim to create mechanisms of consultation has been determined at the constitutional level.[12] The ruling confirmed the constitutionality of the mechanism established by the Executive Branch in this sense: it was a consultation about consultation.

The process of creating a Mechanism for Consultation with Indigenous Peoples (CMGDPI)

The CMGDPI process is based on three orders by the government: Executive Directive 042-MP (Executive Branch, 2016), which initiated the process, and Executive Order No. 40932-MP-MJP (Executive Branch, 2017), and Directive No. 101-P (Presidente de la República de Costa Rica, 2018), which are the result of this consultation process over consultation itself.[13] All of the regulations were issued by the Ministry of the Presidency and signal the beginning and ending of the process to create policy.

With respect to the Executive Branch's determination to issue an Executive Order instead of promoting a bill, it is worth taking several factors into account. First, the Constitutional Court has made a few references to the Executive Branch regarding the way in which consultation processes were being carried out. Since 1994, the legislative branch has been considering the Bill for the Autonomous Development of Indigenous Peoples and – up until now – there has been no progress whatsoever in relation to its approval. This has encouraged the idea that the legislative branch is not to be deemed as particularly efficient at acknowledging and guaranteeing Indigenous Peoples' rights.

Executive Directive 042-MP

The CMGDPI process began with the issuing of Executive Directive 042-MP (Executive Branch, 2016). It established the need to carry out a consultation process on the issue of consultation itself, in accordance with the recommendations issued by the UN Special Rapporteur on the Rights of the Indigenous Peoples (UNSR, 2009; Human Rights Council, 2011; 2012). Indeed, the UN Special Rapporteur highlighted that the need to carry out a consultation on consultation itself derives not only from international obligations but also from the need to create an environment of mutual trust and respect, which is why the consultation process itself should be a result of consent (Human Rights Council, 2011). In his report of 2011, the UN Special Rapporteur warned that the consultation regarding consultation should consist of an open and thorough dialogue between the parties on several aspects of consultation that need to be established, which include defining the different stages of consultation, the corresponding timeframes, and the specific mechanisms of participation. This dialogue should not be initiated if predetermined stances on these matters have been adopted (Human Rights Council, 2011).

In essence, the Directive is an Order directed at the public institutions of the Executive Branch in three ways: (1) it promotes internal organisation to initiate dialogue, (2) it suggests minimum regulations in terms of Indigenous Peoples' Rights (Block of Constitutionality) and (3) summons them to a first Territorial Meeting. Thus, it was suggested that territorial meetings should be held in the 24 Indigenous territories to start the dialogue. The aim was to construct a joint, participative, gradual, exclusive, and final consultation between the government and Indigenous Peoples on the creation of a General Mechanism of Consultation with Indigenous Peoples (Executive Branch, 2016). The Directive incorporated the minimum standards of the Constitutional Block and defined certain meanings, participants, invitations, exchanges of information, governing principles, observers, methodologies, and regulations for the first Territorial Meeting.

Although the Directive planned the first round of meetings in the 24 territories, it was considered that future stages of the process should be agreed upon together with the Indigenous Peoples, and this was respected (Executive Branch, 2016). This decision was established as one of the key elements of the process, as it shows that the government was willing to negotiate, which is key to building trust.

For its part, the Directive did not specify a fixed timeframe for the process, which ultimately went on for 24 months.[14] In total, the process included: over 120 community events (involving at least four visits to each territory); two national meetings of Indigenous delegates[15]; four regional meetings with Indigenous delegates; two drafting sessions between the government and the Indigenous Drafting Commission[16]; a national meeting for the Indigenous youth, a meeting with the Indigenous Women's Forum of the Women's National Institute; territorial meetings prior to the preparation of the activities; and 20 working sessions between public officials and specialised bodies of the United Nations System.[17]

It is important to highlight certain aspects of the Executive Directive that are particularly relevant in relation to international standards on this issue. The first aspect is the incorporation of international treaties and Inter-American case law as a regulatory framework that offers better protection of the right to consultation, in accordance with the case law related to the hierarchy of international treaties.

Second is the creation of a general mechanism of consultation according to international standards which refer to coordinated, systematic actions of protection[18] and the general duty to adapt internal regulations so that rights and freedoms can be made effective.[19] Likewise, the decision to carry out a process of consultation on consultation itself, which stems from recommendations made by the UN Special Rapporteur on the Rights of Indigenous Peoples (Human Rights Council, 2011) and the 2011 constitutional judgement of the HP Diquís case (CCSCJ, 2011b), was particularly apt, given that it gave much-needed legitimacy to the final instrument of consultation.

Another appropriate decision was to seek the involvement of Indigenous Peoples from the early stages of the process. In this sense, the Executive Directive made public the government's intention of initiating a consultation process on consultation itself, so as to reach an agreement or consent on the consultation mechanism. The methodology aimed to garner information for a first draft that would be subject to discussion, revision, and consultation by the Indigenous Peoples on several occasions during the process[20] (Executive Branch, 2016).

It is important to note that 22 of the 24 Indigenous Territories granted their consent for the issue of the Executive Order 40932-MP-MJ (Executive Branch, 2017). The territories that did not grant the said consent were the Indigenous territories of Talamanca Bribri and Alto Chirripó. In the case of Talamanca Bribri, although no court proceeding was held, peculiarities in the way in which the process was carried out meant that the final stages were left unfinished. In the case of Alto Chirripó, a writ of *Amparo* was filed, and the Constitutional Court found that certain standards had not been fulfilled during the consultation process (CCSCJ, 2017). After the proceeding was filed, the Ministry of the Presidency requested clarification from the Constitutional Court, leading to a new decision. This second decision finds that, in spite of the fact that the "order has already been approved and published for the other Indigenous communities [...] it cannot be imposed in the territory of the party issuing the proceeding, except, of course, if it is so required by said party" (CCSCJ, 2018, para. 4). Likewise, as can be understood from the ruling, those territories which have not granted their consent to the text, can do so once the requirements indicated by the aforementioned Decision No. 14,522 (CCSCJ, 2017) – which establishes standards on consultation – have been met.

Table 10.1 summarises the main stages in this process.

Executive Order 40932-MP-MJ

Executive Order 40932-MP-MJ is a norm collectively created by Indigenous Peoples together with the government (Executive Branch, 2017). It is the result of the process known as the "consultation on the consultation", whereby 22 Indigenous territories granted their consent for the order to be adopted. It was formally published on 6 March 2018.

The purpose of the order is to regulate the Executive Branch's obligation to carry out prior consultations with Indigenous Peoples and obtain their Free, Prior and Informed Consent (FPIC), whenever measures carried out by the legislative branch, the public administration, and the private sector that may affect their collective rights are considered. In a similar way to the Directive, the Order is based on Costa Rica's constitutional block.[21] Specifically, it refers to two norms: ILO 169, which mentions the State's obligation to carry out consultation processes in its article 6; and the ACHR, which – via the interpretation of the IACtHR – has established the same obligation. Likewise, it includes elements contained in the UNDRIP.

The Order contains rules related to the scope of consultation within the institutions of the Executive Branch, the binding nature of the agreements, the roles of consent, the dimensions of consultation (local, territorial, regional, or national), and the levels of impact (positive or negative) (Executive Branch, 2018). For its part, the obligation to carry out consultations is duly established: consultations are a right of Indigenous Peoples; the State's obligation is unavoidable; and there must be suitable ways of informing, summoning and communicating the process. Furthermore, based on the constitutional block, the Order provides a list of cases where consultation is required, as well as a broad definition of unforeseen cases (Executive Branch, 2018).

Table 10.1 Stages in the creation of a prior consultation mechanism in Costa Rica (2015–2019)

	Stage	Content	Outcome
Stages of the mechanism			
1	Preliminary discussions	General debates about the standards	Proposals for an Executive Guideline
Stages proposed for the Executive Guideline			
2	Executive Guideline 042-MP	Minimum rules for initiating the process	Initiation of the process
3	Informative Workshops (22)	Information about the HR standards and the Executive Guideline	Distribution and information. Debate between Indigenous peoples
4	I Territorial Meeting (24)	First debates about the consultation standards with Indigenous peoples	Drafting of the document (one per territory) on how they see HR standards
Stages negotiated with Indigenous Peoples			
5	Visits for planning	Coordinating the performance of activities	Agreed upon dates and activities
6	II Territorial Meeting (24)	Returning the Territorial document and delivering the proposal of the steps to be taken for the Mechanism	Debate on the draft of steps for the Mechanism and selection of delegates
First draft of the Order			
7	Regional Meetings (4)	Explanation to the delegates of the content of the first draft and planning of the national meeting	Information delivered to delegates, and agreements for carrying out the national meeting
8	I National Meeting	Debate and modifications to the first draft	Creation of the first Drafting CommissionFirst national modifications to the draft
9	I Session of the Drafting Commission	Debate and modifications to the first draft	Approval of the Second Draft of the Order
Second draft of the Order			
10	III Territorial Meeting (22)	Debate and modifications to the second draft in each Territory	Introducing modifications per territory to the draft,Appointment of alternate delegates
11	II Session of the Drafting Commission	Debate and modifications to the second draft with the Drafting Commission	Consensus between the government and the Drafting Commission
12	II National Meeting	Debate and modifications to the second draft with the delegates	Granting of free, prior and informed consent to the text of the Order
Final text of the Order			
	Executive Order	The Order regulates the consultation process for the 22 Indigenous Territories	

Source: own elaboration

In relation to the parties involved, the Order established the powers of the different authorities created therein: the Technical Unit for Consultation (UTCI) and the Territorial Instances of Indigenous Consultation (ITCI), with the latter being a requirement articulated by the Indigenous Peoples in the first national meeting of delegates. Moreover, the roles of the opposing parties interested in the consultation processes (either public or private) are regulated (Executive Branch, 2018).

Finally, the Order regulates eight stages of the consultation process; namely, (1) request; (2) admissibility; (3) preparatory agreements; (4) exchange of information; (5) internal assessment made by the Indigenous Peoples; (6) dialogue, negotiation and agreements; (7) completion of the process and (8) fulfilment and monitoring of the agreements (Executive Branch, 2018).

Executive Order No. 40932-MP-MJP (Executive Branch, 2017) is currently in its implementation stage. From the institutional perspective, the Ministry of Justice is undertaking actions for the establishment of a Technical Unit of Indigenous Consultation (*Unidad Técnica de Consulta Indígena*), such as estimating the budget and calculating the human resources necessary for its approval. Unofficially, it is known that to date at least five requests for consultation have been submitted to the Ministry of Justice, which is waiting for indications on how to proceed.

From the territorial point of view, some Indigenous territories have begun to carry out the first regional and territorial discussions on the establishment of the ITCI and its respective regulations, making commendable progress since the issuing of the Executive Order. This is the case of the Indigenous Territories of Boruca, Curré, Cabagra and Tayní, which have formed the TIIC and are taking steps at a territorial level to initiate the consultation processes.

Some conclusions

The identification of introductory rules for International Human Rights Law is a step forward in the implementation of an international obligation, given that it clarifies the integration and hierarchy of norms between the international and national levels. In the case of Costa Rica, the identification of introductory constitutional rules and case law has allowed us to carry out a regulation of the right to prior consultation of Indigenous Peoples on a solid legal base that is both national and international. This brings about legal certainty for the parties, and coherence between the Costa Rican General Mechanism of Consultation Mechanism for Indigenous Peoples (CMGDPI) and the block of constitutionality.

Consultation over consultation is a recent standard that has begun to be promoted by the UN Special Rapporteur on the Rights of Indigenous Peoples and which is included within Costa Rica's constitutional case law. Its importance does not only derive from its inclusion in these standards but also because it is directly related to trust-building among parties. The Costa Rican case demonstrates that, in appropriate conditions, it is possible to obtain the FPIC of Indigenous Peoples over the issuance of measures that affect their rights. In the consultation process studied here, gaining Indigenosu FPIC becomes more relevant, since it is the norm that will regulate all prior consultation processes from now on.

Although the Executive Order (Executive Branch, 2017) is the main rule in this process, it does not consider all of the particularities of the 22 Indigenous territories. Thus, the territorial regulation of the mechanism needs to incorporate the necessary adjustments for the proper functioning of the CMGDPI in each territory, and will be applied jointly with the Order. Taking into account the particularities of each territory means that the Order can be adapted to the distinctive elements of each territory and people, thereby leaving behind the application of unique mechanisms or protocols that reveal the differences in each territory.

Naturally, there are unresolved challenges in relation to how the General Mechanism of Consultation Mechanism for Indigenous Peoples is implemented. On the one hand, a dialogue between the executive and legislative branches must be generated, because currently it is not foreseen that existing bills being dealt with by the legislative branch will be consulted (rather, only those which have been promoted by the executive). Likewise, the application of the consultation mechanism to local governments or administrative measures issued by the Judicial Branch is still pending. Other implementation challenges include the financial resources required to carry out consultation processes, the continual training and capacity building of public institutions and civil servants, the decentralisation of the mechanism from the capital city to areas with a greater presence of Indigenous peoples, and the territorial regulation in each of the 22 Indigenous territories.

Notes

1 Introductory clauses refer to provisions or resolutions through which the State integrates IHRL into the national legal system, affecting and transforming the essence of the constitution. It may be due to constitutional reforms or constitutional case law, which establish the constitutional or supra-constitutional hierarchy.
2 In particular, articles 7 (superior hierarchy to law of international treaties) and 48 (habeas corpus and *amparo* proceedings regarding rights included in international treaties) of the Costa Rican Constitution.
3 Through a constitutional reform, in 1989, the Constitutional Court of the Supreme Court of Justice was created with competent jurisdiction to declare the unconstitutionality of the regulations of any nature and that of actions subject to Public Law (judicial review); resolve the competence conflicts between the branches of the State; to be informed of consultations regarding constitutional reform projects, approval of international agreements or treaties and other bills, pursuant to law; and to guarantee the rights and freedoms set

forth in the Constitution and the human rights acknowledged by the International Law in force by means of the *habeas corpus* and writs of *Amparo*.

4 Although the ICE stated that the cancellation of the project was due to the lesser demand for energy production, it is also true that PH Boruca, predecessor of PH Diquís, was abandoned as an initiative due to the opposition of the Indigenous communities. Although the consultation was never carried out, the obligation to carry out a consultation process had the effect of containing the project.

5 In this regard, decision No. 14,522 of 2017 reviews the main judgments related to Indigenous Peoples' rights, particularly, regarding the right to consultation (CCSCJ, 2018).

6 The mentioned judgments stated that the Foreign Affairs Ministry, Costa Rican Institute of Aqueducts and Sewage System, the Ministry of Housing, and the Ministry of the Presidency – all of which are institutions of the Executive Branch – should carry out the consultation processes.

7 The Court resolved that no unconstitutionalities were found in the public interest order (CCSCJ, 2011). However, this judgement was subject to a consultation process on the HP Diquís to be carried out within six months. However, this consultation was not carried out, which encouraged Decision No. 15711–2016, stating that the order was partially unconstitutional (CCSCJ, 2016).

8 Since 1956, recognition of dozens of Indigenous territories has begun via an Executive Order. Such territories were legally recognised with the Indigenous Law.

9 This consultation process is fostered by the decision of the Constitutional Court in decision No 6,655–2013 and has been extended for over four years waiting for a consultation mechanism.

10 The Ministry of the Presidency was created in 1961 with the aim of providing support to the President of the Republic and to serve as a link between the State branches, actors of the civil society, and the different ministries. In general terms, this ministry works as a coordination agency between the Republic and such actors.

11 The Indigenous territory of Alto Chirripó filed a writ of *Amparo* against the General Mechanism of Consultation. This Indigenous People claimed that this process did not comply with international standards (CCSCJ, 2017).

12 The Constitutional Court had already established in its decision No. 12,975–2011 (on HP Diquís) the obligation of determining the methodologies for consultation with the affected peoples (CCSCJ, 2011b).

13 The Executive Order No. 40932-MP-MJP (Executive Branch, 2017) and Directive No. 101-P (Presidente de la República de Costa Rica, 2018) are rules with the same content. The first one is aimed at the Central Administration and the second one at the Decentralised Administration.

14 Considering at least six months of planning and 24 months between the Directive being issued and the Order being issued. The Directive was created with the support of the United Nations, public institutions, the office of public defines and external Indigenous consultants.

15 The Government proposed the selection of four representatives per territory and with conditions of gender equality. For the First National Meeting, 96 delegates of 23 territories were assigned and for the Second National Meeting, 94 delegates of 22 Territories were assigned. All together, they formed the National Assembly of Indigenous Delegates. As shall be seen, it worked as a national instance for the representation of Indigenous Peoples with regard to the creation of the Mechanism, and it was aimed at negotiating the details of the Order with the Government.

16 The Drafting Commission was an instance required and created by the delegates of the Indigenous Peoples during the First National Meeting. Its aim was to negotiate certain stages of the process directly with the Government.

17　See Table 12.1.
18　Article 2 of ILO 169 fosters the development of coordinated and systematic actions which seek the protection of Indigenous Peoples' rights.
19　Both article 2 of the ACHR and the case law of the IACtHR foresee the State's duty to adopt the necessary measures for guaranteeing the effectiveness of conventional rights and freedoms.
20　Article 8, related to the informed process; article 10, which deals with the agreement on methodologies; article 12, concerned with the summoning to territorial meetings; article 13, which suggests a series of goals for the first territory meetings, and article 14, which sets out agreements on the stages that take place after the process.
21　See the "considering" (*considerando*) section of the Directive and of the Order, where the rules and case law of form and substance that serve as the basis of this process are summarised.

References

Acuña Ortega, V. (2002). La invención de la diferencia costarricense, 1810–1870. *Revista Historia*, 45, 191–228.
Botero Marino, C. (2008). *Los retos del juez constitucional en un Estado multicultural: el caso de Colombia*. Madrid: Centro de Estudios Políticos y Constitucionales.
CCSCJ. (1990). Decision No. 1,147, 13 January 1990 [on file with author].
CCSCJ. (1992). Decision No. 3,003, 7 October 1992 [on file with author].
CCSCJ. (2007a). Decision No. 1,682, 9 February 2007 [on file with author].
CCSCJ. (2007b). Decision No. 4,276, 27 March 2007 [on file with author].
CCSCJ. (2011a). Decision No. 3,084, 11 March 2011 [on file with author].
CCSCJ. (2011b). Decision No. 12,975, 23 September 2011. Retrieved from https://vlex.co.cr/vid/-499615202.
CCSCJ. (2013). Decision No. 6,247, 9 May 2013 [on file with author].
CCSCJ. (2014). Decision No. 18,643, 12 November 2014. Retrieved from https://www.poder-judicial.go.cr/salaconstitucional/index.php/control-convencionalidad/684-14-0 18643.
CCSCJ. (2016). Decision no. 15,711, 26 October 2016. Retrieved from https://nexuspj.poder-judicial.go.cr/document/sen-1-0007-875804.
CCSCJ. (2017). Decision No. 14,522, 16 June 2017 [on file with author].
Chacón Castro. R. (2005). *El sistema jurídico indígena en Costa Rica: Una aproximación inicial*. San José: Instituto de Investigaciones Jurídicas de la Universidad de Costa Rica.
Consulta Indígena Costa Rica. (n.d.). Consultas en Costa Rica. Retrieved from http s://www.consultaindigena.go.cr/consultas-costa-rica/.
Executive Branch. (1978). Executive Order No. 8487-G. Reglamento a la Ley Indígena. Retrieved from http://www.pgrweb.go.cr/scij/Busqueda/Normativa/Normas/nrm_texto_completo.aspx?param1=NRTC&nValor1=1&nValor2=56355&nValor3=61774&strTipM=TC.
Executive Branch. (2013). Executive Order No. 37801. Reforma del Subsistema de Educación Indígena. Retrieved from http://www.pgrweb.go.cr/scij/Busqueda/Normativa/Normas/nrm_texto_completo.aspx?param1=NRTC&nValor1=1&nValor2=10147&nValor3=10857&strTipM=TC.
Executive Branch. (1993). Executive Order No. 22072-MEP. Crea Subsistema Educación Indígena. Retrieved from http://www.pgrweb.go.cr/scij/Busqueda/Normativa/

Normas/nrm_texto_completo.aspx?param1=NRTC&nValor1=1&nValor2=10147& nValor3=10857&strTipM=TC.

Executive Branch. (2016). Executive Directive No. 042-MP. Construcción del Mecanismo de Consulta a Pueblos Indígenas. Retrieved from http://www.pgrweb.go.cr/ scij/Busqueda/Normativa/Normas/nrm_texto_completo.aspx?param1=NRTC&nValor1=1&nValor2=81277&nValor3=103560&strTipM=TC.

Executive Branch. (2017). Executive Order No. 40932-MP-MJO. Mecanismo General de Consulta a Pueblos Indígenas. Retrieved from https://www.imprentanacional.go. cr/pub/2018/04/05/ALCA70_05_04_2018.pdf.

Góngora Mera, M. (2014). La Difusión del Bloque de Constitucionalidad en la jurisprudencia latinoamericana y su potencial en la construcción del Ius Constitutionale Latinoamericano. In F. Fierro, H. Bogdandy, & M. Morales (Eds.), *Ius constitutionale commune en América Latina. Rasgos, potencialidades y desafíos* (pp. 301–327). Ciudad de México: UNAM/Max-Planck-Institut/Instituto Iberoamericano de Derecho Constitucional.

Human Rights Council. (2009). Special Rapporteur on the Rights of Indigenous Peoples, the Situation of Indigenous Peoples in Chile: Follow Up the Recommendations made by the Previous Special Rapporteur. UN Doc. A/HRC/12/34/ Add.6, 5 October 2009.

Human Rights Council. (2011). Special Rapporteur on the Rights of Indigenous Peoples, the Situation of the Indigenous Peoples Affected by the El Diquís Hydroelectric Project in Costa Rica. UN Doc. A/HRC/18/35/Add.8, 11 July 2011.

IACtHR. (2007). *Saramaka People v. Suriname*, Judgment of November 28, 2007 (Preliminary Objections, Merits, Reparations, and Costs). Inter-Am. Ct. H.R., (Ser. C) No. 172(2007).

Legislative Assembly. (1977). Law No. 6172. Ley Indígena. Retrieved from http:// www.pgrweb.go.cr/scij/Busqueda/Normativa/Normas/nrm_texto_completo.aspx? param1=NRTC&nValor1=1&nValor2=38110&nValor3=66993&strTipM=TC.

Legislative Assembly. (2015). Law No. 9305. Reforma constitucional del artículo 1 para establecer el carácter multiétnico y pluricultural de Costa Rica. Retrieved from http://www.pgrweb.go.cr/scij/Busqueda/Normativa/Normas/nrm_texto_comp leto.aspx?param1=NRTC&nValor1=1&nValor2=80269&nValor3=101779&strTip M=TC.

Mesa Nacional Indígena de Costa Rica. (2007). Informe alternativo presentado por los pueblos indígenas al informe presentado por el Estado de Costa Rica al Comité contra la Discriminación Racial de la Convención Internacional sobre la eliminación de todas las formas de discriminación. Retrieved from https://tbin ternet.ohchr.org/Treaties/CERD/Shared%20Documents/CRI/INT_CERD_NGO_ CRI_71_8460_E.pdf.

Ministry of Housing and Human Settlements. (2013). Agreement No. 2. Requisitos para la postulación del bono familiar de vivienda dentro del programa indígena casos individuales y proyectos colectivos. Retrieved from http://www.pgrweb.go.cr/ scij/Busqueda/Normativa/Normas/nrm_texto_completo.aspx?param1=NRTC&nValor1=1&nValor2=75965&nValor3=94667&strTipM=TC.

Presidente de la República de Costa Rica. (2018). Directriz N° 101-P. Implementación del mecanismo general de consulta a pueblos indígenas, 6 March 2018. Retrieved from http://www.pgrweb.go.cr/scij/Busqueda/Normativa/Normas/nrm_texto_completo.aspx ?param1=NRTC&nValor1=1&nValor2=86192&nValor3=111716&strTipM=TC#dd own.

Soto Quirós, R. (2008). Imaginando una nación de raza blanca en Costa Rica: 1821–1914. *Amérique Latine Histoire et Mémoire Les Cahiers ALHIM*, 15. Retrieved from http://journals.openedition.org/alhim/2930.

UNSR. (2009). Comentarios del Relator Especial sobre los derechos de los pueblos indígenas en relación con el documento titulado: "Propuesta de gobierno para nueva normativa de consulta y participación indígena de conformidad a los artículos 6° y 7° del Convenio N° 169 de la Organización Internacional del Trabajo". Retrieved from http://unsr.jamesanaya.org/docs/special/2012-11-29-unsr-comentarios-a-propuesta-reglamento-consulta-chile.pdf.

11 The construction of a national mechanism of prior consultation in Honduras

Irati Nahele Barreña

Introduction

The right of Indigenous Peoples to be consulted has been the subject of many studies and research projects. This chapter aims to review the background to and development of the process of dialogue to construct a national mechanism for the prior consultation of Indigenous Peoples[1] in Honduras between 2014 and 2018.

In most countries around the world, relations between States and Indigenous Peoples have been characterised by the lack of trust between both groups. This is also the case in Honduras. In addition, exercising the right to consultation is one of the main sources of tension between Indigenous Peoples and the State. In Honduras, there have been many human rights mechanisms and bodies that have emphasised the need for some type of instrument that regulates the right to prior consultation of Indigenous Peoples, and this is the issue we intend to analyse in this chapter.

In the pages that follow, the issue of Indigenous Peoples in Honduras will be contextualised and the existing national regulatory framework on the subject will be analysed. Subsequently, the background to the construction of a national mechanism of prior consultation will be examined, analysing several events that gave rise to the process. Next, the different stages in the process of dialogue to construct the mechanism will be reviewed. Finally, some conclusions are offered, including an analysis of the difficulties encountered during the process, and the identification of future challenges surrounding the implementation of the right to prior consultation and the mechanism itself in Honduras.

Contextualisation and status of Indigenous Peoples

Since 1994, through a Presidential Order (La Gaceta Diario Oficial de la República de Honduras, 1994a), the Honduran State has recognised the multi-cultural and multi-linguistic nature of Honduran society, and assumes this diversity as a resource for internal development, in particular, for the comprehensive development of national communities.

There is a broad consensus (RENUDPI, 2016, p. 3) on the lack of up-to-date or accurate data[2] on the Indigenous population in Honduras. However,

DOI: 10.4324/9781351042109-15

it is believed that Indigenous Peoples make up between 10% (INE, 2013) and 20% (BID, 2007) of the total population. Currently, the existence of nine culturally distinct Peoples is recognised: the Lenca People, the Maya Chortí People, the Tolupán People, the Garífuna People, the Habla Inglesa [English-speaking] People, the Nahua People, the Pech People, the Tawahka People, and the Miskitu People.

Regarding Indigenous representativeness, there is a Confederation of Autonomous Peoples of Honduras (CONPAH), which is an effort to unite the Federations of the different Peoples and defend their rights (ACNUDH/ América Central, 2011, p. 277); however, it suffers from serious internal divisions (RENUDPI, 2016, p. 16). This division and fragmentation of Indigenous organisations is one of the greatest challenges both for the peoples themselves and for the State when it comes to establishing processes of dialogue and the full and effective participation of Indigenous Peoples.

Honduras is organised into 18 departments, of which 15 are inhabited by Indigenous Peoples, who own 1,322,774.50 hectares of land. While more than 500 title deeds have been issued to them (RENUDPI, 2016, p. 9), only 10% of Indigenous Peoples have a property title (IWGIA, 2010). Furthermore, 80% of the Indigenous Peoples still live in their traditional territory and the remaining 20% live in urban areas,[3] mostly as a result of migration in search of better living conditions (Fondo de las Naciones Unidas para la Infancia, 2012, p. 31).

Although Indigenous Peoples in Honduras are currently in a critical situation regarding their land, territory, and natural resources specifically, they are also affected by the fact that Honduras has the highest levels of poverty and social and economic inequality in Latin America, with a Gini index of 0.52 (PNUD, 2015) and 62.8% of households living in poverty (INE, 2015). Likewise, the human development indicators relating to Indigenous Peoples fall well below the national average, with a median monthly income equivalent to 36.8% of the national average, while 72% of households are unable to cover the costs of basic consumer goods, in contrast to 41.6% of households nationwide. Another of the main challenges facing Indigenous Peoples is the violence that results from the "imposition of development and investment plans and projects and licences for the extraction of natural resources in their ancestral territories" (Comisión Interamericana de Derechos Humanos, 2015, p. 40).

Regulatory framework

The 1982 Constitution of the Republic of Honduras includes and recognises principles relating to human dignity and both rights and responsibilities. At the constitutional level, Honduras recognises itself as a multi-ethnic and multi-cultural country, committing itself to preserving and promoting native cultures and protecting the rights and interests of Indigenous communities (Asamblea Nacional Constituyente, 1982, article 348). This is the only specific mention of the rights of Indigenous Peoples in the Constitution; however, two constitutional provisions (Asamblea Nacional Constituyente, 1982, articles 172 and 173) refer to the State's obligation to promote its cultural, anthropological, and folkloric wealth.

Although the Constitution does not mention peoples, but rather communities, the rights of Indigenous Peoples are protected by the Constitution insofar as it recognises the principles of international law. The Constitution (Asamblea Nacional Constituyente, 1982, article 15) establishes that Honduras endorses the principles and practices of international law and that international treaties are part of domestic law once they have been duly ratified by the State. Similarly, it states that in the event that a national law comes into conflict with an international treaty or convention, the treaty will prevail (Asamblea Nacional Constituyente, 1982, article 18). However, the Office of the UN High Commission for Human Rights (OHCHR) concludes that there is limited awareness of international law in the judiciary and that this rarely leads to invoking international law in Honduran courts (ACNUDH, 2017).

Also, the Honduran State explicitly recognises the rights of Indigenous Peoples, having ratified ILO Convention No.169 Concerning Indigenous and Tribal Peoples in Independent Countries of 1989 (ILO 169), in force in Honduras since March 1995 (La Gaceta Diario Oficial de la República de Honduras, 1994b). Additionally, it voted affirmatively both on the 2007 United Nations Declaration on the Rights of Indigenous Peoples (UNDRIP) and, nine years later, in 2016, on the American Declaration on the Rights of Indigenous Peoples (ADRIP) of the Organization of American States (OAS). As indicated by the former Special Rapporteur on the Rights of Indigenous Peoples, James Anaya (RENUDPI, 2012, p. 8), the right to consultation is not only regulated in ILO 169 or UNDRIP but it is also included in other international instruments that Honduras is also signatory to. In addition, the Inter-American Court of Human Rights (IACtHR) has concluded that the obligation to consult with Indigenous Peoples, "in addition to constituting a conventional standard, is also a general principle of International Law" (IACtHR, 2012, p. 49).

It is important to point out that, although the Honduran State has adopted secondary legislation establishing some of the rights of Indigenous Peoples and some forms of citizen participation in general, 23 years after ILO 169 entered into force, Honduras has not yet created a mechanism to develop and operationalise the right to consultation of Indigenous Peoples. The OHCHR has concluded that Honduran legislation contains no recognition of the rights of Indigenous Peoples, and neither does the Constitution (ACNUDH, 2017, p. 10). In any case, the Constitution contains the peremptory rule of the inapplicability of secondary rules that diminish, restrict, or distort the exercise of the declarations, rights, and guarantees established in it (Asamblea Nacional Constituyente, 1982, article 64).

Despite all this, the State has made some efforts and has taken steps to adapt its regulatory framework to include the right to prior consultation of

Indigenous Peoples. However, these measures and the processes to implement them have not been exempt from difficulties. Both within the Executive Branch, where several national policy instruments have been developed, and civil society, with initiatives for public policy proposals, the right to consultation is on the agenda.

In 2013, the First Public Policy on Human Rights and the National Human Rights Action Plan (La Gaceta Diario Oficial de la República de Honduras, 2013) was approved, which includes the action of agreeing with Indigenous Peoples on a participatory mechanism for holding prior consultations (OIT/CEARC, 2016a). In addition, at the same time, Honduras appeared before the Universal Periodic Review in 2015, and the "Policy against Racism and Racial Discrimination for the Comprehensive Development of Indigenous and Afro-Honduran Peoples" was officially approved by Executive Decree (La Gaceta Diario Oficial de la República de Honduras, 2016) with participation, consultation, and consent as guiding principles.

Within the Executive Branch, in 2014, the Secretariat of State for Indigenous (and Afro-Honduran) Peoples (SEDINAFROH), which had been created in 2010, became the Directorate of Indigenous and Afro-Honduran Peoples (DINAFROH),[4] the governing body of programmes and policies for Indigenous (and Afro-Honduran) Peoples. The institution's change of rank, from Department to Sub-Department, was perceived by the representatives of Indigenous Peoples as reflecting a decrease in support and resources for Indigenous affairs (RENUDPI, 2016, p. 5).

Within the Judicial Branch, in 1994, the "Special Prosecutor's Office for Ethnic Groups and Cultural Heritage" was created, a specialised body of the Public Ministry responsible for hearing complaints on violations of the rights of Indigenous Peoples and ensuring the application of international standards on the matter. As of June 2016, this Special Prosecutor's Office, with technical support from UNDP, has developed a "Manual of Procedures for the Investigation and Protection of the Rights of Indigenous and Afro-Honduran Peoples", which is mandatory for all personnel at the institution.

Construction of the mechanism of prior consultation in Honduras

Background

Several factors converged to pave the way for a mechanism of prior consultation in Honduras, which has been promoted and constructed by a range of different actors. These factors include recommendations from international human rights mechanisms (which are discussed in detail below), commitments acquired by the State in national public policies and their respective action plans, initiatives such as the European Union's Voluntary Partnership Agreement on Forest Law Enforcement, Governance and Trade (FLEGT) (RENUDPI, 2017, pp. 7–8), claims made by Indigenous movements, and/or pressure from the private sector regarding the need for a mechanism that guarantees legal certainty for its investments in the country.

Of the above factors, it is important to highlight the recommendations made by international human rights mechanisms regarding the importance of developing a legal instrument that regulates and implements the right to consultation.

For instance, the ILO Committee of Experts on the Application of Conventions and Recommendations (CEACR) has made several Observations and Direct Requests to the Honduran State regarding the right to consultation. In both 2015 and 2016, employer and worker organisations expressed their concern to the ILO about the process that the government was proposing to create the consultation mechanism. In the 2015 Observation (OIT/CEACR, 2016a), the CEACR asked the government to report on the initiatives underway, in order to establish the appropriate procedures to carry out consultation. The CEACR (OIT/CEARC, 2016a) also noted that the Honduran Council of Private Enterprise (COHEP) reaffirms the need to adopt a national law on prior consultation. In parallel, through a Direct Request, the CEACR requested the government to prepare the next report on the process to develop the mechanism, in consultation with social actors and Indigenous Peoples (OIT/CEACR, 2016b). In a following Observation (OIT/CEACR, 2017a), the CEACR asked the government to report on ongoing initiatives to establish appropriate consultation and participation mechanisms, and urged it to promptly adopt regulations on the issue, in consultation with social actors. Also, in the subsequent Direct Request (OIT/CEACR, 2017b), the CEACR requested the government to provide information on the procedures established to consult with Indigenous Peoples before undertaking or authorising any project to explore or exploit natural resources on their lands.

In the framework of special procedures, and under article 22 of the ILO Constitution, in November 2015, the Confederation of Honduran Workers (CUTH), presented an alternate report alleging non-compliance by the Honduran government with the provisions of ILO 169. The information sent to the CEACR was related to the case of the Tolupán People, with additional information on the Lenca and Garífuna Peoples, denouncing that several members of these peoples have been victims of threats and murders related to defending the environment. This situation has been stated not only by the ILO[5] but also by the current Special Rapporteur on the Rights of Indigenous Peoples, Victoria Tauli-Corpuz (hereinafter, the Special Rapporteur) (RENUDPI, 2016). After reviewing the information provided by all the actors and the commentaries from the CEACR in 2015, the ILO agreed to call on Honduras to provide more information at the 105th session of the 2016 International Labour Conference, where the government committed itself to promoting an inclusive dialogue with Indigenous Peoples, workers, and employers.

In the framework of the most recent Universal Periodic Review (UPR) of Honduras,[6] there were eight specific recommendations on Indigenous rights, and one particularly urging improved participation and consultation of Indigenous Peoples over public policies that affect them (CDH, 2015). All of these recommendations were voluntarily accepted by the State. For its part, the Committee on the Elimination of Racial Discrimination (CERD, 2014) also advised on this matter and, likewise, reminded the State of its obligation to observe the right to consultation and urged it to develop practical mechanisms for the implementation of this right, respecting the consent of Indigenous Peoples.

Similarly, the Committee on Economic, Social and Cultural Rights (CDESC, 2016) welcomed information on the construction of the consultation mechanism, but showed concern about the information received on the lack of adequate participation of Indigenous Peoples in the process. It recommended initiating a broad process of consultation and participation with regard to the construction of the mechanism, with reference to the highest international standards in the matter. In addition, the ACNUDH (2017) concluded that one of the main sources of tension between Indigenous Peoples, State authorities and private companies is the exercise of the right to consultation, specifically in the context of hydroelectric projects.

Finally, it is also important to note that the State of Honduras, having accepted the jurisdiction of the IACtHR, was found responsible for not having consulted Indigenous Peoples on matters that affected them, thus violating the American Convention on Human Rights (IACtHR, 2015).

Process[7]

As of 2012, Honduras has established a process to establish a national mechanism to regulate and operationalise the right to consultation of Indigenous Peoples. With this aim in mind, various Indigenous movements and the government have been carrying out a series of activities with a view to developing a national consultation mechanism.

The process carried out by the government was channelled through the Interinstitutional Technical Commission (CTI),[8] composed of 19 State institutions and constituted in May 2015 by the Minister of Labour and Social Security (STSS) and the director of DINAFROH, who serve as Secretary and President, respectively. The CTI was established in order to respond to the Observations and Direct Requests made by the ILO and to continue the task of constructing a consultation mechanism. A regulation was adopted on January 2016 to lay out, primarily, the structure, functions, and attributions of the body. The CTI invited the ILO, UNDP, and OHCHR to participate as observers in its meetings and, at the same time, established continuous communication with Indigenous Peoples – primarily with the federations affiliated with CONPAH – to agree on the process to construct the mechanism.

In July 2015, DINAFROH received a proposal for a draft of the consultation and FPIC Law that CONPAH and its Federations had been developing since 2012. Based on this document, and from November 2015 to May 2016, the CTI reviewed the proposals made by CONPAH and DINAFROH, and included observations and comments from various government entities, including the legal analysis of the Secretariat of General Government Coordination (SCGG). During this period, the CTI received training on the right to consultation. In this way, by May 2016, the CTI had a first draft approved by the 19 State institutions that constitute the CTI, which it would share with Indigenous Peoples to start the dialogue process. In the pre-preparation phase, it was always understood that the draft proposed by the CTI was a proposal, as its name indicated – "First draft of the Legal Bill on consultation and FPIC (Secretaría de Trabajo y Seguridad Social, 2016)" – as a preliminary document to make headway[9] in the dialogue process with Indigenous Peoples on a possible future mechanism.

In this process, CONPAH seemed to assume the representation of the nine Indigenous Peoples, and has been the main counterpart of the government, particularly in planning workshops and initial discussions to choose dates, locations, the number of meetings, etc. However, although CONPAH accompanied the process, over 100 grassroots organisations – some of which were affiliated to CONPAH and others not – participated.

When the process began, the majority of the Lenca People were represented through the National Lenca Indigenous Organisation of Honduras (ONILH) as a member of CONPAH. However, during the process – and continuing to the present – together with 25 other Lenca organisations, they formed the Lenca Sectorial Committee, and disaffiliated from CONPAH in February 2017, expressing their will to establish direct dialogue with the government in developing the mechanism.

Negotiations to agree on the plan for dialogue workshops lasted more than four months, and the plan was changed and adapted according to the agreements reached between Indigenous Peoples and the government. For example, the first government proposal – in January 2016 – was to hold three regional meetings to gather input on the proposal; however, after the initial talks, a second plan was made to hold a meeting to consult separately with each Indigenous People, and ended finally with a proposal to hold between one and three meetings with each Indigenous People at the request of the representatives and according to each Indigenous People's characteristics and number of members.

From May to October 2016, 18 dialogue workshops were held, with an open invitation to all IP organisations and institutions to participate in the process. More than 1,400 representatives of organisations, federations, territorial councils, councils of elders and Garífuna government, etc. participated. It is important to note that neither COPINH nor OFRANEH participated in this process, despite having been invited by the Government, although the reasons for their absence remain unknown (RENUDPI, 2016).

The first draft was translated into Miskitu, Tawahka, and English and was interpreted during the meetings with Tawahka, Miskitu, Garífuna and English-speaking peoples at the request of these Indigenous Peoples. In addition, several organisations brought to the table proposals that they had previously worked on with their members in community meetings. All of the input from the 18 dialogue workshops was systematised and published (Secretaría de Trabajo y Seguridad Social, 2016) and, at the same time, a new draft document was constructed with this input as well as the international standards on the subject by a team of international experts provided by UNDP, at the government's request. The team of experts recommended modifying the articles (PNUD, 2016a) of the first draft to incorporate Indigenous Peoples' proposals and also adapt them to international legal documents adopted by Honduras.

In addition, within the framework of ILO Convention No. 144 on Tripartite Consultation, the CTI has been working with representatives of workers and employers. Thus, in October 2016, the government held a tripartite consultation workshop to gather input from social actors on the process of constructing the mechanism, in which COHEP and the three worker unions (CGT, CTH, and CUTH) participated (Secretaría de Trabajo y Seguridad Social, 2016). In addition, in January 2017, bipartite meetings were carried out again, in order to report on progress.

In July 2016, the Resident Coordinator of the United Nations System in Honduras (UNS) sent a formal petition for support to the Special Rapporteur, at the government's request, in order to review and comment on the first draft that the government was developing. However, the Ministry of Development and Social Inclusion reported that this document had already been submitted to the Special Rapporteur during her visit to Honduras in late 2015 (PNUD, 2016b).

Following the request for technical assistance from the government and the Resident Coordinator of the UNS in Honduras, the Special Rapporteur conducted a preliminary analysis of the documentation received by different actors regarding the procedure. Thus, the Special Rapporteur provided comments in relation to the first draft, in which she emphasises that there are problems in relation to its content, (it is not in keeping with international standards on the matter) and its development and socialisation process (especially with respect to the timeframe for finalising the document and submitting it for approval in the National Congress). In addition, the Special Rapporteur concluded that, in the midst of the serious situation of violence and social conflict that Honduras faced due to energy, extractive, or investment projects on Indigenous lands, the effective implementation of the right to consultation was necessary but that other measures to protect and guarantee Indigenous rights were also needed (RENUDPI, 2016, p. 24).

Later, invited by the government, the Special Rapporteur carried out a mission in Honduras in April 2017 with the purpose of advising and guiding the process to construct the consultation mechanism. During her visit, she met with different actors involved in the process: Indigenous Peoples, Government, Public Ministry, National Congress, Private Sector, Trade Unions, Civil Society Organisations, and UNS. The Special Rapporteur emphasised the main challenges encountered during her visit: the situation regarding human rights, land titling, inclusion, and representativeness (RENUDPI, 2017).

After her mission, the Special Rapporteur provided additional commentaries on the process to regulate consultation in Honduras (RENUDPI, 2017). In these observations, she expressed concern about two recurring issues: first, that the socialisation process of the first draft had not been sufficiently inclusive, with several Indigenous organisations not participating in the process, and, second, that the dichotomy between consultation and consent[10] may become a stumbling block in the dialogue process (on this, see also the chapter by Herrera in this volume). In this regard, the Rapporteur expressed her concern after observing the link some actors made between consent and veto, which hindered genuine dialogue, and urged actors not to lose sight of the spirit of the principles of consultation and consent according to the international standards on the matter (RENUDPI, 2017, p. 10). The Rapporteur also noted during her visit that her previous recommendations had not yet been implemented (RENUDPI, 2017, p. 2).

At the same time, another initiative was put forward by several Indigenous organisations – including the Indigenous and Black Peoples of Honduras Observatory (ODHPINH) (BIP, 2016, p. 37), made up of the Honduran Black Fraternal Organization (OFRANEH) and the Civic Council of Popular and Indigenous Organisations of Honduras (COPINH) – which, via the representative of the Liberty and Refoundation Party (LIBRE), Rafael Alegría, in April 2016 presented the initiative of a Framework Law on the Consultation and Free, Prior and Informed Consent of Indigenous Peoples before the National Congress (COPINH, 2017). To analyse this initiative, an Opinion Committee was formed within the National Congress in May 2016, comprising seven representatives. Therefore, at the national level, there are two proposals: one supported by the Executive Branch with the Federations of CONPAH, and another supported by ODHPINH and the opposition in the National Congress.

Returning to the main proposal for a consultation mechanism in Honduras, there is no public final document even though the project has been sent to the National Congress by the Executive Branch. Thus, it is impossible to analyse or evaluate it, or indeed to find out whether the input from Indigenous Peoples or the recommendations of the Special Rapporteur and other mechanisms have been taken into account.

Challenges and conclusions

The internal regulation of the duty of consultation and its implementation is a relatively recent development for countries in the Americas (Hartling, 2017, p. 19); thus, finding points of reference and good practices is fundamental.

However, the characteristics and contexts of each country are different, so it is vital to understand the context in which there is an effort to implement this right. For its part, the Honduran context is both critical and hostile, especially for Indigenous Peoples (RENUDPI, 2016, p. 1). There are several challenges to deal with before this right can be correctly observed, in terms of the creation of policy or legislation, as well as its practical implementation in the territories.

The process adopted by the Honduran government was not technically a "consultation about consultation" in the terms of former Special Rapporteur on the Rights of Indigenous Peoples, James Anaya (RENUDPI, 2012, p. 4). However, it can be said that for the first time in the history of Honduras, a draft proposal of the consultation mechanism was socialised among the communities by the Executive Branch, where more than 100 representative organisations of Indigenous Peoples become involved and offered very valuable input to the government. The great challenge of the process is ensuring that organisations that have not felt the confidence to participate, or for various reasons have not wanted to do so, can do so if they decide to, in a climate of trust and good faith between the government and Indigenous Peoples. This process of dialogue and regulation of consultation requires considerable effort, so that the proposed mechanism is constructed in the most participatory manner possible and with the greatest consensus.

The conditions necessary to build such a climate of trust between Indigenous Peoples and the government will not be easy. The general context of Honduran society's polarised climate is a considerable challenge that hinders any dialogue process.[11] In addition, State-Indigenous Peoples relations have historically been characterised by a deep-rooted lack of trust. In this regard, for example, the Special Rapporteur recommended that the State should demonstrate its good faith by ordering an

> extension[12] for the approval and operation of development or investment projects or other similar activities that could affect the rights of Indigenous Peoples, until a new dialogue process has been satisfactorily completed and a prior consultation law, as a result of that dialogue, enters into force and can be applied to those licences or projects.
>
> (RENUDPI, 2016, p. 6)

Another obstacle to building confidence in this process has been the participation of Indigenous Peoples in the previous dialogue for the preparation of meetings, which, since their inception, should have had a broader participation, beyond the presence of CONPAH and its Federations. However, as the Confederation pointed out, as a first effort, it can be considered positive, since it was the first time that Indigenous Peoples and the government sat together to plan a process of this magnitude. Between the government and the Federations that participated in both the preliminary and final meetings, a climate of trust and mutual respect was generated, and most of the subsequent meetings were the result of that consensus.[13] However, this was not the case with the organisations that did not participate in these previous preparations.

Another considerable stumbling block in this process – which is also an obstacle to observing Indigenous Peoples' rights in Honduras in general – is the lack of capacity-building with respect to consultation and Indigenous rights among different actors. Indeed, there is no standard training plan for officials of the judicial branch or civil servants working in different departments.

It would be important for the final text of the Law to be the result of the consensus reached in a broad process of consultation about consultation, for it to respect all the international standards in the matter, and for it to be built in a climate of trust between all parties. Otherwise, the consequence of adopting a final version that does not have input from Indigenous Peoples as a result of a consultation process, and that does not respect international standards, is that it will lack legitimacy in the eyes of important sectors of Indigenous Peoples, and its implementation will be complicated. In addition, and at the same time, it is important to move forward in generating the conditions necessary to overcome the lack of trust existing between the government and Indigenous Peoples, beset for years by unresolved conflicts both in terms of natural resources and human rights violations.

Similarly, it is important to emphasise that the process described here was carried out by the government; however, the legislative power must finally approve this law. As indicated by the Special Rapporteur in her observations, it will be important

> to pay attention to the stage of legislative debate once the bill of prior consultation is submitted to the National Congress. Like the other branches of government, Congress also has the obligation to comply with Honduras' international obligations regarding the human rights of Indigenous Peoples.
>
> (RENUDPI, 2016, p. 12)

Notes

1 This chapter focuses on Indigenous Peoples but the author is well aware that all the considerations drawn here may be equally applied to all other groups in Honduras, such as Afro-Hondurans.
2 Until 2016, the reference used was from the 2001 Census.
3 Figures from the self-census of 2007, included in the Rapporteur's report (RENUDPI, 2016, p. 3).
4 DINAFROH became part of the Secretary of State in the Offices of Development and Social Inclusion (SEDIS) as of 2014, as established in Executive Decree PCM-03–2014.
5 Issue discussed within the ILO (OIT/CEACR, 2016b).
6 The last UPR conducted on Honduras was in May 2015. Honduras's next review will be in May 2020.
7 In addition to the cited references, the information narrated in this section was collected through participant observation and accompaniment in the process.

8 Initially called the "Interinstitutional Technical Board".
9 As was the case of Chile and noted by the former Special Rapporteur on the Rights of Indigenous Peoples, James Anaya (RENUDPI, 2012, p. 4).
10 This was the most debated issue in the dialogue process between the PIAH and the Government.
11 Zeid Ra'ad Al Hussein, the United Nations High Commissioner for Human Rights, has spoken of deep political and social polarisation in Honduras (ACNUDH, 2018)
12 This recommendation was made by CIPRODEH (2016).
13 Thus, in January 2017, CONPAH, in a note signed by its president representing all its federations, explained to the Special Rapporteur on the Rights of Indigenous Peoples how they had participated in previous dialogues and that it was a participatory process in which the majority of representative organisations have taken part.

References

ACNUDH/América Central. (2011). *Diagnóstico sobre la situación de los derechos humanos de los pueblos indígenas en América Central. Tomo II.* Panamá: OACNUDH.
ACNUDH. (2017). Informe anual del Alto Comisionado de las Naciones Unidas para los Derechos Humanos sobre la situación de los derechos humanos en Honduras. UN Doc. A/HRC/34/3/Add.2, 9 February 2017.
ACNUDH. (2018). *Informe: Las violaciones a los derechos humanos en el contexto de las elecciones de 2017 en Honduras.* Tegucigalpa: OACNUDH.
Asamblea Nacional Constituyente. (1982). Constitución Política de 1982. Retrieved from http://www.poderjudicial.gob.hn/CEDIJ/Leyes/Documents/ConstitucionRepublicaHonduras.pdf.
BID. (2007). *Plan Estratégico para el Desarrollo Integral de los Pueblos Autóctonos.* Tegucigalpa: BID.
BIP. (2016). *Boletín PBI Honduras.* Tegucigalpa: PBI.
CDESC. (2016). Observaciones finales sobre el segundo informe periódico de Honduras. Aprobadas por el Comité en su 58o período de sesiones. UN Doc. E/C.12/HND/CO/2, 11 July 2016.
CDH. (2015). 30° período de sesiones, Examen periódico universal. Informe del Grupo de Trabajo sobre el Examen Periódico Universal- Honduras, UN Doc. A/HR/30/11, 15 July 2015.
CERD. (2014). Observaciones finales sobre los informes periódicos primero a quinto de Honduras aprobadas por el Comité en su 84° período de sesiones. UN Doc. CERD/C/HND/CO/1–5, 13 March 2014.
CIPRODEH. (2016). *Análisis comparativo de las propuestas de ley de consulta y consentimiento previo, libre e informado para pueblos indígenas y negros de Honduras: insumo para la negociación con el Estado.* Tegucigalpa: CIPRODEH.
Comisión Interamericana de Derechos Humanos. (2015). Informe sobre la situación de derechos humanos en Honduras. OEA Doc. CIDH/OEA/Ser.L/V/II. Doc. 42/15, 31 December 2015. Retrieved from http://www.oas.org/es/cidh/informes/pdfs/Honduras-es-2015.pdf.
COPINH. (2017) Proyecto de Decreto elaborado por OFRANEH y COPINH "Ley marco de consulta, consentimiento previo, libre e informado a los Pueblos Indígenas". Retrieved from http://copinhonduras.blogspot.com/2017/04/proyecto-de-decreto-elaborado-por.html.

Fondo de las Naciones Unidas para la Infancia. (2012). *Niñez Indígena y Afro-hondureña en la República de Honduras.* Tegucigalpa: UNICEF.

Hartling, J. (2017). *Guía de buenas prácticas para la consulta previa en las Américas.* La Paz: Konrad Adenauer Stiftung (KAS).

IACtHR. (2012). *Kichwa Ingenous People of Sarayaku v. Ecuador,* Judgment of June 27, 2012 (Merits and reparations). Inter-Am. Ct. H.R., (Ser. C) No. 245(2012).

IACtHR. (2015). *Caso Comunidad Garífuna de Punta Piedra y sus miembros v Honduras,* Judgment of October 8, 2015 (Merits, Reparations, and Costs). Inter-Am. Ct. H.R., (Ser. C) No. 305(2015).

INE. (2013). *Censo nacional de población y vivienda.* Tegucigalpa: Gobierno de Honduras.

INE. (2015). *Encuesta Permanente de Hogares de Propósitos Múltiples 2014.* Tegucigalpa: Gobierno de Honduras.

IWGIA. (2010). *El mundo indígena 2010.* Copenhagen: IWGIA.

La Gaceta Diario Oficial de la República de Honduras. (1994a). Approval of Presidential Agreement 0719. AP, 0719-EP-94, 3 August 1994.

La Gaceta Diario Oficial de la República de Honduras. (1994b). Decreto Ejecutivo Número 26–1994. Ratification of ILO Convention 169 in Honduras.

La Gaceta Diario Oficial de la República de Honduras. (2013). Decreto Ejecutivo Número PCM-003–2013. Política Pública y Plan Nacional de Acción en Derechos Humanos.

La Gaceta Diario Oficial de la República de Honduras. (2016). Decreto Ejecutivo Número PCM-027–2016. Política Pública Contra el Racismo y la Discriminación Racial para el Desarrollo Integral de los Pueblos Indígenas y Afrohondureños.

Ministerio Público de Honduras & Fiscalía Especial de Etnias y Patrimonio Cultural. (2016). *Manual de procedimientos de investigación y protección de los derechos de los pueblos indígenas y afrohondureños.* Tegucigalpa: PNUD/Ministerio Público.

OIT/CEACR. (2016a). Observación-CEACR – Adopción: 2015, Publicación: 105ª reunión CIT. Ginebra.

OIT/CEACR. (2016b). Solicitud Directa (CEACR) – Adopción: 2015, Publicación: 105ª reunión CIT. Ginebra.

OIT/CEACR. (2017a). Observación-CEACR – Adopción: 2016, Publicación: 106ª reunión CIT. Ginebra.

OIT/CEACR. (2017b). Solicitud Directa (CEACR) – Adopción: 2016, Publicación: 106ª reunión CIT. Ginebra.

PNUD. (2015). *Informe sobre desarrollo humano 2015.* New York: PNUD.

PNUD. (2016a). Matriz de insumos y comentarios recogidos en talleres comunitarios sobre el primer borrador de propuesta de anteproyecto de ley de consulta previa. Retrieved from http://www.hn.undp.org/content/dam/honduras/docs/ddhh/pnud_hn_nuevas%20propuestas%20del%20articulado%20-%20version%20completa.pdf.

PNUD. (2016b). El PNUD acompaña proceso de diálogo entre el Gobierno de Honduras y los nueve Pueblos Indígenas y Afrohondureños para la construcción conjunta de la Ley de Consulta Previa, Libre e Informada (Convenio 169 de la OIT). Retrieved from http://www.hn.undp.org/content/honduras/es/home/presscenter/articles/2016/11/17/el-pnud-acompa-a-proceso-de-di-logo-entre-el-gobierno-de-honduras-y-los-nueve-pueblos-ind-genas-y-afrohondure-os-para-la-construcci-n-conjunta-de-la-ley-de-consulta-previa-libre-e-informada-convenio-169-de-la-oit-.html.

RENUDPI. (2012). Comentarios del Relator Especial sobre los derechos de los pueblos indígenas en relación con el documento titulado: "Propuesta de gobierno para nueva normativa de consulta y participación indígena de conformidad a los artículos 6° y 7° del Convenio N°169 de la OIT". Retrieved from http://unsr.jamesanaya.org/special-reports/comentarios-a-la-propuesta-del-normativa-de-consulta-chile.

RENUDPI. (2016). Comentarios de la Relatora Especial de las Naciones Unidas sobre los derechos de los pueblos indígenas en relación con el Anteproyecto de Ley Marco de consulta libre, previa e informada a los pueblos indígenas y afrohondureños (Honduras). Retrieved from http://unsr.vtaulicorpuz.org/site/images/docs/special/2016-honduras-unsr-comentarios-anteproyecto-ley-consulta-sp.pdf.

RENUDPI. (2016). Informe de la Relatora Especial sobre los derechos de los pueblos indígenas sobre su visita a Honduras. UN Doc. A/HRC/33/42/Add.2, 21 July 2016.

RENUDPI. (2017). Observaciones adicionales de la Relatora Especial sobre los derechos de los pueblos indígenas sobre el proceso de regulación de la consulta previa en Honduras. Retrieved from http://unsr.vtaulicorpuz.org/site/images/docs/special/2017-06-09-honduras-unsr-additional-observations.pdf.

Secretaría de Trabajo y Seguridad Social. (2016). Informe de avances del proceso de Consulta previa, libre e informada y Convenio 169 de la OIT. Retrieved from http://www.trabajo.gob.hn/?s=convenio+169.

12 Towards an effective prior consultation law in Paraguay

Sara Mabel Villalba Portillo

Introduction

Over the last decades, progress has been achieved by Latin American States regarding the rights of Indigenous Peoples in general and the right to prior consultation in particular. However, there is a clear implementation gap and, in some cases, a lack of mechanisms to facilitate their effective application. Indeed, there is little clarity about the extent of Indigenous rights, as well as a lack of appropriate secondary legislation (Rowlands, 2013; Yrigoyen, 2009, p. 17).

There appears to be a tendency on the part of governments not to generate clear mechanisms to implement prior consultation. One of the critical points in this sense is the unwillingness to provide Indigenous Peoples with a tool to avoid the extraction of natural resources from their territories, considering that such extraction is encouraged by governments themselves. This has even led to a contradictory situation, taking into account that the Latin American States have entered into international obligations regarding Indigenous Peoples' rights, at the same time as they promote a model of development based on the extraction of natural resources, which is generally carried out in ancestral Indigenous lands (Rowlands, 2013).

In Paraguay, prior consultation with Indigenous Peoples has been poorly implemented by different government authorities, despite the constitutional recognition of Indigenous Peoples' rights and the ratification of the ILO Convention No.169 Concerning Indigenous and Tribal Peoples in Independent Countries of 1989 (ILO 169). One of the main difficulties in the past was the absence of a law to regulate this process, or a PC protocol. Several proposals on this issue have been presented by Indigenous and government authorities, and – at last – a protocol has recently been adopted.[1]

For a long time, the poor implementation of prior consultation has hindered the right to participation of Indigenous Peoples, which is set forth in the National Constitution. Also, this situation has opposed self-determination in the selection of models of development, participation, prior consultation, and Free Prior and Informed Consent (FPIC), which are part of both international and national legislation. In fact, this new set of rights goes against the protective tradition of the States, which considered themselves to be the

DOI: 10.4324/9781351042109-16

owners of Indigenous Peoples' territories, and, therefore, the exclusive decision makers over these lands. The new set of rights is based on the principle that Indigenous Peoples have the same dignity and capacity as other peoples to control their institutions and freely establish their ways of living (Yrigoyen, 2011).

The aim of this chapter is to present a general view regarding the implementation of prior consultation with Indigenous Peoples in Paraguay. Despite the absence of a protocol of prior consultation for several years, some participation processes have been encouraged, such as the addition of chapters that acknowledge Indigenous rights in the National Constitution, as well as the passing of the Law on Indigenous Health, and the Law on Indigenous Education. In these cases, a broad participation process was carried out, which involved working with several Indigenous groups.

This chapter is based on qualitative and exploratory research, taking into account that the subject has not yet been studied in the case of Paraguay. The main methodological tool used was the review of documents such as institutional reports from Indigenous organisations, journalistic publications, and audio-visual material, in addition to specialised references on Indigenous participation and prior consultation and the corresponding legal framework.

The chapter is structured as follows: first, it includes basic information about Indigenous Peoples in Paraguay and the national and international legislation regarding prior consultation and Indigenous Peoples' participation. Next the background to the process of adopting an executive decree to establish the "General Guidelines for the Free, Prior and Informed Consultation Process with Indigenous Peoples" in Paraguay is described, as well as some conflicts that have arisen during this process. Also, some examples are referenced of Indigenous participation encouraged by private organisations. For illustrative purposes, the creation and enactment of the Law on Indigenous Health and the Law on Indigenous Education is described. Last of all, some final considerations about the situation regarding prior consultation in Paraguay are presented.

Indigenous Peoples in Paraguay

According to the National Population and Housing Census for Indigenous Peoples carried out in 2012 (DGEEC, 2012), there are currently 112,848 Indigenous individuals in Paraguay, constituting approximately 1.8% of the total population. As can be seen in Table 12.1, there are 19 Indigenous Peoples, grouped according to five different language families.

With regard to their demographic distribution, the Indigenous Peoples living in the Western Region of the country represent a high proportion of the total population. According to data collected from the Third National Population and Housing Census for Indigenous Peoples, specifically in the Boquerón, Alto Paraguay and Presidente Hayes Departments, the Indigenous population constitutes 40%, 37% and 24%, respectively, of the entire

Table 12.1 Population and language families in Paraguay

Family	Peoples	Population
Guaraní	Païi Tavyterä	15.097
	Mbyá Guaraní	21.492
	Avá Guaraní	17.697
	Aché	1.942
	Guaraní Ñandeva	2.393
	Western Guaraní	2.379
Maskoylanguage	Toba Maskoy	2.817
	North Enlhet	8.632
	South Enxet	5.740
	Sanapaná	2.833
	Angaité	6.638
	Guaná	86
Mataco-Mataguayo	Nivaclé	16.350
	Maká	1.892
	Manjui	385
Zamuco	Ayoreo	2.481
	Ybytoso	1.824
	Tomaraho	183
Guaicuru	Qom	2.057
Total		112.848

Source: Author's own elaboration, according to data collected from DGEEC (2012).

population of the Western Region. In the Eastern Region, there are important minorities of the Indigenous population in the Departments of Amambay (12.1%), Canindeyú (11.1%) and Caaguazú (8%) (DGEEC, 2012).

The socioeconomic indicators of this census also reveal the level of exclusion of the Indigenous Peoples: in formal education, on average, this population only attended the first three years of school; 41% of children under the age of five suffer from chronic malnutrition; illiteracy affects 37.6% of the Indigenous population older than 15 years old; the labour force participation rate, in comparison with the total population old enough to work, is just 52.6%; only 64.0% have an identity card; only 2.5% of Indigenous individuals have access to safe drinking water (most of them only have access to dykes or rivers); and 31.2% have access to electricity (DGEEC, 2012).

Over the last decades, pressure over the territories of Indigenous Peoples, including the seizure of traditional lands and the eviction of communities, has increased. In most cases, the purpose is to clear the areas, to grow soy or other cereals in the Eastern Region and for livestock farming in the Western

Region. In such a context of rapid expansion of the agricultural border, attempts have been made to amend Law No. 904 of 1981 (Congreso de la Nación Paraguaya, 1981), which guarantees the right to land for Indigenous Peoples (FAPI, 2017a, p. 10).

Legal framework

The most important legal instrument on prior consultation is the ILO 169, followed by the United Nations Declaration on the Rights of Indigenous Peoples (UNDRIP), adopted in 2007, and the American Declaration on the Rights of Indigenous Peoples (ADRIP), which was adopted in 2016.

To facilitate intercultural dialogue with the State, the aforementioned legal instruments establish the following fundamental rights for Indigenous Peoples: the right to participation in all processes that may affect their lives; the right to Consultation oriented towards reaching an agreement or consent; and FPIC (especially in case of any situation that may involve displacement or relocation of their territory) (Rowlands, 2013, p. 75). The consultation process should be carried out with the aim of obtaining FPIC and should involve the complete participation of the Indigenous Peoples.

With regard to consultation, ILO 169 establishes that the State should be obliged to implement procedures and mechanisms allowing Indigenous Peoples to exercise the right to be consulted, wherever there are administrative and legislative measures that may affect them (articles 6, 15, 17, 22, 27 and 28). In particular, the Convention states that prior consultation should be carried out before any authorisation or commencement of the exploitation of natural resources in their territories (Linares, 2013, p. 99).

This way, the Convention established a new framework for relations between States and Indigenous Peoples. Additionally, by means of the right to participation, Indigenous Peoples may intervene in all phases of constructing and applying national or local policies, plans and development programmes that may affect them (Yrigoyen, 2011). And specifically, consultation is the main instrument to exercise this participation.

Prior consultation is considered to be more than just a right, since it is also a key mechanism to guarantee the right to self-determination, allowing for the integration of concepts and strategies for the development of Indigenous Peoples within public policies. Therefore, it constitutes a critical tool that may help to reduce the asymmetry of power, mitigate conflicts, improve decision-making and potentially reduce costs (Camacho Nassar, 2017, p.8; Weitzner, 2011, p. 65).

In Table 12.2, we can observe some of the legal instruments that apply to Paraguay and which safeguard the right of Indigenous Peoples to participate in public policies that directly affect them. However, the national legal instruments fail to expressly refer to prior consultation.

Paraguay has ratified ILO 169 by means of Law No. 234 of 1993 (Congreso de la Nación Paraguaya, 1993). Nevertheless, there is still no

Table 12.2 The legal framework that guarantees the right to participation of Indigenous Peoples in Paraguay

Instrument	Nature	Scope	Year
National Constitution	Binding	National	1992
Convention No. 169 concerning Indigenous and Tribal Peoples in Independent Countries	Binding	International	1989
Declaration on the Rights of Indigenous Peoples	Non-binding	International	2007
Law 904/81 *Estatuto de las Comunidades Indígenas* (Statute of Indigenous Communities)	Binding	National	1981
Law 234/93 by means of which the Paraguayan government ratifies the Convention No. 169	Binding	National	1993
Law No. 3231/07 "creating the Directorate General for Indigenous Schooling"	Binding	National	2007
Law 5469/15 "creating the Indigenous Health System and the National Directorate for the Health of Indigenous Peoples and the National Council on Indigenous Peoples' Health"	Binding	National	2015

Source: Author's own elaboration.

law to regulate the prior consultation process. However, as the convention has been ratified, the State is obliged to facilitate Indigenous Peoples' participation in all matters that may affect them (Quiroga, 2013, p. 63). The laws that effectively mention prior consultation in Paraguay are Law No. 3,231 of 2007, which creates the Directorate General for Indigenous Schooling (Congreso de la Nación Paraguaya, 2007), and Law No. 5,469 of 2015, of Indigenous Health (Congreso de la Nación Paraguaya, 2015).

The path towards a protocol on prior consultation

For many years now – 29 in total, since the fall of the Stroessner dictatorship – there has been no law or protocol for the effective application of the right to prior consultation, in line with Law No. 234 (Congreso de la Nación Paraguaya, 1993). Nevertheless, despite this lack of regulation, Indigenous Peoples have been consulted regarding specific measures that may concern them, such as the law on Indigenous Health and the Law on Indigenous Education, among others.

There have been several attempts to regulate prior consultation, and some of the main initiatives have been made in the last ten years. As a matter of fact, in 2008, with the election of former bishop Fernando Lugo to the Presidency of the Republic,[2] the promise to pay more attention to Indigenous Peoples was a key feature of the pre- and post-election government discourse. This was translated into a series of public policies addressed to this group.

One of the first measures adopted in 2008 by Lugo's government was the implementation of an Inter-Institutional Plan of Action for Vulnerable Sectors, where attention was paid to Indigenous Peoples. Also, the National Program for Indigenous Peoples (PRONAPI) was created. The aim of PRONAPI was to design a policy for Indigenous Peoples and to create institutional mechanisms to offer comprehensive assistance through the participation of the affected peoples by means of constant consultation (Ayala Amarilla, 2009, p. 233).

In this context, between 2009 and 2010, a project – the Strengthening of Institutional Capacities for the Implementation of Public Policies Addressed to Indigenous Peoples – was developed by the Social Affairs Cabinet of the Executive, Indigenous organisations, civil society, and UN agencies.[3] Within this project, technical workshops were carried out for 18 State institutions whose work is linked to Indigenous Peoples. One of these workshops was on prior consultation, during which it was noted that the State of Paraguay is obliged to consult and that such consultation should reach all Indigenous Peoples, through their organisations (Secretaría de la Función Pública/Instituto Paraguayo del Indígena, 2011, p. 35).

The project also gave rise to a draft document – Basic Notions for a Civil Servant's Work regarding Indigenous Peoples – which established the main working guidelines between governmental institutions and Indigenous Peoples. As a starting point, it highlighted the "Free, Prior and Informed Consultation" (Secretaría de la Función Pública/Instituto Paraguayo del Indígena, 2011, p. 33) for the adoption of measures involving Indigenous Peoples.

Prior to this, in 2009, the Federation for the Self-Determination of Indigenous Peoples[4] (FAPI) created a document – Proposals for Public Policies aimed at Indigenous Peoples – supported by the United Nations Development Programme (UNDP), wherein the fundamental nature of prior consultation as a right of Indigenous Peoples is referred to repeatedly:

> All State measures adopted within the framework of Public Policies addressed to Indigenous Peoples must take into account: [...] that prior, free and informed consultation and the right to grant their Free, Prior and Informed Consent or not regarding any kind of activity affecting Indigenous communities, members and organisations are compulsory and binding; that decisions must be made at the meetings attended by legitimate, representative leaders, after inter-community meetings; that projects must be coordinated and created together with Indigenous communities and organisations, rather than present them in a haphazard way, as the national government.
>
> (CAPI, 2009, p. 21)

Likewise, the document stipulates that whenever public policies deal with the infringement of the rights of Indigenous Peoples – as well as the "Free, Prior and Informed Consultation" – "consent for carrying it out within the framework of respect for free determination guaranteed in the United Nations Declaration on the Rights of Indigenous Peoples" (CAPI, 2009, p. 11) shall be required as well.

In 2011, as the result of an internal process and with the support of the UNDP, the FAPI made and published a Protocol Proposal for the Process of Consultation and Consent with the Indigenous Peoples of Paraguay (FAPI, 2011).

On the other hand, in 2013 and during President Horacio Cartes' term of office (2013–2018), there was a setback regarding human rights in general and, specifically, the rights of Indigenous Peoples. As well as the limited budget devoted to the Paraguayan Indigenous Institute (INDI) for the purchase of claimed lands, and both medical and educational assistance, environmental licenses were granted that promote deforestation, agrobusiness and the use of agro-toxic products which affect the communities. Violent evictions of Indigenous communities also took place, together with their forced relocation to Asunción so that they could make claims over their rights (Ayala Amarilla, 2014; Barreto, 2017; Mendieta & Cabello, 2016; Servín, 2017).

In this context, in 2013, the Ombudsman's Office submitted a bill to regulate Indigenous Peoples' right to prior consultation, which was supported by representatives of the lower house belonging to the party in office, the ANR (ABC, 2013). This bill was drafted with the cooperation of the Indigenous Peoples Department of the Ombudsman's Office.

The documents referred to when drawing up the bill included the National Constitution, the Statute of Indigenous Communities (Congreso de la Nación Paraguaya, 1981), ILO 169, UNDRIP, Peru's Law No. 29785 on the right to consultation (see the chapters by Doyle and Flemmer in this volume) and Bolivia's Law No. 222/2012 (see the chapter by Schilling-Vacaflor in this volume) (Defensoría del Pueblo, 2014a, p. 69). However, there is no record of any participation of Indigenous Peoples in this process. In fact, a report from the UN Rapporteur points out that information was received on the existence of several bills on prior consultation and consent.

> The Special Rapporteur would like to point out that any measure designed to guarantee the fulfillment of the State's duty to consult Indigenous Peoples must be consistent with ILO's Covenant 169 and the Declaration, and jointly drafted with the Indigenous Peoples.
>
> (UNSR, 2015, p. 7)

Afterwards, Indigenous organisations also requested the House of Representatives to file the draft law No. D-1326050, on the Right to Prior Consultation of Indigenous Peoples acknowledged in ILO 169 submitted by the Ombudsman's Office (Defensoría del Pueblo, 2014b). It was claimed that this

bill was not drafted with the participation of representative Indigenous organisations (FAPI, 2014a).

In August 2014, the First Workshop on FPIC and Consultation of Indigenous Organisations was held. The event was organised by the INDI and the
National Joint Program on Reducing Emissions from Deforestation and
Forest Degradation (REDD+) in Paraguay. The latter entity is made up of
the FAPI, the Secretary of the Environment (SEAM) and the National Forestry Office (INFONA). The aim of this event was the joint revision of the
proposed protocol on consultation and consent for future REDD+ projects.
A team was established with the purpose of consolidating the bill on consultation and FPIC (FAPI, 2014c). Finally, the Indigenous leaders approved
the draft bill on a Protocol for the Process of Consultation and Consent for
the Indigenous Peoples of Paraguay (FAPI, 2014b).

Afterwards, in July 2017, a conflict arose because Indigenous organisations
claimed that the INDI was engaged in parallel work on the consultation
protocol. It was pointed out that the document – "The Means by which the
General Guidelines for the Free, Prior and Informed Consultation Process
with Indigenous Peoples in Paraguay are Established" – did not include the
comprehensive content of the bill put forward in November 2016 by the
Indigenous organisations. It was demanded that the original document –
"The Legal Framework that Establishes a Protocol for the process of Consultation and Consent of the Indigenous Peoples and Communities" – should
be taken into consideration (FAPI, 2017b).

In April 2018, FAPI urged the candidates from all political parties to consider Indigenous Peoples a priority within their governmental programme,
should they win the elections. They claimed that the institutionalisation of the
right to consultation and FPIC should be deemed a priority. They further
added that FAPI, together with Indigenous organisations in Paraguay, had
actively prepared the aforementioned bill.

The practice of consultation

With respect to the implementation of prior consultation in Paraguay, the
document entitled "Proposals for Public Policies Aimed at Indigenous Peoples"
(CAPI, 2009) affirms that when dealing with Indigenous communities, the State
has implemented projects in a unilateral way. Specific reference is made to the
process of regularisation of the lands being claimed, or the provision of drinking
water, implemented with funds from the World Bank. "No consultation was
made either before, during, or after the project was carried out. No records exist
that the State has coordinated its actions together with truly representative
Indigenous organisations" (CAPI, 2009, p. 17).

Likewise, during the Third Latin American Meeting of Local Governments
in Indigenous Territories, which took place in Asunción in 2009, 150 representatives from Indigenous organisations gathered to hold a debate on the
issues of Indigenous self-government and prior consultation. The participants

emphasised, once again, the lack of consultation over the relocation of communities in the case of large-scale construction works (FAPI, 2010).

With respect to the question of whether there had been any processes of consultation made by governments to the communities, those participating in the conference also pointed out that:

> No consultations have been made to Indigenous Peoples. San Cosme [an Indigenous community] was relocated by the Yacyreta binational entity. The communities from Itapúa are seldom consulted regarding NGO's projects, which are carried out without any consultation being made with Indigenous Peoples.

> (FAPI, 2010, p. 64)

In 2013, the Commission on Indigenous Peoples within the House of Representatives suggested the amendment of Law No. 904 of 1981, the Statute of Indigenous Communities (Congreso de la Nación Paraguaya, 1981), arguing that having an updated framework law was a necessity. This bill, likewise, was not the result of prior consultation. Regarding this issue, the National Coordinator of Organisations of Working, Rural, and Indigenous Women (CONAMURI)[5] made a statement in which it expressed that "the rumour about the amendment to Law 904/81 without the active participation of all the peoples by means of Free, Prior and Informed Consultation" (Viveros & Ramírez, 2013, p. 81) was a threat to their self-determination and the sovereignty of their territories.

The lack of prior consultation with Indigenous Peoples over administrative or legislative measures that affect them has even been confirmed by the UN Special Rapporteur on the Rights of Indigenous Peoples in her report on the situation of Indigenous Peoples in Paraguay. In this document, she observes that "in Paraguay, there is a generalised lack of fulfilment of the State's duty to consult before adopting legislative, political, and administrative measures that directly affect Indigenous Peoples and their lands, territories and natural resources" (UNSR, 2015, p. 7).

Likewise, in the same report, it is stated that no consultation has taken place over the majority of institutional programs and projects directed towards Indigenous Peoples, including those carried out by the Paraguayan Secretariat for Social Action (SAS). Furthermore, the Special Rapporteur adds that the same occurs with the granting of environmental licenses for deforestation in areas that affect Indigenous Peoples, as well as with the establishment of protected areas[6] and other conservation initiatives without consultation with or the consent of the affected Peoples.

Nonetheless, despite the lack of a prior consultation protocol and formal processes, significant participative processes have been carried out, the aim of which has been to let Indigenous Peoples' voices be heard by the State. In general, these processes have been promoted by non-governmental organisations, especially by the National Indigenous Ministries Coordinating Body

(CONAPI).[7] To illustrate this, reference is made to two experiences in which there was a high degree of Indigenous participation in the processes of drafting and enacting two specific laws on education and health policies for the Indigenous Peoples (Gaska & Ferreira, 2012; DGEEI, GSEI & CONAPI, 2013). This legislation comprises the abovementioned Law No. 3,231 of 2007, creating the Directorate General for Indigenous Schooling (Congreso de la Nación Paraguaya, 2007) and Law No. 5,469 of 2015, creating the National Directorate for the Health of Indigenous Peoples (Congreso de la Nación Paraguaya, 2015).

The involvement of Indigenous Peoples in creating the Indigenous law of schooling and Indigenous educational policies, in general, is a long-standing practice in Paraguay. In 1990, a National Commission for Educational Reform was created to elaborate a diagnosis and establish alternatives to cover the deficiencies in the existing educational system. Paraguay's Indigenous Peoples asked to participate in the National Consultation, and the Ministry of Education and Science (MEC) agreed to their request. Meetings, conferences, and workshops were arranged for parents and political and religious leaders to discuss proposals regarding Indigenous education (Gaska & Rehnfeldt, 2017).

During the 1990s, proposals were made for Indigenous school education, and in 2001 the First National Gathering for Indigenous Education, promoted by the MEC, was held. In 2003, a Meeting for Indigenous Education was held to discuss a bill and establish a law for Indigenous Education, and a Monitoring Group for Indigenous Education (GSEI) was created. It was made up of Indigenous representatives who took part in workshops to study and develop legal strategies. In 2004, a Department for Indigenous Education was created within the MEC, and in 2005 and 2006, several monitoring entities for the education of Indigenous Peoples were created (CONAPI, 2009).

At the same time, progress was made in relation to the preparation of the bill for the law on Indigenous Peoples Education, and finally, in 2007, Law No. 3,231 was enacted (Congreso de la Nación Paraguaya, 2007), thereby creating the Directorate General for Indigenous Schooling. A debate ensued between the GSEI and the MEC, which elaborated the draft for this law's regulation, which was enacted in 2011 (Gaska & Rehnfeldt, 2017).

With regard to the law concerning Indigenous health, its approval implied a series of prior activities in which Indigenous Peoples fully participated and which was also encouraged by CONAPI. In the 1990s, workshops and conferences were held with the aim of analysing the context surrounding the health of Indigenous Peoples, as well as their cultural diversity, which demanded a special kind of medical attention. In the year 2000, upon the request of Indigenous organisations and with the support of UNICEF, an inter-institutional group was formed in order to draft a bill on the Indigenous Health System, which would be incorporated into the National Health System. The bill was drafted on the basis of consultations with Indigenous Peoples (Gaska & Ferreira, 2012). Several workshops were held in the course of these years, and the law was finally passed and enacted in 2015 (Congreso de la Nación Paraguaya, 2015).

In conclusion, both laws were enacted after lengthy processes which involved the permanent intervention of Indigenous organisations – by means of meetings and training workshops – as well as the intervention of other non-governmental organisations that supported these undertakings.

Conclusions

In this chapter, I have provided a general overview of the right to prior consultation in Paraguay and its actual and effective application in the country. This should be placed into a context where for many years there has been no regulatory law or protocol for the effective enforcement of this right, despite the ratification of ILO 169, and other national laws.

Several attempts have been made by Indigenous organisations to regulate prior consultation, especially over the last ten years. The Federation for the Self Determination of Indigenous Peoples (FAPI) has had a key role throughout this period. In 2009, it even encouraged an internal process of consultation that resulted in the preparation and publication of a "Proposed Protocol for the Process of Consultation and Consent with the Indigenous Peoples of Paraguay" (FAPI, 2011), supported by the UNDP. Moreover, some State bodies have drafted bills for the regulation of prior consultation, albeit without the participation of Indigenous organisations, which has led to conflicts and prevented the regulation from being passed.

Nonetheless, broad processes involving the participation of Indigenous Peoples have been carried out in Paraguay, especially for drafting and enacting Law No. 3,231 of 2007 creating the Directorate General for Indigenous Schooling (Congreso de la Nación Paraguaya, 2007) and Law No. 5,469 of 2015 creating the National Directorate for the Health of Indigenous Peoples (Congreso de la Nación Paraguaya, 2015). These processes have been supported by non-governmental organisations, particularly by the National Coordination of Indigenous Ministry (CONAPI).

Although some consultation processes have been carried out, the main disadvantage of not having a prior consultation protocol has been the lack or deficit of consultation processes with Indigenous Peoples in most of the activities or large-scale development projects that affect them. This has been a recurrent claim by Indigenous organisations and has involved different State bodies. One of the reasons for this situation might be the lack of inter-institutional collaboration, which is an impediment to achieving the necessary continuity for effectively creating public policies in Paraguay. This situation affects several sectors of society and is a longstanding problem in relation to the State's administration of the country.

Lastly, there is a conflict of interests with regard to the preservation of ancestral Indigenous territories that goes against the progress of agro-business in Paraguay. A law on prior consultation that regulates its effective and real implementation requires providing Indigenous Peoples with a tool of empowerment for defending their right to territory and free access to natural resources.

Notes

1 The "Protocolo para el proceso de consulta y consentimiento libre, previo e informado con los pueblos indígenas que habitan en el Paraguay" (Protocol for the process of consultation and free, prior and informed consent with Indigenous Peoples who live in Paraguay) was adopted during the process of editing this book. The norm is based on Note No. 211 of 2018 presented by Paraguay's Indigenous Institute (INDI) and was confirmed by Presidential Decree No. 1039 of 2018 (Presidente de la República del Paraguay, 2018). The Federation for the Self-Determination of Indigenous Peoples (FAPI) has identified as an achievement the passing of the legal instrument and thanked the Presidency of the Republic (FAPI, 2019).
2 In 2008, there was a change in power after 60 years of dominance by the National Republican Association (ANR), which had supported Alfredo Strossner's dictatorship (1954–1989).
3 Three networks of Indigenous organisations were involved; two organisations made up of Indigenous women; a network of private entities working with Indigenous peoples and four agencies of the United Nations' System (ILO, UN WOMEN, UNDP and UNICEF) (Secretaría de la Función Pública/Instituto Paraguayo del Indígena, 2011, p. 7).
4 FAPI is made up of 12 Indigenous organisations of the Western and Eastern regions. Its goal is to defend the collective and individual rights of Indigenous Peoples.
5 CONAMURI is an organisation of Indigenous women farmers, created to defend their rights. It is divided into committees which are, at the same time, categorised into production committees and different associations. They are currently present in 12 departments all over the country.
6 In accordance with the FAO's report (2008), 9 out of 28 National Protected Areas (NPA) established in the country overlapped with ancestral Indigenous lands (32%). According to data provided by the Department of Environment, this situation affects the Ayoreo peoples (in voluntary isolation), as well as the Mbya Guaraní, the Guaraní Ñandeva and the Aché peoples. In all of these cases, these NPAs have been created without any kind of consultation with the affected Indigenous communities (Villalba, 2015, p. 447).
7 CONAPI is a body of the Paraguayan Episcopal Conference (CEP) and deals with Indigenous issues in Paraguay.

References

ABC. (2013). Los indígenas deben ser consultados. *ABC*, 19 April. Retrieved from http://www.abc.com.py/nacionales/indigenas-deben-ser-consultados-562616.html.
Ayala Amarilla, O. (2009). Una política pendiente. Derechos de los pueblos indígenas. In CODEHUPY (Ed.) *Derechos humanos en Paraguay 2009* (pp. 395–408). Asunción: CODEHUPY.
Ayala Amarilla, O. (2014). Los derechos de los pueblos indígenas en tiempos de una impronta empresarial para el Estado. In CODEHUPY (Ed.), *Derechos humanos en Paraguay 2014* (pp. 65–78). Asunción: CODEHUPY.
Barreto, V. (2017). Sintomatología de la agudización neoliberal en agravio a los pueblos indígenas en el Paraguay. In CODEHUPY (Ed.), *Derechos humanos en Paraguay 2017* (pp. 59–73). Asunción: CODEHUPY.
Camacho Nassar, C. (2017). Guía de acción para la consulta a pueblos indígenas: conceptos y prácticas a considerar. Retrieved from http://www.agter.org/bdf/_docs/camacho_guia-consulta-comunidades-indigenas.pdf.
CAPI. (2009). *Propuestas de Políticas Públicas para Pueblos Indígenas*. Asunción: FAPI-PNUD.

CONAPI. (2009). *Antecedentes y alcances de la Ley 3231/07*. Asunción: AGR.

Congreso de la Nación Paraguaya. (1981). Ley N° 904/81. Estatuto de las Comunidades Indígenas, 18 December 1981.

Congreso de la Nación Paraguaya. (1993). Ley N° 234 que aprueba el convenio N° 169 sobre pueblos indígenas y tribales en países independientes, adoptado durante la 76ª. Conferencia internacional del trabajo, celebrada en Ginebra el 7 de junio de 1989, 19 July 1993.

Congreso de la Nación Paraguaya. (2007). Ley N° 3231 Que crea la Dirección General de Educación Escolar Indígena, 29 June 2007.

Congreso de la Nación Paraguaya. (2015). Ley 5469 De Salud Indígena, 29 July 2015.

Defensoría del Pueblo. (2014a). *El Derecho a la Consulta Previa de los Pueblos Indígenas: El Rol de los Ombudsman en América Latina. Memoria del Encuentro Extraordinario de la Federación Iberoamericana de Ombudsman, Lima, Perú, 25 y 26 de abril del 2013.* Lima: Defensoría del Pueblo en el Perú, Programa de Pueblos Indígenas.

Defensoría del Pueblo. (2014b). Expediente N° D-1326050. Rechazo del Proyecto de Ley "De Derecho a la Consulta Previa a los Pueblos Indígenas, Reconocido en el Convenio 169 de la Organización Internacional del Trabajo", 1 October 2014.

DGEEC. (2012). III Censo Nacional de Población y Viviendas para Pueblos Indígenas. Asunción.

DGEEI, GSEI & CONAPI. (2013). *Plan de implementación del modelo educativo plurilingüe desde los pueblos indígenas en el Paraguay 2013–2018.* Asunción.

FAPI. (2010). *Tekoháre. Por nuestro territorio. III Encuentro Latinoamericano de Gobiernos Locales en Territorios Indígenas.* Asunción: FAPI.

FAPI. (2011). *Propuesta de protocolo para un Proceso de Consulta y Consentimiento con los Pueblos Indígenas del Paraguay.* Asunción: FAPI-PNUD.

FAPI. (2014a). Culmina taller sobre Consulta y Consentimiento Libre Previo e Informado con la posición de las Organizaciones Indígenas. Retrieved from http://www. fapi.org.py/culmina-taller-sobre-consulta-y-consentimiento-libre-previo-e-informado -con-la-posicion-de-las-organizaciones-indigenas.

FAPI. (2014b). Organizaciones indígenas aprueban documento que establece un protocolo para un proceso de consulta y consentimiento. Retrieved from http://www.fapi.org.py/ organizaciones-indigenas-del-paraguay-aprueban-documento-que-establece-un-protoco lo-para-un-proceso-de-consulta-y-consentimiento.

FAPI. (2014c). Seguimiento al Primer Taller de Consulta y Consentimiento Libre, Previo e Informado en el Marco del Programa ONUREED. Retrieved from http://www.fapi.org. py/seguimiento-al-primer-taller-de-consulta-y-consentimiento-libre-previo-e-informado-e n-el-marco-del-programa-onureed.

FAPI. (2017a). Organización Nacional de Pueblos Indígenas (ONPI) solicita al INDI considerar documento original de CLPI. Retrieved from http://www.fapi.org.py/orga nizacion-nacional-de-los-pueblos-indigenas-solicita-al-indi-considerar-documento-orig inal-de-clpi.

FAPI. (2017b). *Plan Estratégico Quinquenio 2017–2021.* Asunción: FAPI-PNUD.

FAPI. (2018). La FAPI insta a los candidatos que ganen las Elecciones Generales cumplir siete puntos prioritarios para los Pueblos Indígenas. Retrieved from http://www.fapi. org.py/la-fapi-insta-a-los-candidatos-que-ganen-las-elecciones-generales-cumplir-siete- puntos-prioritarios-para-los-pueblos-indigenas.

FAPI. (2019). President of FAPI highlights the institutionalisation of the Consultation and Consent protocol during the presentation of Decree 1039. Retrieved from http://www.

fapi.org.py/presidente-de-la-fapi-destaca-la-institucionalizacion-del-protocolo-de-consul
ta-y-consentimiento-durante-la-presentacion-del-decreto-1039/?Fbclid=IwAR1AvNUn
O3o6N0hmSNxb0tiCsQi0T2QfRgiNR6exIrPxDknfuZiEXPWlr4Y.

Gaska, H. & Ferreira, S. (2012). *Presencia misionera junto a los pueblos indígenas: cuatro décadas de Pastoral Indígena de la Conferencia Episcopal Paraguaya (CEP).* Asunción: Coordinación Nacional de Pastoral Indígena (CONAPI).

Gaska, H. & Rehnfeldt, M. (2017). *Construyendo la educación intercultural indígena: una propuesta para formación docente.* Asunción: Centro de Estudios Antropológicos de la Universidad Católica (CEADUC).

Linares, S. (2013). Derecho de consulta indígena e innovación democrática: un debate complejo. In S. Martí i Puig, C. Wright, J. Aylwin & N. Yáñez (Eds.), *Entre el desarrollo y el buen vivir. Recursos naturales y conflictos en los territorios indígenas* (pp. 99–129). Madrid: Ediciones La Catarata.

Mendieta, M. & Cabello, J. (2016). Discriminación estructural del Estado paraguayo contra los pueblos indígenas. Políticas neoliberales y acciones ilegales como instrumento de violaciones de los derechos territoriales (pp. 55–65). In CODEHUPY (Ed.), *Derechos humanos en Paraguay 2016.* Asunción: CODEHUPY.

Presidente de la República del Paraguay. (2018). Decreto N° 1039 por el cual se aprueba el "Protocolo para el proceso de consulta y consentimiento libre, ¿previo e informado con los pueblos indígenas que habitan en el Paraguay, 28 December 2018.

Quiroga, L. (2013). ¡Arriba las manos! Derechos de los pueblos indígenas. In CODEHUPY (Ed.), *Derechos humanos en Paraguay 2013* (pp. 59–74). Asunción: CODEHUPY.

Rowlands, J. (2013). Estados Latinoamericanos frente a la consulta indígena: tensión cultural en la implementación de los derechos territoriales. In S. Martí i Puig, C. Wright, J. Aylwin & N. Yáñez (Eds.). *Entre el desarrollo y el buen vivir. Recursos naturales y conflictos en los territorios indígenas* (pp. 68–98). Madrid: Ediciones La Catarata.

Secretaría de la Función Pública/Instituto Paraguayo del Indígena. (2011). Fortalecimiento de capacidades institucionales para la implementación de políticas públicas orientadas a los pueblos indígenas. Principales actividades y lecciones aprendidas. Retrieved from http://www.py.undp.org/content/dam/paraguay/docs/ Libro%20fortalecimiento.pdf.

Servín, J. (2017). *Historia Sauce. Informe antropológico Comunidad Avá Guaraní. Tekoha Sauce.* Asunción: Centro de Estudios Antropológicos de la Universidad Católica Nuestra Señora de la Asunción.

UNSR. (2015). Informe. Situación de los pueblos indígenas en el Paraguay. UN Doc. A/HRC/30/41/Add.1, 13 August 2015.

Villalba, S. (2015). Áreas protegidas en territorios indígenas. El caso del Pueblo Aché en el Paraguay. *Suplemento Antropológico,* 50(2), 417–492.

Viveros, D. & Ramírez, E. (2013). Los pueblos indígenas y el viejo rumbo de la discriminación estructural. In CODEHUPY (Ed.), *Derechos humanos en Paraguay 2013* (pp. 75–85). Asunción: CODEHUPY.

Weitzner, V. (2011). *Inclinando la Balanza del Poder. Logrando que el Consentimiento Libre, Previo e Informado Funcione: Lecciones y orientaciones políticas obtenidas en 10 años de investigación acción sobre actividades extractivas con pueblos indígenas y afrodescendientes en las Américas.* Ottawa: Instituto Norte-Sur.

Yrigoyen, R. (2011). El derecho a la libre determinación del desarrollo. Participación, consulta y consentimiento. In M. Aparicio (Ed.), *Los derechos de los pueblos indígenas a los recursos naturales y al territorio: conflictos y desafíos en América Latina* (pp. 103–146). Barcelona: Icaria.

Part IV

Avoiding prior consultation

13 The failure to consult Indigenous Peoples and obtain their free, prior and informed consent in Ecuador

The Yasuní ITT case

Malka San Lucas Ceballos

Introduction

The Yasuní National Park (YNP) is the largest protected area in continental Ecuador and one of the world's most genetically diverse areas.[1] At the same time, it is home to several Indigenous Peoples, including those in voluntary isolation (IPVI).[2] However, it is also an area subject to oil extraction, since it is considered strategic for Ecuador's energy sector. Therefore, in its effort to meet its energy needs and sustain the national economy, the State encourages the use of non-renewable natural resources such as oil, which has not only led to environmental degradation but has also had effects on the rights of Indigenous Peoples, such as the right to consultation and to Free, Prior and Informed Consent (FPIC).

This chapter analyses the degree of success that prior consultation and FPIC have had in the oil extraction project carried out in the ITT oil block in the Yasuní, after the Yasuní-ITT Initiative (hereinafter, also the Initiative) was abandoned in 2013. Even though, according to international and national legislation, Ecuador is obliged to apply a consultation process in order for Indigenous Peoples to give their consent, the implementation of the process in this case has clearly failed, since operations began on Indigenous Peoples' lands without them being properly consulted first. This omission went against not only the existing regulations but also the very public demand for a consultation process.

In this sense, the aim of this chapter is to argue that Ecuador has implemented legislation that requires FPIC, but in practice continues to support extractivist development strategies, without observing the regulations and the real will of Indigenous Peoples. Hence, more political commitment to the process of FPIC is required by the national government in order for it to serve its intended purpose of protecting Indigenous Peoples and the entire Ecuadorean population. The impacts of oil exploitation in the ITT remain to be seen, since the operations began in September 2016. However, based on similar activities in other blocks of the Yasuní, it is clear that there will be serious economic, social, and environmental impacts, as well as life-threatening risks for IPVI.

The following chapter is the result of a legal study from the perspective of human rights on the Yasuní-ITT case, including an analysis of the legislation

DOI: 10.4324/9781351042109-18

in force and its degree of implementation, as well as its failures, in this specific case. The struggle for participation in the decision-making process on this project reflects the important challenges in the construction of national and socio-environmental policies, and that is why it is a fundamental case to study regarding Indigenous Peoples' right to consultation and their FPIC.

Ecuador's legal framework for Indigenous Peoples' consultation and FPIC

The legal framework on Indigenous Peoples' right to consultation and their FPIC in Ecuador has been clearly established within national and international legal instruments. Internationally, Ecuador is a signatory to and has ratified the main international instruments on Indigenous Peoples' rights. Indeed, Ecuador ratified ILO Convention No.169 Concerning Indigenous and Tribal Peoples in Independent Countries of 1989 (ILO 169) in May 1998.[3] This Convention contains specific articles establishing the right of Indigenous Peoples to prior consultation regarding development projects in their lands (see its articles 6, 7 and 15). Consultation and participation have been defined as the cornerstones of ILO 169, since they underscore the importance of Indigenous Peoples' involvement in decision-making processes and also establish the requirement for information to be provided prior to consent (Calí Tzay, 2014; see also other chapters in this book). However, despite the fact that these provisions have been incorporated into Ecuador's domestic law, many Indigenous territories in Ecuador have been drilled for oil, without any prior consultation process.

Ecuador is also party to the United Nation Declaration on the Rights of Indigenous Peoples (UNDRIP) of 2007, which not only grants Indigenous Peoples the right to consultation, but also determines the ways in which their land and resources will be used, according to its article 32. Furthermore, article 19 specifies that States shall consult Indigenous Peoples in order to obtain their FPIC before adopting and implementing legislative or administrative measures that may affect them.

For its part, the American Convention on Human Rights is of particular importance. Even though it does not deal specifically with Indigenous Peoples, it contains provisions on individual rights that can and have been invoked by Indigenous Peoples to defend their rights. Specific protection for consultation and FPIC has been granted by the Inter-American Court for Human Rights (IACtHR), which ruled against Ecuador in the Case of the Kichwa Indigenous People of Sarayaku (IACtHR, 2012), which, significantly, is located in the Yasuní. The Sarayaku claimed that oil operations had started in their territories without them being consulted first. The Court ruled in favour of the Sarayaku and, furthermore, set FPIC standards even higher than those established in ILO 169 (Khatri, 2013).

There are other international instruments worth mentioning regarding Indigenous Peoples' rights that are part of Ecuador's legal framework on the subject. However, since they are more recent, they were not applicable when the decision was taken to exploit the ITT. These instruments are the American Declaration on the Rights of Indigenous Peoples from 2016 and the Regional Agreement on Access to Information, Public Participation and Justice in Environmental Matters in Latin America and the Caribbean from 2018.

At the national level, the main instrument for Indigenous Peoples' rights is the Ecuadorean Constitution of 2008,[4] which was established as the legal basis for a far-reaching change in understanding and managing relations between Ecuadorean society and nature, as well as between all its different peoples, nationalities, and social groups. In this sense, the Constitution of 2008 establishes three main legal pillars to construct this new paradigm: (1) the recognition of nature as a subject of law[5]; (2) the welfare regime[6]; and, (3) the recognition of a plurinational State.[7] In this context, the Yasuní-ITT Initiative was presented as the first effective step towards an economy focusing less on extraction and a change in the production matrix (Dávalos & Silveira, 2017).

Article 57 of the Constitution includes a broad catalogue of Indigenous rights, including prior consultation. However, according to the same article, consent is not mandatory, and it establishes that if consent is not reached the Government must proceed according to the Constitution and the law. Therefore, there is no possibility to veto a project. The same provision recognises the right to consultation before the adoption of any legislative measure that may affect any collective right.

In addition, there are several legal norms that develop and regulate the contents of the rights granted in the Constitution. However, these laws do not reflect the constitutional mandate. This is either because they were enacted prior to the adoption of the Constitution but remain in force, or – in the case of those subsequent to the Constitution – they do not develop the standards enshrined in the Constitution efficiently. These include the Citizens' Participation Law of 2010; the Regulations for the Implementation of the Pre-Legislative Consultation Process of 2012; and the Environmental Management Act of 1999, which includes consultation as a social participation mechanism (CONAIE, Fundación Pachamama, Fundación Regional de Asesoría en Derechos Humanos-INREDH, Fundación Centro Lianas, & Centro por la Justicia y el Derecho Internacional-CEJIL, 2009). The latter, however, relates to an environmental consultation process and does not refer to the specific right to consultation of Indigenous Peoples.

Thus, the legal provisions for consultation and FPIC in Ecuador are clearly established within different legal instruments. However, we must highlight that they contain limitations and contradictions that do not allow for the effective implementation of the protection of Indigenous Peoples and the environment (Aparicio Wilhelmi, 2011).

The Yasuní and the ITT Initiative

Ecuador's economy is heavily oil dependent; indeed, oil has been the mainstay of Ecuador's national development for more than forty years. Notwithstanding a recent decrease in production, the country continues to drill within one of the most ecologically complex and fragile areas, the YNP (Rival, 2011). The Yasuní was declared a National Park in 1979 (Ministry of Agriculture and Livestock and Ministry of Industries, Trade and Integration, 1979) in order to protect its biological richness, and it is also home to great ethnic diversity. Many different Indigenous Peoples inhabit the park, such as the Huaorani, the uncontacted voluntary isolated Tagaeri and Taromenane, the Kichwa, the Shuars, and other local communities.

In 1989, the YNP and a significant part of its adjacent territory that constitutes the Huaorani Ethnic Reserve became a Biosphere Reserve under the UNESCO's Man and the Biosphere Programme (MAB, 1989). In 1990, due to social pressure and criticism, the government created the "Huaorani Territory" (Ministry of Agriculture and Livestock, 1990) in order to protect this Indigenous People. However, rather than protecting them, the declaration contributed to the development of oil-related activities in the area. In 1999, the southern part of the YNP was declared a *Zona Intangible* (Intangible Zone), because of its exceptional natural and biological importance, and therefore extractive activities were banned, as well as any other human activity that might jeopardise its cultural and biological integrity. The Executive Decree (Presidency of the Republic of Ecuador, 1999) that creates this zone declares the living and development lands of the Huaorani people known as Tagaeri and Taromenane, and others that remain in complete isolation, as an 'intangible conservation zone', thus permanently banning all types of extractive activities.

Notwithstanding all the protection mechanisms and the prohibition on extraction in protected areas established in article 407 of Ecuador's Constitution, the YNP has been subject to controversy, since its creation, due to the oil fields found within its territory. Indeed, its territory is divided into oil blocks. Therefore, there are two legal and institutional frameworks that should have safeguarded the YNP: the conservation of protected areas in Ecuador; and the rights of Indigenous Peoples established at both international and national level illustrated above. However, this has not been sufficient to stop exploration and exploitation in tropical forests and protected fragile areas, and the YNP case has been no exception. With the expanding oil frontier, the Yasuní territory has been divided into another six blocks for exploitation: 14, 15, 16, 17, 31 and 43, (Ministry of Environment, 2011), which is also known as ITT (Ishpingo-Tambocoha-Tiputini) and which formed part of the Yasuní-ITT Initiative.

The Yasuní-ITT Initiative was a proposal officially put forward by civil society in 2007.[8] In essence, it consisted of banning oil drilling within the ITT block, which represents 20% of the oil reserves that have been discovered so far in the country. In return, the country was supposed to receive the economic cooperation of the international community, which would provide at least half of the profit that Ecuador would earn if exploiting the crude or through carbon emissions trading. The main objectives of the Yasuní-ITT Initiative were: (1) to respect the territory of the Indigenous communities from the Yasuní; (2) to fight climate change by avoiding the emission of 407 million metric tons of carbon dioxide (CO_2) into the atmosphere; and, (3) to protect the park and its biodiversity (Acosta Espinosa, 2014).

In an era where the rush for natural resources deeply affects social justice, the proposal to leave oil in the Yasuní underground represented an innovative action. The Yasuní-ITT Initiative offered a way to protect the Amazon, its biodiversity, and Indigenous territories, as well as to combat climate change. In a broader context, it was also meant to be part of a major attempt to break Ecuador's dependency on oil and avoid exacerbating social and environmental tensions that had already been caused by oil operations in the Ecuadorean rainforest, where Indigenous Peoples are particularly affected.

However, after years of using the Yasuní-ITT Initiative as the country's environmental flagship, in August 2013 the Ecuadorean government publicly abandoned it after not meeting the established economic goals required for its implementation (Presidency of the Republic of Ecuador, 2013). The international community failed to support it and the proposal was abandoned due to inconsistencies and contradictions within the government. In Ecuador, decisions are often taken to assure the inflow of cash into the State's coffers. In addition, since the State is the one and only owner of non-renewable natural resources and has exclusive competence over them in accordance with article 1 of the 2008 Constitution (National Constituent Assembly, 2008), the government announced that it would proceed to exploit the ITT, setting aside its public commitment to protecting the environment and diversifying the energy matrix. Even worse, this decision was made without any consultation process and despite the opposition of some Indigenous Peoples living in the area and a large part of the Ecuadorean population (Lang, 2014).

Therefore, it could be said that although it was ultimately a failure, during its brief existence the Yasuní-ITT Initiative was considered a valid mechanism to guarantee respect for Indigenous Peoples' rights, and it was praised both at the international and national level. It became a public issue that attracted considerable attention not only from the international community, States and multilateral organisations, but also within Ecuador's civil society. Indeed, the author of this chapter senses that Ecuadoreans looked at it as an opportunity to participate and exert influence over policy related to oil extraction and the protection of Indigenous Peoples' rights.

The Yasuní-ITT: the failure to consult and obtain FPIC

Oil has been considered a pillar of the Ecuadorean national economy but, at the same time, it is clear that this resource has become a source of conflict by promoting the deterritorialisation of Indigenous Peoples, affecting their cultural identity, and jeopardising their survival. The fact that the country's economy depends on oil, in addition to constant institutional changes and limited national resources, has meant that legal dispositions are rarely implemented; as a consequence, Indigenous rights continue to be unobserved.

Although Ecuadorean government policies have included a change in the productive matrix and a transition to a post-oil economy that will enable the nation to achieve a welfare regime, the State continues to seek the population's prosperity through the economic resources obtained from oil: a decision that threatens the survival of Indigenous Peoples, especially of those in voluntary isolation. Decisions such as exploiting the ITT block, as mentioned above, confirm that extractive industries not only persist in Ecuador, but are becoming more intense.

In this sense, constitutional progress regarding the protection of nature, Indigenous Peoples in general, and those in voluntary isolation are invoked to defend such rights. Notwithstanding this recognition and the above-mentioned clear legal provisions regarding the right to consultation and the FPIC of Indigenous Peoples, Ecuador has not complied with either the international or the national legal framework in the case of the Yasuní-ITT. Indeed, no consultation was carried out when deciding to proceed with the exploitation in 2013, and the consultation carried out for the licensing process also failed since it was not taken into consideration before oil production activities began in 2016 (El Universo, 2016).

Not necessarily all the Indigenous Peoples from the Yasuní opposed the oil extraction of the ITT block. Some members of the Huaorani Indigenous People supported the government's proposal of exploiting petroleum from the ITT block. They actually considered it an opportunity to leave poverty behind, not only for them but for the entire nation. Others backed the project too, especially those involved with environmental organisations such as Fundación Pachamama (Lang, 2014).

No matter what the different positions were, in 2013, the Indigenous Peoples of the Yasuní were not consulted before the decision to exploit the ITT was taken. Their right to be consulted before the adoption of any legislative measure that may affect their collective rights, as mentioned, is recognised in international instruments and the Constitution itself. Based on these provisions, any law or legislative measure adopted without a previous consultation process would be considered unconstitutional (Due Process of Law Foundation & OXFAM, 2011). Furthermore, consultation and participation rights are not subject to the discretion of the authorities: they are rights that must be respected and that impose correlative duties (Potes, 2013).

The decision to proceed with the exploitation of the ITT, which declared the exploitation of the oil from the ITT block as a matter of national interest, was implemented through a Resolution of the National Assembly, the national legislative body (Asamblea Nacional de la República del Ecuador, 2013). The adoption of this resolution without carrying out a consultation process *per se* violates this right since it should have been carried out prior to the adoption of the decision itself (Salazar, 2013). The National Assembly ruled in favour of authorising activities in Indigenous Peoples' territories, thus contravening the pre-legislative consultation requirement established in article 57 (para. 17) and that of prior consultation of article 57 (para. 7) of the Constitution (National Constituent Assembly, 2008).

According to the National Assembly, this Resolution was a legislative act prior to the prospection, exploitation, and commercialisation of non-renewable resources. Therefore, it did not require prior consultation, since it was not a law itself but a resolution from a legislative body on a specific issue (Asamblea Nacional de la República del Ecuador, 2013, p. 20). However, we must highlight that all the legal instruments that refer to pre-legislative consultation do not necessarily mention 'laws' as the subject of this kind of consultation, but they rather refer to 'legislative measures' (above all, see article 6 of ILO 169). Thus, there is no legal basis for this distinction.

As explained before, as well as the omission of the pre-legislative consultation, the consultation process set out in article 57 (para. 7) of the Constitution was also set aside. Even though there was a consultation process within the environmental licensing process,[9] it was not a prior consultation as such. It was a consultation prior to exploitation but not prior to the decision to exploit the resource. Regardless of the nature of the decision (administrative or legislative), consultations must be prior, that is to say, carried out within the early stages of the decision-making process, not when the decision has already been made, i.e., when the potentially affected peoples can no longer influence the decision (Potes, 2013).

Furthermore, in this case, not only was consultation required, but FPIC should have been a requisite too. The expansion of oil and mining activities overlooks the right to FPIC, but in a project of the magnitude of the ITT, FPIC should have been required before determining whether to continue with the exploitation (Melo, 2013). According to the IACtHR, in specific circumstances, such as large-scale projects or projects likely to cause major impacts in a territory, the right to be consulted implies the State's obligation to obtain FPIC according to Indigenous Peoples' traditions (IACtHR, 2007, paras. 134 & 137). Despite the fact that consent may not be required by domestic law, as in the Ecuadorean case, FPIC is recognised by international human rights laws and State parties must respect this specific aspect set out in those instruments (Tauli-Corpuz, 2015).

Regarding the specific situation of the Tagaeri and Taromenane, as IPVI, the right to consultation to obtain their FPIC must be interpreted taking into account their decision to remain isolated and their need for greater protection given their vulnerability (UN Office of the High Commissioner for Human Rights, 2012, para. 22 & 14). Therefore, as a result of respecting their right to self-determination and their choice to remain isolated, IPVI do not intervene through traditional participation mechanisms (UN Office of the High Commissioner for Human Rights, 2012). The only option would be for States to coordinate actions with the Indigenous organisations representative of those parts of the same population in contact. Thus, paying particular attention to the principles of *pro personae* and non-contact, the main factors to consider when analysing whether IPVI consent or not to the presence of people who do not belong to their groups within their ancestral territories would be the expression of clear rejection to outsiders and their decision to remain in isolation regarding other populations and people in general (CIDH, 2013).

A missed opportunity for broader social participation

Not only Indigenous Peoples but also activists and Ecuador's civil society in general opposed the oil extraction from the ITT, demanding a consultation process that should have been free, informed and prior to any action. However, the Government assumed the central role in defining the entire process and set aside the affected communities and civil society. Since the beginning of the initiative until its end, Indigenous organisations were excluded from its management. Civil society actors who tried to participate and aimed to influence the decision-making process regarding oil exploitation were left without any opportunity to participate. In a short period, the entire country passed from the dream of the oil moratorium[10] to the nightmare of collective rights (Dávalos & Silveira, 2017).

It is important to recall that civil society was fundamental for the origins of the Yasuní-ITT Initiative, but over the years it fell behind, and its participation in forums was significantly reduced. This became clear when former President Correa decided to end the Initiative and the National Assembly permitted the exploitation of the ITT. This decision attracted the general interest of Ecuadorean citizens, signatures were collected, and the National Assembly could have called for a popular referendum, but in the end nothing happened (Dávalos & Silveira, 2017).

In 2014, in response to the decision to exploit the ITT, a group of activists called "*Yasunidos*" began a campaign to collect signatures to demand the longed-for national referendum on keeping the oil underground. However, while they claimed to have collected over 750,000 signatures, the National Electoral Council only considered about 360,000 valid (National Electoral Council, 2014), a number that did not meet the minimum required for the referendum.

This all goes to show that not only directly affected Indigenous Peoples opposed the ITT project, but also civil society in general. Nevertheless, the government did not listen to either of them, thereby demonstrating its lack of will to effectively implement the constitutionally-guaranteed rights of consultation and FPIC of Indigenous Peoples. This in turn demonstrates that there was never a real intention to keep the oil underground or promote a new paradigm and energy matrix. The government just moved forward with its neo-developmental model, relying on a constitutional exception, that of the national interest (Asamblea Nacional de la República del Ecuador, 2013).

However, this exception was not applicable to this case, since IPVI were involved. Hence, Ecuador violated international provisions on the consultation and FPIC of Indigenous Peoples, as well as human rights and environmental, national, and international provisions.

Indigenous Peoples are one of the most vulnerable groups of the population, and their consent should be required before any development project can be implemented, as clearly stated by the IACtHR (2007, paras. 134 & 137). Even though there is legislation that recognises the right to consultation and FPIC of Indigenous Peoples, governments continue to implement extractivist development strategies, which is why a greater political commitment to the process of FPIC is required for it to serve its intended purpose of protecting Indigenous Peoples. Oil exploitation in the Yasuní may be costly in monetary terms, but the cost of human rights violations and ecological damage is incalculable.

Conclusions

It is clear that the implementation of the right to consultation and FPIC of Indigenous Peoples failed in the Yasuní-ITT case. In reality, there was never an intention of implementing a consultation process that could have led to FPIC. Ecuador is a country with considerable legislation on the subject. However, it seems that consultation rights have only been included within its legal instruments in order to bolster the country's appearance as sustainable and plurinational.

Ecuador's Constitution of 2008 was recognised as progressive, incorporating a broad catalogue of Indigenous rights and even granting rights to nature, and these provisions were the basis for the Yasuní-ITT Initiative. However, the lack of real political commitment has led to the current situation, not only at the Yasuní but all over the country. It seems that Indigenous rights have been intentionally set out with gaps, contradictions, and limitations to empower the government over the final decision of continuing under the same extractive regime. A consultation process was not just considered unnecessary by the National Assembly, but was never even considered as an option, as mentioned above. Moreover, the exploitation process can be attributed to the lack of political commitment to the environmental and Indigenous protection enshrined in the Constitution. The decisions taken, and the actions carried out, not only in the Yasuní case, but also regarding other extractive and development projects in Ecuador, show that the government so far has not been interested in resorting to alternative development methods other than extractive industries and activities.

Real commitment is required for consultation and FPIC to be fully applied and work as effective tools for Indigenous Peoples to demand a fundamental change in the way development is understood and implemented. The problem lies in the extractivist development model and the complex relationship between natural resource extraction and Indigenous Peoples. It is evident that violations of Indigenous rights occur if they are not consulted about the projects that take place in their traditional lands, and that is why consultation and FPIC are tools for them to have a voice. Therefore, we can conclude that the greatest impediment to the effective implementation of consultation and FPIC in Ecuador is the lack of governmental commitment to the protection that would be granted by consultation processes through national and international legislation. These processes could also benefit from being more specific and from the elimination of contradictions and limitations within the legislation.

Beyond the failure of all the efforts from Indigenous Peoples and different social groups to participate in the decision-making process regarding the Yasuní-ITT, the fight for protecting human rights and the environment must continue, and new mechanisms should be explored in order to effectively integrate the different actors involved and their interests.

Notes

1 Many scientific studies have recognised the Yasuní as one of the most biologically and culturally diverse places on Earth as well as its importance for global conservation efforts. See Bass et al. (2010), Bravo & Yánez, (2005) and Puyol, Ortiz, Inchausty, & Yépez (2010).

2 Ecuador, as defined by the Constitution, is a plurinational country, which implies the existence of different nationalities and peoples which legally and politically make up the Ecuadorean State. According to the Ecuadorean System of Social Indicators from the Ministry for the Coordination of Social Development (n.d.), there are 13 different indigenous nationalities and 14 peoples, with different traditions and social models. Some peoples are part of some nationalities. Nationalities maintain their own language and culture. Indigenous Peoples in Ecuador have defined themselves as nationalities with traditional roots. Having said that, the author – who is well aware of the complexity of Ecuador's social make-up – will use the internationally accepted terminology of Indigenous Peoples.

3 The Convention was approved by Ecuador's National Congress on 14 April 1998 and ratified by the Government on 15 May 1998. It was published in Registro Oficial No. 304, of 24 April 1998.

4 The text of the Constitution of the Republic of Ecuador was agreed by the National Constituent Assembly, approved by popular referendum on 28 September 2008 and published in Registro Oficial No. 449 of 20 October 2008 (National Constituent Assembly, 2008).

5 See articles 71–74 of the Ecuadorian Constitution (National Constituent Assembly, 2008).

6 Title VII of the Ecuadorean Constitution (National Constituent Assembly, 2008).

7 According to article 1 of the Constitution (National Constituent Assembly, 2008), Ecuador is a Constitutional, Plurinational and Intercultural State governed by rights and justice.

8 The Initiative was officially and internationally launched by former President Rafael Correa on 28 September 2007 at the United Nations General Assembly.

9 All the documents regarding the exploitation of the ITT blocks are available at the official website 'Yasuní Transparente'. Regarding prior consultation there is only information on the social participation process carried out within the additional environmental impacts study of block 43 (Gobierno de la República del Ecuador, & Ministerio del Ambiente, 2018).

10 An oil moratorium is the suspension of hydrocarbon exploration and production activities for a specific period of time or indefinitely. The suspension may be due to different reasons, e.g., to evaluate the suitability of areas for sustainable exploration and exploitation.

References

Acosta Espinosa, A. (2014). La Iniciativa Yasuní ITT. Una crítica desde la economía política. *Coyuntura*, 16, 39–49.

Aparicio Wilhelmi, M. (2011). Nuevo Constitucionalismo, Derechos y Medio Ambiente en las Constituciones de Ecuador y Bolivia. *Revista General de Derecho Público Comparado*, 9. Retrieved from http://www.iustel.com/v2/revistas/detalle_revista.asp?id_noticia=410617&texto.

Asamblea Nacional de la República del Ecuador. (2013). Resolución que declara de Interés Nacional la explotación de los Bloques 31 y 43, en una extensión no mayor al uno por mil de la superficie actual del Parque Nacional Yasuní, 4 October 2013. Published in Registro Oficial Segundo Suplemento No. 106 of 22 October 2013.

Bass, M., et al. (2010). Global conservation significance of Ecuador's Yasuní National Park. *PloS One*, 5(1), 22. Retrieved from http://www.plosone.org/article/fetchObject.action?uri=info:doi/10.1371/journal.pone.0008767&representation=PDF.

Calí Tzay, J. (2014). Notas sobre el Convenio 169 y la lucha contra la discriminación. In J. Aylwin & L. Tamburini (Eds.), *Convenio 169 de la OIT. Los desafíos de su implementación en América Latina a 25 años de su aprobación* (pp. 28–44). Copenhagen: IWGIA.

CIDH. (2013). Pueblos Indígenas en Aislamiento Voluntario y Contacto Inicial. OEA Doc. OEA/Ser.L/V/II. Doc. 47/13, 30 December 2013.

CONAIE, Fundación Pachamama, Fundación Regional de Asesoría en Derechos Humanos-INREDH, Fundación Centro Lianas, & Centro por la Justicia y el Derecho Internacional-CEJIL. (2009). Derechos de los pueblos indígenas en Ecuador. Audiencia - 137 período ordinario de sesiones CIDH. Retrieved from http://www.inredh.org/archivos/pdf/informe_pueblosindigenas.pdf.

Dávalos, J. & Silveira, S. (2017). La iniciativa Yasuní-ITT: Del sueño de la moratoria petrolera a la pesadilla de los derechos colectivos. *ARACÊ Direitos Humanos Em Revista*, 5, 346–364.

Due Process of Law Foundation & OXFAM. (2011). The Right of Indigenous Peoples to Prior Consultation: The Situation in Bolivia, Colombia, Ecuador and Peru. Retrieved from http://www.oxfamamerica.org/explore/research-publications/the-right-of-indigenous-peoples-to-prior-consultation-the-situation-in-bolivia-colombia-ecuador-and-peru/.

El Universo. (2016). Ecuador inicia fase de producción en bloque petrolero ITT, 7 September. Retrieved from https://www.eluniverso.com/noticias/2016/09/07/nota/5787927/ecuador-inicia-fase-produccion-bloque-petrolero-itt.

Gobierno de la República del Ecuador & Ministerio del Ambiente. (2018) Yasuní Transparente. Retrieved from http://yasunitransparente.ambiente.gob.ec/inicio.

IACtHR. (2007). *Saramaka People v. Suriname*, Judgment of November 28, 2007 (Preliminary Objections, Merits, Reparations, and Costs). Inter-Am. Ct. H.R., (Ser. C) No. 172(2007).

IACtHR. (2012). *Kichwa Ingenous People of Sarayaku v Ecuador*, Judgment of June 27, 2012 (Merits and reparations). Inter-Am. Ct. H.R., (Ser. C) No. 245(2012).

Khatri, U. (2013). Indigenous Peoples' Right to Free, Prior, and Informed Consent in the Context of State-Sponsored Development: The New Standard Set by Sarayaku V; Ecuador and its Potential to Delegitimize the Belo Monte Dam. *American University International Law Review*, 29(1), 165–207. Retrieved from https://digita lcommons.wcl.american.edu/cgi/viewcontent.cgi?referer=https://www.ecosia.org/&ht tpsredir=1&article=1805&context=auilr.

Lang, C. (2014). Ecuador's Continued Conflict Over Oil Drilling, Indigenous Rights and Biodiversity. *Redd Monitor*, 3 June. Retrieved from http://www.redd-monitor.org/2014/ 06/03/ecuadors-continued-conflict-over-oil-drilling-indigenous-rights-and-biodiversity/.

MAB. (1989). Report of the Meeting of the MAB Bureau. Paris, France. UNESCO Doc. SC-89/Conf.227/14, 28 August 1989. Retrieved from http://unesdoc.unesco. org/images/0008/000850/085095eb.pdf.

Melo, M. (2013). Consulta previa en el Ecuador: la mirada de los organismos internacionales de Derechos Humanos. Retrieved from https://correismo.wixsite.com/ elcorreismoaldesnudo/mariomelo.

Ministry of Agriculture and Livestock and Ministry of Industries, Trade and Integration. (1979). Inter-Ministerial Agreement No. 0322 of July 26, 1979. Creation of the Yasuní National Park. Registro Oficial No. 69 of 20 November 1979.

Ministry of Environment. (2011). Management Plan for the Yasuní National Park. Quito, Ecuador.

Ministry of Agriculture and Livestock. (1990). Ministerial Agreement No. 19. Published in Registro Oficial No. 408 of 2 April 1990.

Ministry for the Coordination of Social Development. (n.d.). Listado de nacionalidades y pueblos indígenas del Ecuador. Retrieved from http://www.siise.gob.ec/siise web/PageWebs/glosario/ficglo_napuin.htm.

National Constituent Assembly. (2008). Constitution of the Republic of Ecuador Approved by Popular Referendum on September 28, 2008. Published in Registro Oficial No. 449 of 20 October 2008.

National Electoral Council. (2014). Resolution PLE-CNE-2-8-5-2014. Record No. 031-PLE-CNE, 8 May 2014.

Pérez, M., Bravo, E., & Yánez, I. (Eds). (2005). *Asalto al paraíso: Empresas petroleras en áreas protegidas*. Quito: Oilwatch & Manthra Editores.

Potes, V. (2013). Artículo 57 y Decisiones Estatales. *Gkillcity*, 7(2). Retrieved from http s://gk.city/2013/09/16/articulo-57-y-decisiones-estatales/.

Presidency of the Republic of Ecuador. (1999). Executive Decree No. 552 of January 29, 1999. Published in Registro Oficial Suplemento No. 121 of 2 February 1999.

Presidency of the Republic of Ecuador. (2013). Executive Decree No. 74 of August 15, 2013. Published in Registro Oficial Suplemento No. 72 of 3 September 2013. Modified by Executive Decree No. 84 of August 17, 2013. Published in Registro Oficial Suplemento No. 77 of 10 September 2013.

Puyol, A., Ortiz, B., Inchausty, V., & Yépez, O. (2010). Género, alternativas productivas y soberanía alimentaria: Estrategias políticas para lograr cambios positivos y disminuir la cacería comercial en Yasuní. Retrieved from https://cmsdata.iucn.org/ downloads/informe_avance_caceria_fauna_silvestre_2_.pdf.

Rival, L. (2011). Planning development futures in the Ecuadorian Amazon. The expanding oil frontier and the Yasuní-ITT initiative. In A. Bebbington (Ed.), *Social Conflict, Economic Development and Extractive Industry. Evidence from South America* (pp. 153–171). London and New York: Routledge.

Salazar, D. (2013). Los Derechos Humanos y la Explotación Petrolera en Yasuní. Asuntos del Sur (Discussion Paper), November. Retrieved from http://www2.con greso.gob.pe/sicr/cendocbib/con4_uibd.nsf/39212BE70B0C957F05257D7300761310/ $FILE/Documento-de-debate-yasun%C3%AD-itt.pdf.

Tauli-Corpuz, V. (2015). Victoria Tauli-Corpuz: "Indígenas deben ser consultados". *El Universo*, 30 August. Retrieved from http://www.eluniverso.com/noticias/2015/08/30/ nota/5092145/indigenas-deben-ser-consultados.

UN Office of the High Commissioner for Human Rights. (2012). *Guidelines on the Protection of Indigenous Peoples in Voluntary Isolation and in Initial Contact of the Amazon Basin and El Chaco and the Eastern Region of Paraguay.* Geneva: UNOHCHR.

14 The right to consultation and free, prior and informed consent in Argentina

The case of Salinas Grandes-Laguna de Guayatayoc

Marzia Rosti

Introduction

Argentina is a federal country made up of 23 provinces with a population of over 40 million people, of whom 955,032 (or 2.4%) identify as Indigenous. This sector of the population lives mainly in the provinces of Chubut (8.7%), Neuquén (8%), Jujuy (7.9%), Río Negro (7.2%), Salta (6.6%), Formosa (6.1%) and La Pampa (4.5%) (INDEC, 2010). There are 35 different, officially recognised Indigenous Peoples[1] who have been granted specific rights in the federal and provincial constitutions. These rights are supplemented by those listed in international treaties which have been ratified by the country. Despite a favourable legal framework, an increase in territorial disputes was recorded in 2017 because the State has guaranteed neither the territorial rights of Indigenous Peoples (in order to avoid the demarcation of their territories) nor their right to consultation and exercise of their Free, Prior and Informed Consent (FPIC) (Ramírez, 2018). Both rights are yet to be regulated by a national law since they are incompatible with the neo-developmental economic model adopted by the latest governments, which is based on extractive activities (including agriculture, mining, and hydrocarbons) that have been carried out in Indigenous territories over the last few years.

In this context, the case of Salinas Grandes-Laguna de Guayatayoc is interesting as the region is divided between the provinces of Salta and Jujuy, which are inhabited by the Kolla and Atacama Indigenous Peoples. Here lithium was discovered in 2010, giving rise to a process of exploration by multinational companies and promotion by national and provincial governments. The case involves the violation of three Indigenous rights. First, the right to communal ownership of the land, since, although a survey has been carried out, the title deed of the land has not yet been granted. Second, the right to consultation and exercise of FPIC, given that the government took measures related to lithium exploration and exploitation without the consultation and consent of the affected communities. Third, this is a case of interest because it involved an unprecedented process of Indigenous mobilisation which has led to the *Kachi Yupi/Huellas de la Sal* protocol, drawn up

DOI: 10.4324/9781351042109-19

by the communities themselves and which represents the model of how the rights to consultation and FPIC of Indigenous Peoples should be exercised.

Current legislation

The constitutional reform made in 1994 introduced a wide variety of Indigenous rights to article 75, para. (*sección*) 17, which deals with the powers of Congress. Federal Law No. 23,302 on Indigenous Policy and Support for the Aboriginal Communities was adopted in 1985. This law defined

> Indigenous communities as groups of families that recognise themselves as such, as a result of being descendants of the peoples that inhabited national territory at the times of the conquest or colonisation and the members of these communities are defined as indigenous or natives.
>
> (article 2, para.2)

By means of registration in the Indigenous Communities Registry (*Registro de Comunidades Indígenas*), their 'legal personality' would be acknowledged (article 2, para. 3). This legal personality is necessary for them to claim rights and access certain benefits, such as the adjudication of lands "without charge" (article 7; see also articles 8–9). Law No. 23,302 also created the National Institute of Indigenous Affairs (INAI, for its Spanish acronym), which is responsible for the implementation of Indigenous policies in coordination with the provinces. In 2010, the National Register of Indigenous Peoples' Organisations (*Registro Nacional de Organizaciones de Pueblos Indígenas*) was created by means of Federal Resolution No. 328/2010 (Gobierno de Argentina, 2010), which frames such organisations as those that "represent the majority of the indigenous communities of one or several Indigenous Peoples at the provincial, regional or national level" (Gobierno de Argentina, 2010, article 4).

During the governments of Néstor Kirchner and Cristina Fernández de Kirchner (2003–2015), the Indigenous Participation Council (2004) and the Coordination Council (2008) were created. Nonetheless, these councils have not been able to guarantee the effective participation and presence of the Indigenous Peoples in the INAI's programmes. In 2016, Mauricio Macri's government incorporated the Consultation and Participative Council for the Indigenous Peoples within the Human Rights and Cultural Pluralism Secretariat of the Ministry of Justice and Human Rights by means of Decree No. 672 of 2016 (Gobierno de Argentina, 2016). This Council was created without consulting the Indigenous Peoples and was therefore rejected through the Parliament of the First Nations (Parlamento de Naciones Originarias et al., 2016, p. 7). The purpose of this body was both to promote respect for the rights foreseen by the Argentine National Constitution, ILO 169 and the United Nations Declaration on the Rights of Indigenous Peoples [UNDRIP] (Gobierno de Argentina, 2016, article 1) and promot[e] the reform of Law No. 23,302, thus adjusting it to international standards, proposing a plan for the regulation of the right to Free, Prior and Informed Consultation, as established by Convention 169 [...] and a plan for the regulation of indigenous communities.

(Gobierno de Argentina, 2016, article 3, section a)

However, to this day, there is no news about the Council's works and projects.

Since the 1980s, and within the framework of a federal structure, provinces with significant Indigenous population have issued constitutional rules and laws on the Indigenous property issue, both in general or in relation to specific details (Carrasco, 2000; Anaya, 2012). This is a concurring competence at the federal level, as established by the Constitution (article 75, para. 17), and set out by the Argentine Supreme Court of Justice (2013).

Finally, in 2001, ILO 169 came into force via Law No. 24,071 of 1992 (Gomiz & Salgado, 2010).[2] In 2007, Argentina endorsed the UNDRIP and, in 2016, the American Declaration on the Rights of Indigenous Peoples.

Territorial disputes: Law 26,160 of 2006

Since the constitutional acknowledgement of Indigenous Peoples and their rights in 1994, territorial disputes in the country have increased dramatically. This is because, on the one hand, Indigenous Peoples began to claim their rights over the ancestral lands from which they were evicted and which they often occupied, being well aware of Argentina's constitutional and international obligations. On the other hand, there were Argentine or foreign individuals or companies that purchased these lands long ago and defend their rights on the basis of the legal system and national legislation (Rosti, 2010; 2016).

Along with these developments, there have been more recent conflicts in relation to the extractivist model (Giarracca & Teubal, 2013), which now appears in the form of progressive neo-extractivism (Gudynas, 2015) or developmental extractivism (Svampa, 2016; Svampa & Viale 2014; Göbel, 2015). Promoted by the Kirchner governments and reaffirmed – and perhaps even reinforced – by Macri's administration, this has led to the penetration of the so-called 'extractive frontier' (*frontera extractivista*) in the agriculture, mining, and hydrocarbon sectors, together with the involvement of government authorities in areas of the country that had – up until recently – been practically excluded from exploitation projects. This situation led to an increase in tension and disputes given that other activities had been undertaken in these lands or because there were Indigenous communities living there that claimed their ownership rights. These situations have often turned into violent incidents which have led to the eviction of communities and, most of the time, the presence of different types of companies in these areas, with the help of the State and without any consultation process or consent granted by the Indigenous communities.

The unfolding of territorial disputes has, first of all, given rise to initiatives at the provincial level aimed at regulating Indigenous possession of the land. These initiatives have included Court orders in which the territorial rights of Indigenous communities were acknowledged on the basis of their "ethnic and cultural pre-existence" established in the Constitution in its above-mentioned article 75 (para. 17).[3] Other disputes have also arisen which are more complex due to the passivity and complicity of the authorities, whose main goals for the economic growth of the country were – and still are – based on the aforementioned extractive industries, which affect Indigenous rights, including land, natural resources, food, health, development, as well as consultation and FPIC. In November 2006, the increase in territorial claims and disputes led the government to enact Law No. 26,160 on the "emergency of indigenous communities' property" (Congreso, 2006). This law suspended evictions from ancestral lands (article 2) and made the INAI responsible for undertaking "technical, legal, and cadastral surveys on the status of the lands inhabited by indigenous communities in the three subsequent years" (article 3) (Congreso, 2006).

In 2007, the INAI established the Regularisation and Land Allocation Programme for Indigenous Communities (RETECI, for its Spanish acronym), and – due to the lack of results – the timeline foreseen by Law No. 26,160 was extended three times: in 2009, in November 2013 and, later on, in November 2017 (on this last occasion, until 2021) despite fears that Macri's government would not extend the law and Indigenous Peoples would be left without protection and defence (Amnesty International, 2017). It is worth pointing out that this law is the only national measure for the protection of the territorial rights of Indigenous communities, although its scope is limited and its implementation is scarce.

As a matter of fact, the data provided by INAI in 2016 reveal that 759 of the 1,532 communities identified by the RETECI programme have initiated the process of surveying (49%). Only 459 of these processes have been resolved, meaning that their report has been concluded and the data collected can be found in a cartographic report in a technical file, together with practical instructions for legalising the ownership of the lands. However, the destination of these documents is often unknown due to a lack of interest and response from government authorities. In the meantime, dozens of communities continue to be evicted and their right to consultation and information is still denied (Amnesty International, 2017).

As to the limited legal extent of Law No 26,160, the aforementioned RETECI programme constitutes only a part of the State's obligations towards Indigenous Peoples, as it does not include the possibility of territorial demarcations and title granting. Likewise, this law does not offer a solution to the disputes that arise as a result of claims over the land from which Indigenous Peoples were evicted.

On this matter, it is worth highlighting the official request made in 2011 by the then UN Special Rapporteur on the Rights of Indigenous Peoples, James Anaya, who visited the most critical provinces (Neuquén, Río Negro, Salta, Jujuy, and Formosa) during his mission in Argentina. His report showed that the State must develop a consultation mechanism or procedure in accordance with international standards to increase the participation of Indigenous Peoples in decisions that affect them (Anaya, 2012, p. 34). This point of view was reaffirmed by the Committee on the Elimination of Racial Discrimination in its Concluding Observations on Argentina's periodic reports (CERD, 2017).

Right to consultation and Free, Prior and Informed Consent

In Argentina, there is a direct relation between the legal personality of Indigenous communities, territorial claims, consultation, and FPIC, since acknowledging the first element (legal personality) is a requirement to exercise the right to consultation and FPIC. These are two instruments which, up to this day, are not foreseen by the Argentine legislation. They are not considered when assessing investment and exploitation plans that affect Indigenous lands or for passing bills or issuing regulations that involve them (CERD, 2017). This is of particular concern if we take into account that the Argentine Constitution guarantees Indigenous Peoples "their participation in issues relating to their natural resources and other interests affecting them" (article 75, para. 17) and that Argentina has ratified ILO 169 and signed the UNDRIP. Both of these instruments set out international standards with respect to the right to consultation and FPIC, which are part of the Argentine legal system and to which all provinces shall be bound according to the Constitution (article 31). Moreover, some specific Indigenous rights have been inserted into the provinces' constitutions, i.e., in Jujuy (article 50); Rio Negro (article 42); Buenos Aires (article 36, para. 9), Chaco (article 37); Chubut (article 34); La Pampa (article 6, para. 2); Salta (article 15); Formosa (article 79); Neuquén (article 53); Tucumán (article 149); and Entre Ríos (article 33) (Instituto Nacional de Asuntos Indígenas, n.d.).

However, only a small number of consultations can be identified, and those that have taken place have "done so with vested interest and fail to observe the international standards" (CERD, 2017, p. 4). Moreover, the decision on which projects to carry out, which Indigenous Peoples to consult, and which modality to adopt is made in an entirely discretionary way. Finally, it is often the case that the Environmental Impact studies or files are not translated into the native tongue of the Indigenous People affected by said projects. In 2016, several Indigenous organisations[4] pointed out before the Human Rights Committee that

it is common to find practices characterised by coercive decisions, pressure, manipulation of indigenous leaders, corruption, fake organisations, forged documents; [...] in order to yield in favour of the interest of corporations and their economic interests, without taking into account vital interests at stake, and the development of the communities and Indigenous Peoples.

(Parliament of the First Nations et al., 2016, p. 6)

In these circumstances, the cases of the provinces of Salta, Jujuy, and Catamarca stand out. In these provinces, licences were granted for lithium exploration and exploitation without consultation and FPIC of local peoples, thus leading to disputes raised not only between companies, the government, Indigenous communities and farmers but also among the communities themselves due to the imaginaries generated by lithium (Argento & Zicari, 2017; Svampa & Viale, 2014), as illustrated below.

Lithium: white gold

Lithium has gained special importance among natural resources in the past few years, as it is an essential component in the manufacturing of Ion-Lithium rechargeable batteries, which are used in electric cars and some medicines. It is estimated that approximately 80% of the world's reserves can be found in the so-called 'Lithium Triangle' (also known as the 'White Saudi Arabia') that encompasses Northern Chile, the South of Bolivia, and Northwest Argentina (Fornillo, 2018).

With regard to Argentina, lithium exploration and exploitation projects have been part of the State policy to promote mining as a pillar of the national economy since the Kirchner administration. Today, the country is the world's second producer of lithium and has the potential to become the top producer in five years' time, thanks to a very favourable regulatory framework which can lure foreign investments. This framework is based on the Argentine Constitution that acknowledges that provinces shall "have the ownership of the natural resources existing in their territory" (article 124), the Mining Code of 1997, and Law No. 24,196 on Mining Investments of 1993. Both of the latter texts promote investment projects by international companies without any reference being made to the territories and Indigenous Peoples, although the Code of 1997 requires the permission of the landowners for the extraction of minerals. For its part, the recent Law No. 25,675 on General Aspects of the Environment of 2002 does not make any reference to consulting Indigenous Peoples either. However, what it does establish is the necessity of an Environmental Impact study on these projects, access to environment-related information, and the right of any individual to be consulted, although it further adds that said opinion or objection shall not be binding (article 20).

With respect to lithium, in the several salt flats located in the provinces of Salta, Jujuy, and Catamarca – i.e., an area of over 300,000 hectares – licences have been granted for lithium exploration and exploitation for about two decades (Fornillo, 2018; Nacif, 2017). The national government has adopted different stances on this issue. A legal bill of 2014 declared lithium a strategic asset and suggested the creation of a State-owned company and of a National Committee of Lithium Exploitation to regulate the extraction of this mineral in salt flats, but made no mention of Indigenous rights (Aranda, 2015; Fornillo, 2015; Ferradás Abalo, Lobo & Lucero, 2016). In 2015, another legal bill declared lithium a strategic natural resource but deemed necessary the consultation and FPIC of Indigenous communities that may be affected by exploration and exploitation. A similar bill introduced in March 2017 is still under discussion but, since 2016, the government has already found foreign countries and multinational companies that are interested in investing in lithium extraction (Fornillo, 2018).

Although, from an environmental point of view, lithium mining is not as risky and destructive as other types of open-pit mining or non-conventional hydrocarbon extraction techniques, it does have an impact on the environment. This is because huge quantities of water are required, which is scarce in the Puna de Atacama region, as it is all over Northern Argentina. Therefore, lithium mining competes with the farming and grazing activities of local Indigenous communities. It may also affect tourism, as the region's visitors are attracted by nature and the moon-like landscapes of the salt flats. In Göbel's words, "the concessioned areas are not empty spaces; instead, lithium mining *lands* in territories with their own history, specific practices, cultural significance, and dynamics of social organisation. These areas overlap with lands for grazing, indigenous territories and natural reserves" (Göbel, 2014, p. 174). Additionally, Indigenous organisation and titling of community lands varies greatly in these territories and, in this context, lithium mining has led to different reactions, dynamics, and results.

The case of Salinas Grandes-Laguna Guayatayoc

The case of Salinas Grandes-Laguna Guayatayoc has been chosen to illustrate the consequences of the lack of compliance with the right to consultation and FPIC of Indigenous Peoples in Argentina. This case is important due to the reactions of the communities involved in order to defend their rights.

Salinas Grandes is a 212 square kilometre salt flat located in the Argentine Puna of Atacama and part of the Laguna Guayatayoc sub-basin. This area is located in the provinces of Jujuy and Salta, where the Kollas and Atacama Indigenous communities live (about 7,000 inhabitants in total). They are engaged in subsistence crop and livestock farming, while some of them work in salt extraction by means of a reasonable use of both salt and water, based on respect for the salt flat cycles, which has been the key to the survival of Salinas Grandes (Schiaffini, 2013). As part of the Andean culture, these communities aim at living well (*Buen Vivir*), rather than strategic development or individual economic growth. These communities have been evicted from their lands and their rights violated since colonial times, and these actions have been repeated by the Argentine State. In 1946, during Juan Perón's administration, the communities drew national attention when they organised the *Malón de la Paz*, in which 150 natives from the Puna area marched to Buenos Aires to claim possession and ownership of their lands and to meet with the President. Despite their efforts, no significant results were achieved, and in 1949 the national government expropriated 58 latifundiums in the Puna. These territories were transferred to the province of Jujuy as public land to be sold. Some families that had the economic means acquired the individual possession and ownership of their lands, but most of them were not granted a title deed (Solá, 2016; Valko, 2007). After the Constitutional reform of 1994, the province signed the abovementioned RETECI programme. However, so far no title deeds have been granted to any Indigenous community in Salinas Grandes and Laguna Guayatayoc due to a hurdle regarding the territorial boundaries of the different families that are collectively established among the community and passed down through generations orally, instead of by means of written deeds. In other words, even though it is neither written nor registered, "each person knows the boundaries of their own territory" (Puente & Argento, 2015, p. 128). However, that is not enough.

At the beginning of 2010, when lithium was discovered in the underground brines that feed the salt flat, multinational companies, the national government and the provinces of Salta and Jujuy undertook a rapid process of exploration in the area. This process was carried out under the notion of 'development' and triggered a conflict with the Indigenous communities inhabiting their ancestral territories.

In the same year, the province of Salta declared the project carried out by the private company Bolera Minera S.A. to be of 'public interest'. The company's aim was to explore, exploit, and industrialise lithium in seven mines of the Salinas Grandes salt flat by means of Executive Order No. 3860 of 2010. At the same time, the government of Jujuy declared lithium to be a strategic natural resource generated by the socio-economic development of the province by means of Executive Order No. 7592 of 2011. This allowed the government of Jujuy to take a more interventionist position and become a business partner in mining projects, in contrast to the government of Salta, which took the role of a private investment facilitator. Both these legislative measures were adopted without any kind of consultation with the affected Indigenous communities. The Environmental Impact reports carried out by the National Agricultural Technology Institute (INTA, by its Spanish acronym) and by the Council of Indigenous Organisations of Jujuy were not considered either (Solá, 2016). The same happened with the licences granted to two companies which are currently focused on the exploration phase: the Australian company Orocobre and the Canadian company Daijin Resources Corp, which have partnered with Toyota and Mitsubishi, respectively (Puente & Argento, 2015).

At first, without acknowledging the right to consultation, the companies were able to obtain the signature of some Indigenous community members on agreements regarding the assignment of territorial rights in exchange for a one-off payment of 25,000 Argentine pesos. Such agreements were presented as evidence for apparent acceptance by the community.[5] The reaction was quick: some members of the Santuario Tres Pozos Cooperative obtained an annulment for the agreements (Puente & Argento, 2015), and, as of May 2010, 33 communities[6] affected by the projects and licences began to meet on a monthly basis in the form of a supra-communitarian organisation, the Board of the Indigenous People of the Salinas Grandes Basin and Laguna Guayatayoc for the Defence and Management of Territory, in order to determine the defence strategy for the different areas (Solá, 2016). These 33 communities are legal entities, but the possession and ownership of their lands have still not been granted to them.

In November 2010, a writ of *Amparo* was filed before the Argentina Supreme Court to demand prior consultation regarding the Salinas Grandes licence. The lawsuit was based on the Constitution (article 75, para. 17), the Environment General Law No. 25,675 of 2002, ILO 169 and the UNDRIP (Corte Suprema de Justicia de la Nación, 2011).

On 28 March, 2012, the heads of the Indigenous communities and the representatives of the government of the province of Jujuy were summoned to a public hearing by the CSJN (Schiaffini, 2013). In January 2013, the Court claimed it had no jurisdiction to hear the case and transferred it to the provincial courts (Corte Suprema de Justicia de la Nación, 2012, paras. 7 and 13). As pointed out by Solá (2016), this meant that both the case and the territory were divided between the two provinces and, therefore, access to justice by the correct authority was denied. As a result of this decision, on 13 June 26, the communities resorted to the Inter-American Commission on Human Rights (IACHR). Currently the admissibility of the case is still pending.

At the same time, international attention was drawn to this case after it was flagged by the UN Special Rapporteur on the Rights of Indigenous People and the UN Committee on Economic, Social and Cultural Rights (CESCR) in 2011. During his official visit in December of the same year, the then Special Rapporteur James Anaya met the Ojo de Huáncar community and quoted the Salinas Grandes case specifically in his report (Anaya, 2012). The CESCR (2011), for its part, expressed its concern about the "persistence" of threats, violent displacements, and evictions of Indigenous Peoples from their native lands in several provinces; it regretted the deficiencies in the consultation processes with the affected Indigenous communities which have allowed the exploitation of natural resources in the territories traditionally inhabited or used by these communities through violations of the Constitution and ILO 169; it registered concern about the "negative consequences of the lithium exploitation in Salinas Grandes [...] for the environment, access to water, the way of life and the survival of the indigenous communities (articles 1, 11 and 12)" (CDESC, 2011, p. 3); and recommended the Argentine State to:

adopt the measures required to put an end to the violations of Indigenous Peoples' rights [...] and to carry out effective consultations [...] before granting licences to state-owned companies or to third parties, for the economic exploitation of the lands and territories traditionally inhabited or used by the communities, and to comply with the obligation to obtain Free, Prior and Informed Consent by those affected by such economic activities.

(CDESC, 2011, p. 3)

In addition, other direct actions were taken by those affected, including, for example, a blockade of highway 52, and several protests towards the capital cities of the different provinces and camps. These actions had a strong media impact around the country – so much so that the exploration activities in Salinas Grandes were interrupted, achieving a precautionary measure *de facto* (Ferradás Abalo, Lobo & Lucero, 2016, p. 10).

Finally, on 22 August, 2015, after two years of meetings and workshops throughout the basin (Solá, 2016), the 33 communities gathered at the General Assembly in the Quera and Aguas Calientes Community, and approved the document Kachi Yupi – Huellas de la Sal/Proceeding for the FPIC and Consultation for the Indigenous Communities of Salinas Grandes Basin and Laguna de Guayatayoc (NPKA, 2015). The text seeks to preserve their ancestral culture and offers an exact model of how to perform the consultation and FPIC process. In a Resolution of May 2016, the Argentine Ombudsman acknowledged its importance and made a plea to several national and provincial authorities for both rights to be respected. This resolution would represent an important step, since it legitimises rights to prior consultation and FPIC every time an administrative or legislative measure that may affect one or several communities of the basin is adopted, development plans or programmes are formulated or implemented, and/or the exploitation of existing resources in their territories is authorised (Solá, 2016).

Conclusions

This chapter has focused on the case of Salinas Grandes-Laguna de Guayatayoc since, despite a favourable legal framework, it reflects an "implementation gap" of land and consultation rights and the FPIC of Indigenous communities in Argentina. Additionally, this case shows the mobilisation of the Indigenous communities involved, which, facing a lack of access to justice at the national level, attracted international attention and brought the case before the Inter-American Court of Human Rights, where it is still pending. Finally, the conflict set in motion a process of Indigenous self-organisation and recovery of identity of hitherto unknown proportions in Argentina, and opened up new spaces for social interaction, negotiation, and cooperation (Göbel, 2013). The most specific outcome is the document Kachi Yupi-Huellas de la Sal (NPKA, 2015), which can be understood as "an expression of the exercise of self-determination within the framework of *Buen Vivir* [Good Living] in their territories, defence of their rights, and the relationship with the State in a context of equality and respect" (Solá, 2016, p. 215). It is important to mention that the communities

> do not plainly and simply reject lithium extraction [but they are against] not being taken into account in lithium-related projects, arguing that their presence in the salt flats goes back centuries, and their existence is obviously prior to that of the national State.
>
> (Fornillo, 2018, p. 195)

As regards the national and provincial governments, besides the violation of the rights mentioned, the evident fact is that policies that favour mining by the communities themselves have not been encouraged. Likewise, there has been no approval of legislative measures to regulate exploitation by third parties which would allow for the protection of the environment and the communities, which end up living in a context of generalised legal uncertainty (Anaya, 2012).

Some useful tools to "close the implementation gap" (Economic and Social Council, 2006, para. 5) and find shared solutions in the legal, economic, social, and environmental spheres might be the above-mentioned Consultation and Participative Council for the Indigenous Peoples created in 2016. Moreover, the recent Lithium Board, comprising representatives of the three provinces involved in the case (Salta, Jujuy and Catamarca), could be empowered to create a good practices protocol as regards the exploitation of this mineral (Dinatale, 2018). Likewise, this Board could consider the will of the affected Indigenous communities and not just the interests of the neo-developmentalist economic model adopted by the government, which is based on extractive activities that have been carried out in Indigenous territories over the last few years.

Notes

1 In the Northeast Region (provinces of Chaco, Formosa, Misiones and Santa Fe): Mbya-Guarany, Mocoví, Pilagá, Toba, Vilela and Wichí peoples; in the Northwest Region (provinces of Catamarca, Jujuy, La Rioja, Salta, San Juan, Santiago del Estero and Tucumán): Atacama, Avá-Guarany, Chané, Chorote, Chulupí, Diaguita-Calchaquí, Kolla, Omaguaca, Tapiete, Toba, Tupí-Guarany and Wichí peoples; in the South Region (provinces of Chubut, Neuquén, Santa Cruz and Tierra del Fuego): Mapuche, Ona, Tehuelche and Yamana peoples, and in the Central Region (provinces of Buenos Aires, La Pampa and Mendoza): Atacama, Avá-Guarany, Diaguita-Calchaquí, Huarpe, Kolla, Mapuche, Rankulche, Toba and Tupí-Guarany peoples (INDEC, 2010).
2 The delay between the enactment of the law and its entry into force is due to the late submission of the governmental decree to the ILO, which was enacted on 17 April 2000. In other words, although Argentina did ratify ILO 169 in 1992, it did not transpose it into its legal order until 2000.

3 See further in Gomiz (2015), Rodríguez Duch (2015), and Rosti (2016).
4 Parliament of First Nations; Human Rights Observatory for Indigenous Peoples
   (ODHPI); Attorneys of North Argentina Specialised in Human Rights
   (ANDHES); Agro forestry Network of Chaco Argentina (REDAF); Association of
   Attorneys Specialised in Indigenous Rights (AADI); Social Assistance of the
   Anglican Church of North Argentina (ASOCIANA); Civil Association for the
   Indigenous peoples rights (ADEPI); United Board of Missions (JUM); Centre for
   Legal and Social Studies (CELS); National Secretariat for Indigenous Peoples of
   the Permanent Assembly for Human Rights (APDH); National Native Pastoral
   Team (ENDEPA); Claretian Works for Development (OCLADE); Master's Degree
   in Human Rights at the National University of Salta (Parliament of First Nations
   et al., 2016, p. 1).
5 Indeed, when the mining companies arrived in 2009, local promotion strategies
   were encouraged. These involved specific aid measures (funding for rituals and
   football tournaments, clothes and school material donations, free transportation),
   by means of which they earned the trust of many inhabitants and created patronage
   networks, especially with young people. Between 2011 and 2013, as a result, some
   Indigenous assemblies approved lithium exploration and exploitation projects in
   exchange for a very low annual compensation, such as, for example, a job offer in
   the mining companies, but without a participative process with sufficient information and debates. The absence of the State as guarantor of Indigenous Peoples'
   rights favoured the imposition of the negotiation business logic and the beginning of
   mining projects (Göbel, 2013; 2014).
6 The indigenous communities of Jujuy that participated were: Comunidad Aborigen
   de Santuario de Tres Pozos, Comunidad Aborigen de San Francisco de Alfarcito,
   Comunidad Aborigen del Distrito de San Miguel de Colorados, Comunidad
   Aborigen de Aguas Blancas, Comunidad Aborigen de Sianzo, Comunidad Abori-
   gen de Rinconadilla, Comunidad Aborigen de Lipan, Organización Comunitaria
   Aborigen "Sol de Mayo", Comunidad Aborigen de Pozo Colorado - Departamento
   Tumbaya, Comunidad Aborigen de Santa Ana, Abralaite, Rio Grande y Agua de
   Castilla, Comunidad Aborigen El Angosto Distrito El Moreno, Comunidad de
   Santa Anta. The Saltenean indigenous communities were: Comunidad Aborigen
   Cerro Negro, Comunidad Aborigen de Casa Colorada, Comunidad Esquina de
   Guardia, Comunidad Indígena Atacama de Rangel, Comunidad Aborigen de
   Cobres, Comunidad Likan Antai Paraje Corralitos, Comunidad Aborigen De
   Tipán (Puente & Argento, 2015).

References

Amnistía Internacional. (2017). *Prórroga de la Ley de emergencia territorial indígena
   26.160*. Retrieved from https://amnistia.org.ar/informe-ley-de-emergencia-terri
   torial/.
Anaya, J. (2012). La situación de los pueblos indígenas en Argentina, Informe del
   Relator Especial sobre los derechos de los pueblos indígenas. UN Doc. A/HRC/21/
   47/Add.2, 4 July 2012.
Aranda, D. (2015). YPF del litio: la minería progresista. *Comunicación Ambiental*, 25
   July. Retrieved from http://www.comambiental.com.ar/2015/07/ypf-del-litio-la
   -mineria-progresista.html.
Argentine Supreme Court of Justice. (2013). Confederación Indígena del Neuquén c/
   Provincia del Neuquén, Acción de inconstitucionalidad. 10 December 2013.
   Retrieved from http://www.infojus.gob.ar/jurisprudencia.

Argento, M. & Zícari, J. (2017). Las dísputas por el litio en la Argentina: materia prima, recurso estratégico o bien comun? *Prácticas de Oficios*, 19, 7–49.

Biblioteca del Congreso de la Nación. (2018). Dossier Legislativo. Pueblos Originarios. Legislación e Informes internacionales. Legislación nacional y provincial. *Doctrina y Jurisprudencia*, 5 (155). Retrieved from http://bcn.gob.ar/uploads/DOSSIERlegisla tivo155Pueblosoriginarioslegeinformesint-legnacyprovdoc-juris.pdf.

Carrasco, M. (2000). *Los derechos de los pueblos indígenas en Argentina*. Buenos Aires: IWGIA-Vinciguerra.

CDESC. (2011). Examen de los informes presentados por los Estados partes en virtud de los artículos 16 y 17 del Pacto. Observaciones finales del Comité de Derechos Económicos, Sociales y Culturales. UN Doc. E/C.12/ARG/CO/3, 14 December 2011.

CERD. (2017). Observaciones finales sobre los Informes periódicos 21° a 23° combinados de la Argentina. UN Doc. CERD/C/ARG/CO/21–23, 11 January 2017.

Congreso. (2006). Ley 26.160. Declárase la emergencia en materia de posesión y propiedad de las tierras que tradicionalmente ocupan las comunidades indígenas originarias del país, cuya personería jurídica haya sido inscripta en el Registro Nacional de Comunidades Indígenas u organismo provincial competente o aquéllas preexistentes. Enacted on 1 November and entered into force on 23 November 2006. Retrieved from http://servicios.infoleg.gob.ar/infolegInternet/anexos/120000-124999/ 122499/norma.htm.

Corte Suprema de Justicia de la Nación. (2011). Comunidad Aborigen de Santuario Tres Pozos y otros c/ Jujuy, Provincia de y otros s/ amparo, C.1196.XLVI, 27 December 2011.

Corte Suprema de Justicia de la Nación. (2012). Comunidad Aborigen de Santuario Tres Pozos y otros c/ Jujuy, Provincia de y otros s/ amparo, 18 December 2012.

Dinatale, M. (2018). Avanza la guerra del 'oro blanco' en el norte argentino y toman medidas de control. *Infobae*, 31 March. Retrieved from https://www.infobae.com/p olitica/2018/03/31/avanza-la-guerra-del-oro-blanco-en-el-norte-argentino-y-toman-m edidas-de-control/.

Economic and Social Council. (2006). Human Rights and Indigenous Issues Report of the Special Rapporteur on the Situation of Human Rights and Fundamental Freedoms of Indigenous People, Mr. Rodolfo Stavenhagen. UN Doc. E/CN.4/2006/78, 16 February 2006.

ENDEPA. (2013). Nueva advertencia sobre la inejecución de la Ley 26.160. La brecha entre las declaraciones y la realidad en materia de derechos territoriales indígenas. Retrieved from http://endepa.org.ar/contenido/segunda-advertencia-de-endepa-sobre-la -ley-26160.pdf.

Ferradás Abalo, E., Lobo, A., & Lucero, J. (2016). *Conflicto socioambiental en Salinas Grandes: neoextractivismo, resistencias y nociones de desarrollo en el nuevo escenario político regional*. Villa María: Universidad Nacional de Villa María.

Fornillo, B. (2015). *Geopolítica del Litio. Industria, Ciencia y Energía en Argentina*. Buenos Aires: Editorial El Colectivo-Clacso.

Fornillo, B. (2018). La energía del litio en Argentina y Boliva: comunidad, extractivismo y posdesarrollo. *Colombia Internacional*, 93, 179–201.

Giarracca, N. & Teubal, M. (2013). *Actividades extractivas en expansión: reprimarización de la economía argentina?* Buenos Aires: Antropofagía.

Göbel, B. (2013). La minería del litio en la Puna de Atacama: interdependencias transregionales y disputas locales. *Iberoamericana*, 49, 135–149.

Göbel, B. (2014). La minería del litio en Atacama: disputas sociales alrededor de un nuevo mineral estratégico. B. Göbel & A. Ulloa (Eds.). *Extractivismo minero en Colombia y América Latina* (pp. 167–193). Berlin and Bogotá: Ibero-Amerikanisches Institut and Universidad Nacional de Colombia.

Göbel, B. (2015). Extractivismo y desigualdades sociales. *Iberoamericana*, 58, 161–165.

Gobierno de Argentina. (2010). Resolution (Resolución) No. 328/2010. Créase el Registro Nacional de Organizaciones de Pueblos Indígenas, 10 November 2010. Retrieved from https://ar.vlex.com/vid/registro-organizaciones-pueblos-indigena s-226909365.

Gobierno de Argentina. (2016). Decree (Decreto) No. 672/2016. Consejo Consultivo y Participativo de los Pueblos Indígenas de la República Argentina. Creación, 12 May 2016. Retrieved from http://servicios.infoleg.gob.ar/infolegInternet/anexos/ 260000-264999/261285/norma.htm.

Gomiz M. (2015). El derecho constitucional de propriedad comunitaria indígena en la jurisprudencia argentina. In F. Kosovsky & S. Ivanoff (Eds.). *Dossier propiedad comunitaria indígena* (pp. 119–138). Comodoro Rivadavia: EDUPA.

Gomiz, M. & Salgado J. (2010). *Convenio 169 de la O.I.T. sobre Pueblos Indígenas. Su aplicación en el derecho interno argentino.* Neuquén: ODHPI-IWGIA.

Gudynas, E. (2015). *Extractivismos. Ecología, economía y política de un modo de entender el desarrollo y la Naturaleza.* Cochabamba: CEDIB.

Gutman, N. (2007). La Conquista del Lejano Oeste. *Le Monde Diplomatique*, 95, 12–16.

INDEC. (2010). Censo Nacional de Población y Vivienda 2010. Retrieved from http s://www.indec.gob.ar/indec/web/Nivel4-CensoNacional-3-9-Censo-2010.

Instituto Nacional de Asuntos Indígenas. (n.d.). Normativa sobre Pueblos Indígenas y sus comunidades. Retrieved from https://www.argentina.gob.ar/derechoshumanos/ inai/normativa.

Nacif, F. (2017). El saqueo del litio en el NOA. *Página 12*, 29 December. Retrieved from https://www.pagina12.com.ar/85722-el-saqueo-del-litio-en-el-noa.

NPKA. (2015). Kachi Yupi-Huellas de Sal. Procedimiento de consulta y consentimiento previo, libre e informado para las comunidades indígenas de la Cuenca de Salinas Grandes y laguna de Guayatayoc. Retrieved from http://farn.org.ar/wp -content/plugins/download-attachments/includes/download.php?id=25626.

Parlamento de Naciones Originarias et al. (2016). Evaluación sobre el cumplimiento del Pacto Internacional de Derechos civiles y políticos en Argentina en el marco de la presentación del Quinto Informe periódico ante el Comité de Derechos Humanos 117° Periodo de sesiones, Derechos de los pueblos indígenas en Argentina. Retrieved from http://tbinternet.ohchr.org/Treaties/CCPR/Shared%20Documents/ ARG/INT_CCPR_CSS_ARG_24354_S.pdf.

Puente, F. & Argento, M. (2015). Conflictos territoriales y construcción identitaria en los salares del noroeste argentino. In B. Fornillo (Ed.). *Geopolítica del Litio. Industria, Ciencia y Energía en Argentina* (pp. 123–166). Buenos Aires: Editorial El Colectivo-Clacso.

Ramírez, S. (2018). Argentina. In P. J. Andersen, *et al.*, (Ed.). *El Mundo Indígena 2018* (pp. 199–206). Copenhagen: IWGIA.

Rosti, M. (2010). La terra contesa fra diritto e cultura: Compañía de Tierras del Sur Argentino versus Curiñanco-Rúa Nauhelquir. *THULE Rivista italiana di studi americanistici*, 26, 477–499.

Rosti, M. (2016). El 'modelo extractivista' y los derechos de los pueblos indígenas a los recursos naturales y al territorio en la Argentina de hoy. *DPCE online*, 4, 49–73.

Rodríguez Duch, D. (2015). Apuntes sobre Propiedad comunitaria indígena. In F. Kosovsky & S. Ivanoff (Eds.). *Dossier propiedad comunitaria indígena* (pp. 38–57). Comodoro Rivadavia: EDUPA.

Schiaffini, H. (2013). Litio, llamas y sal en la Puna argentina Pueblos originarios y expropiación en torno al control territorial de Salinas Grandes. *Entramados y perspectivas*, 3, 121–136.

Solá, R. (2016). Kachi Yupi: un ejercicio de autodeterminación indígena en Salinas Grandes. *FARN Informe ambiental anual*. Retrieved from https://farn.org.ar/wp -content/uploads/2016/07/15Solá.pdf.

Svampa, M. (2016). *Debates latinoamericanos. Indianismo, desarrollo, dependencia y populismo*. Buenos Aires: Edhasa.

Svampa, M. & Viale, E. (2014). *Maldesarrollo. La Argentina del extractivismo y el despojo*. Buenos Aires: Katz Editores.

Valko, M. (2007). *Los indios invisibles del Malón de la Paz*. Buenos Aires: Asociación Madres de Plaza de Mayo.

15 Lack of consultation and free, prior and informed consent, and threats to Indigenous Peoples' rights in Brazil

Julia Mello Neiva

Indigenous Peoples and threats to their rights in Brazil

After many years of significant advances in terms of the recognition of human rights in the national and international protection systems, there is a clear global trend of weakening democracies and hence of these rights. The rise of conservative governments is not without consequences. Particularly in Brazil, given the complicity between the government and the private sector, it has led to the weakening of human rights legislation, contributing to an increase in violent land conflicts and forced labour in supply chains, and even the criminalisation of human rights defenders, among other consequences. Unfortunately, the country is considered one of the most dangerous countries in the world for those defending land, environmental rights, and human rights (Business & Human Rights Resource Centre, 2018).

The current political, economic, and social crisis experienced by the country is greatly contributing to the intensification of the exclusion of the most vulnerable groups[1], including Indigenous Peoples. There are currently 305 Indigenous groups, and according to the latest census of 2010 they number 896,000 individuals, 63.8% of whom live in rural areas, and 57.5% on officially recognised Indigenous lands. They speak 274 languages, and about 17.5% of them do not speak Portuguese (IBGE, 2010; FUNAI, n.d.). Indigenous citizens constitute only 0.43% of the Brazilian population, as they were decimated by the colonial powers for over three centuries.

In this sense, the dispute over land is not new. Indeed, since colonial times, the lands of Indigenous Peoples and rural communities have been disputed by governments and companies or other private entities.[2] Over the past 20 years, large infrastructure projects with investments from development banks and private funds, especially in the Amazon region, have further intensified this conflict. As in many other Latin American countries, the impact of large infrastructure projects has driven the extractive industry to supply civil construction needs and meet demand for consumer goods, which are constructed with materials from the extractive sector. The financing of this supply chain is often mixed (public–private), and in this context private interests tend to be privileged to the detriment of the rights of the affected populations, who are

DOI: 10.4324/9781351042109-20

not consulted. Now, the so-called development model is still based on commodities and forced labour conditions, mostly preventing these groups from accessing the wealth produced and participating in the decision-making processes that will directly affect their lives (see Rodríguez-Garavito, 2011).

Entire communities are directly affected by the operations of corporations, often in complicity with the Brazilian State, resulting in combined human rights violations. These violations occur for many reasons, such as gaps in legislation, often conflicting with environmental and human rights norms, the lack of means to hold companies accountable for the human rights violations in which they are involved, the lack of access to justice for those affected, and the chronic inequality and racism in Brazilian society. Often such violations and abuses occur because they involve the intervention of private actors in matters of public interest.

Indigenous and human rights groups such as the *Conselho Indigenista Missionário* – CIMI (2016), as well as the UN Special Rapporteurs on the Rights of Indigenous Peoples, Victoria Tauli-Corpuz (2016), on human rights defenders, Michel Forst, and on the environment, John Knox, and the Inter-American Rapporteur on the Rights of Indigenous Peoples Francisco José Eguiguren Praeli, have spoken out about the high number of Indigenous activists attacked and killed in Brazil. Tauli-Corpuz (2016) expressly mentioned that – over the past 15 years – Brazil has seen an increase in the killing of human rights defenders, and highlighted that Indigenous citizens are especially at risk. She added that these are probably the most dangerous times for Indigenous Peoples since the promulgation of the 1988 Federal Constitution.

The right to consultation and the Free, Prior and Informed Consent of Indigenous Peoples in Brazil

Legislative and administrative institutional framework

The rights of the Brazilian Indigenous Peoples to possess their ancestral lands where they live have been assured in all the country's Constitutions, except for the Charter of 1891 at the beginning of the Republican period. However, it is in the Federal Constitution of 1988 that Indigenous Peoples are recognised as entitled to rights in close cultural connection to their traditional lands, which are necessary for their survival, contrary to the understanding present in previous legal frameworks, which supported their assimilation by Brazilian society.

Article 231 recognises their social organisation, customs, languages, beliefs, and traditions, and their original rights over the lands they traditionally occupy, imposing upon the Union the duty to

> demarcate, protect and enforce all their property. Indigenous lands are those inhabited by them, on a permanent basis, those used for their productive activities, those essential to the preservation of the environmental

resources necessary for their well-being and those necessary for their physical and cultural reproduction, according to their uses, customs and traditions.

(Brazilian Constitution, 1988, article 231, para. 1)

Other parts of this section contain an embryonic form of the right to consultation and participation of Indigenous Peoples in decisions concerning their rights:

Hydric resources, including those with energetic potential, may only be exploited, and mineral wealth in Indian land may only be prospected and mined with the authorisation of the National Congress, *after hearing the communities involved, and the participation in the results of such mining shall be ensured to them, as set forth by law* [emphasis added].

(Brazilian Constitution, 1988, article 231, para. 3.A)

And,

The Indians, their communities and organisations *are legitimate parties to enter in court in defence of their rights and interests*, intervening the Public Ministry in all acts of the process [emphasis added].

(Brazilian Constitution, 1988, article 232)

In an approach that embeds international human rights law, the Constitution of 1988 also recognises that human rights treaties and conventions approved by Congress "shall be equivalent to constitutional amendments" (Brazilian Constitution, 1988, article 5).

At institutional level, the Federal Public Prosecutor's Office is the human rights watchdog institution that works to protect Indigenous rights, especially prosecutors working in the North Region of Brazil, in States such as Pará. For its part, the National Indian Foundation (FUNAI, for its Portuguese acronym), related to the Ministry of Justice, is the national body responsible for coordinating and implementing federal government policies for Indigenous Peoples, as well as protecting and promoting their rights throughout the country.

In addition to national legal instruments, Indigenous Peoples in Brazil are also protected by international treaties. In 1989, Brazil signed the ILO Convention No.169 Concerning Indigenous and Tribal Peoples in Independent Countries (ILO 169). The text was ratified and incorporated into Brazilian legislation through Legislative Decree No. 143 of 20 June 2002, and entered into force one year later, on 20 June 2003. In addition, the country recognises the mandatory nature of the contentious jurisdiction of the Inter-American Court of Human Rights (IACtHR), and, on 10 December 1998, Brazil deposited a document with the Secretary General of the Organization of American States (OAS), pledging to implement the decisions of the body, in recognition of international responsibilities for human rights violations. Furthermore, in 2007, Brazil became a signatory to the United Nations Declaration on the Rights of Indigenous Peoples (UNDRIP) and to the American Declaration on the Rights of Indigenous Peoples, adopted by the OAS in 2016. These documents refer to the State's duty to consult Indigenous Peoples before adopting and implementing administrative or legislative measures that will affect their lives in order to obtain their FPIC, when applicable.

However, even though the country is signatory to these legal instruments, it does not mean that the right to consultation of Indigenous Peoples and their FPIC are fully and effectively implemented in the country. Indeed, legal discussions and the implementation of this right are far behind in Brazil compared to other Latin American countries (Rodríguez-Garavito, 2011; 2018), even considering minimum criteria and standards for its implementation defined by both the IACtHR and the Human Rights Commission.[3]

In view of the aforementioned norms, no specific legislation is required for the implementation of consultation and FPIC in the country. Not only should these rights and standards be immediately effective, but they should also be applied before any measure, decision, project, or government programme affecting Indigenous Peoples, directly or indirectly, is approved. However, referring to an alleged lack of regulation, the Brazilian government has never held a meaningful consultation regarding development and construction projects in Indigenous lands in the country (see below).

This understanding has been challenged by many civil society groups (Rede de Cooperação Amazônica, 2018; Garzón, Yamada & Oliveira, 2016). In 2010, the Labour Union Federation (*Central Única dos Trabalhadores –* CUT), and both Indigenous and *Quilombola* Afro-Descendant organisations denounced the Brazilian State at the ILO mechanism for not fulfilling its international obligation to consult and obtain their FPIC in this type of projects. As a response, the government created in January 2012 an Inter-Ministerial Working Group to produce an administrative proposal on the right to consult. A few months later, the Attorney General's Office (AGU, for its Portuguese acronym) issued Directive 303 (Advocacia Geral da União, 2012), establishing institutional safeguards for Indigenous lands, in accordance with the conditions set out by the Supreme Court's decision on the Case of Raposa Serra do Sol in 2009 (see more below). It allowed activities considered strategic to national defence to be implemented on Indigenous lands without consultation or consent, and it also portrayed a narrow concept of the right to territory. FUNAI (2016) brought a lawsuit against that Directive, as it understood it threatened Indigenous rights, and alleged the AGU could not have used the Supreme Court decision as a reference, since it was not yet a final decision. After this episode, Indigenous organisations such as Brazil's Indigenous People Articulation (*Articulação dos Povos Indígenas do Brasil*) and others decided to withdraw from the government's discussions on regulation, arguing that the government was not acting in good faith. Directive

303 was suspended in 2013, following protests by Indigenous People, human rights groups, and committees opposing that guideline.

In reaction, Indigenous Peoples have started to create their own protocols for consultation and consent.

Judicial decisions

Garzón, Yamada & Oliveira (2016) wrote a thorough study on consultation and FPIC in Brazil that offers both a legal and political perspective on the topic of how to respect such rights. The authors also analysed some Brazilian judicial decisions regarding the implementation of the right to be consulted and to consent, addressing challenges to and opportunities for its fulfilment.

Garzón, Yamada & Oliveira (2016) refer therefore to several decisions in which the Federal Justice ordered the government to (in chronological order): 1) suspend the Teles Pires Dam licensing and consult the Munduruku, Kayabi and Apiaká Indigenous Peoples (Justiça Federal, 2012); 2) consult the Munduruku, Kayabi and Apiaká Indigenous Peoples regarding the construction of the Sao Manoel dam (Justiça Federal, 2014); 3) consult the Munduruku and Sataré-Mawé Indigenous Peoples as well as the Montanha and Mangabal traditional communities regarding the construction of the São Luiz do Tapajós Dam and suspend its licensing process (Justiça Federal, 2015a); and, 4) consult the Awá Guajá Indigenous Peoples before proceeding with the environmental licensing of the duplication of the Carajá railway (Justiça Federal, 2015b).

In almost all judicial decisions regarding consultation and FPIC in Brazil there is a common understanding that they are of immediate applicability. However, as Garzón, Yamada & Oliveira (2016) claim, there have been several challenges and contradictions in the country that prevent the implementation of the right, many of which unfortunately occur within the judiciary itself.

Raposa Serra do Sol case

One of the main threats in this sense is a controversial decision regarding the Raposa Serra do Sol case (Yamada & Villares, 2010). This is a complex case that started in the 1970s with violent land conflicts involving Indigenous Peoples. It is considered an emblematic case decided by the Federal Supreme Court setting standards and clarifications on the demarcation of Indigenous lands in the State of Roraima. The decision recognised that the land actually belonged to the Indigenous Ingaricó, Macuxi, Patamona, Taurepangue, and Uapixana Peoples, confirming the demarcation and homologation of the land that had been challenged by the government of the State of Roraima, jointly by illegal invaders and farmers. The State government requested that the area of land should be smaller, but also challenged the decree that recognised the Indigenous land as ancestral and belonging to the Indigenous Peoples.

The Supreme Court decision in 2009 about this case clarified the provisions of the Federal Constitution regarding the protection of Indigenous lands. In this sense, it was acclaimed by Indigenous Peoples, human rights organisations, academics, and others who hoped that this was an important step in truly protecting Indigenous rights, accepting their own cultures and traditions, and, most importantly, their relationship with their land. This understanding echoed international standards of Indigenous protection such as the provisions enshrined by ILO 169 and the UNDRIP. Nevertheless, the decision also included nineteen controversial conditions regarding its implementation (Yamada & Villares, 2010). These included restricting Indigenous People's use of their own lands, submitting them to public interest, and limiting Indigenous participation and consultation in decisions on the use of these lands.

Another condition that resulted in heated debates and increased the ongoing threats of violence against Indigenous Peoples was the questionable concept of a time limit regarding their land occupation. While restricting the use of lands with the argument of 'public interest', this condition posed a real threat to the implementation of the right to consultation and the FPIC of Indigenous Peoples. They were understood as non-absolute rights, which means that they could be challenged in specific situations when there were other competing constitutional rights (Garzón, Yamada & Oliveira, 2016). This condition also specifically mentioned that military operations would not need Indigenous consultations. This is clearly a controversial and misleading interpretation of article 231 of the Constitution, and disregards Brazilian commitments to international law.

Before arriving at the Supreme Court, the presidents of the higher courts responsible for deciding on such cases have made use of Law No. 8437 of 1992, which suspends an injunction that guaranteed the right to FPIC, alleging one of the possibilities present in article 4, that is:

> to suspend, in a reasoned order, the execution of the injunction in actions brought against the Public Power or its agents, at the request of the Public Prosecution Service or of the legal entity of public interest or flagrant illegality, and to avoid serious damage to public order, health, safety and public.
>
> (Presidência da República, 1992, article 4)

This Law (Presidência da República, 1992) has been used to guarantee large infrastructure and construction projects without consultation or the possibility of being vetoed by Indigenous Peoples or other affected communities, on behalf of so-called development. Again, Garzón, Yamada & Oliveira (2016) claim that:

> Interpretations that restrict the scope of the consultation or establish an exception to the instances of incidence violate Convention 169 / ILO and Article 21 of the American Convention on Human Rights. The 'urgency'

or 'public interest' that supposedly underlies a measure does not authorise the government to stop consulting affected groups, even though these
exceptions are not international standards.

(p. 25)

For his part, Federal Prosecutor Luís de Camões Lima Boaventura said:

Figures collected by the MPF [Federal Prosecutor's Office] show that, just
with respect to the hydroelectric dams in the Teles Pires-Tapajós Basin,
we were victorious in 80 percent of the actions we took, but all of the
rulings in our favor were reversed by suspensions.

(Branford & Torres, 2017)

The UN Special Rapporteur on Indigenous Rights, Victoria Tauli-Corpuz
(2016b), shares this understanding regarding threats within the judiciary with
regard to this case:

All courts have a clear and uniform interpretation of the limitations of
the Raposa-Serra do Sol ruling and its inapplicability to the issuance of
eviction orders for indigenous peoples or the halting of demarcation
procedures. The Federal Supreme Court should continue to accept
requests for the suspension of eviction orders and ensure that future rulings concerning Indigenous Peoples' rights are fully consistent with
national and international human rights standards.

(2016b, para. 97, letter d)

As to more recent developments, in July 2017, the Attorney General's Legal
Opinion (*Parecer Normativo*) No. 001/2017 (Advocacia-Geral da União,
2017) stated that all the bodies of the Federal Administration should follow
the conditions established by the Supreme Court decision in the case of
Raposa da Serra do Sol.

Meanwhile, the Supreme Court (Supremo Tribunal Federal, 2017) decided
on several cases in August 2017, recognising the land rights of Indigenous
groups in Mato Grosso. In its rulings, the Court mentioned that Indigenous
rights have been recognised in all Brazilian Constitutions.

In March 2018, the Federal Prosecutor's Office issued a Legal Opinion
(*Nota Técnica*) No. 02/2018–6CCR (Ministério Público Federal, 2018)
against the above-mentioned Attorney General's Legal Opinion No. 001/
2017, claiming it was unconstitutional and demanding that it be revoked. One
of the arguments was that the Supreme Court had already recognised the
protection of Indigenous rights and that consultation and FPIC had been
violated also in this Legal Opinion, as it did not include consultations with
Indigenous Peoples.

In April 2018, Indigenous People's organisations wrote a protest letter
(Branford, 2018) against the increasing attacks on Indigenous Peoples and

their rights, and had a meeting with justices from the Supreme Court. About 3,500 Indigenous representatives, from over 305 Indigenous Peoples from all over Brazil, demanded, among other issues – besides land demarcation – that consultation and FPIC be applied immediately (Articulação dos Povos Indígenas do Brasil & Mobilização Nacional Indígena, 2018). They also demanded that the Attorney General's Legal Opinion (*Parecer Normativo*) No. 001/ 2017 should be revoked.

Belo Monte case

The second UN Special Rapporteur on the Rights of Indigenous Peoples, James Anaya, in his 2009 report pointed out that Indigenous Peoples in Brazil were not participating in decision-making processes on issues that would affect them (Anaya, 2009). Likewise, the current UN Special Rapporteur Victoria Tauli-Corpuz (2016b) has flagged the lack of effective consultation mechanisms and of implementation of FPIC on big infrastructure projects such as the Belo Monte dam construction. As pointed out by Garzón, Yamada & Oliveira (2016), although the Federal Court suspended the construction of Belo Monte due to a lack of consultation in 2005, and the Inter-American Commission of Human Rights also requested the immediate suspension of the construction of the dam for the same reason in 2011, the project went ahead, and as of 2018 the Indigenous conditions to accept the dam have not yet been met.

Belo Monte is, unfortunately, an illustrative case regarding ongoing human rights violations, as it created difficulties for local communities to access water, due to the Xingu river course being diverted and part of it becoming an artificial lake for the dam. Along with this difficulty, Indigenous Peoples have to cope with floods and droughts, deforestation, division of the communities, co-optation of leaders, and health problems, amongst others. As pointed out by the Brazilian Special Rapporteur Erika Yamada (2017) in her latest report, written jointly with Indigenous groups, this was a very serious episode of the violation of the right to consultation, and should be considered the most emblematic case.

Finally, in 2018, the IACtHR decided upon this case and held Brazil accountable for human rights' violations of the Xukuru Indigenous Peoples for its failure to demarcate their traditional land, offer adequate legal protection, and conduct the process at a reasonable pace (Valente, 2018). This was considered a historical decision given that it was the first time that Brazil had been found guilty of violating Indigenous rights in an international court (Inter-American Court of Human Rights, 2018).

The non-implementation of consultation and FPIC in Brazil

Overall, the consultation and FPIC of Indigenous peoples have not been implemented in Brazil, and when a consultation process occurs, it usually happens through public hearings as required by environmental laws (usually for licensing) but not respecting international standards that oblige the State to consult Indigenous Peoples and allow them to decide on issues that will affect them. The dispute of narratives, with very different perceptions of what consultation and FPIC really mean, and the actual asymmetry of power between, on one hand, Indigenous Peoples, and on the other, the State and companies, has unfortunately affected Indigenous Peoples' enjoyment of their essential rights and standards.

According to the Federal Prosecutor's Office (Ministério Público Federal, 2017), the government has never held any meaningful, respectful consultation in respect of any of its development and construction projects. However, considering plans for the construction of about 40 dams over the next 20 years, especially in the Amazon region, implementation of consultation and FPIC would be more than necessary to ensure the survival of Brazilian Indigenous Peoples, as federal prosecutor Felício Pontes (2016) has stated. The recent developments in governmental policies toward this are a reflection of this attitude.

In January 2018, several federal governmental human rights institutions of the northern State of Pará (where most of the Brazilian Amazon Forest is located and most of the Indigenous Peoples live) issued a joint common institutional recommendation (Ministério Público Federal, Defensoria Pública do Estado do Pará, Ministério Público do Estado do Pará, & Defensoria Pública Da União, 2018) to the Governor and the Chief Federal Prosecutor of the State of Pará calling for the revocation of Decree No. 1969 of 24 January 2018, which had created a working group on FPIC in order to establish a State action plan to address consultation and consent processes.

These federal governmental human rights institutions alleged that this decree is a violation of consultation and FPIC not only because it does not guarantee effective participation of the affected groups and communities, but also because such participation is not culturally adequate. In addition, it does not allow enough time for the Indigenous and other communities to understand, effectively participate, and make decisions in the discussions. Undoubtedly, this decree is also a matter of concern as it does not address the participation of the State authorities that are, by law, responsible for the protection of Indigenous Peoples, as well as others (*Quilombola* Afro-Descendants and other rural communities).

Indigenous and *Quilombola* organisations also opposed this decree and highlighted that it is a contradictory document because it does not allow communities to participate, nor respect their initiatives to develop their own consultation protocols. Finally, the aforementioned authorities think the decree is "a maneuver of the Pará government to validate large projects such as mining, hydroelectric dams, and highways, among others, but representatives of the state government deny it" (Sarraf, 2018).

Despite these recommendations, although the above-mentioned Decree No. 1969 was revoked, a new one, No. 2061, was promulgated with a very similar text on 2 May 2018 (Governo do Estado do Pará, 2018). Indigenous Peoples were not consulted on any of the texts. This is the sort of case the UN Special Rapporteur on Indigenous Rights, Victoria Tauli-Corpuz, highlights in her end of mission statement (Tauli-Corpuz, 2016a) and in her report (Tauli-Corpuz, 2016b) presented to the Human Rights Council. Tauli-Corpuz clearly mentions the lack of meaningful prior consultation regarding the enacting of national legislation that directly impacts Indigenous Peoples.

Indeed, what the government usually considers a consultation is in fact only a public hearing, or a requirement for an environmental licence. Many human rights and environmental groups as well as Indigenous Peoples have claimed that these have not been real instances for fair dialogue and that they are just one item in a 'check list'. In addition, public hearings are opportunities for sharing and gathering information about plans and projects but not for decision-making, and that does not meet the international standards of consultation and consent (Instituto Socioambiental, n.d.). Duprat (2016) has also highlighted that consultation and FPIC really have different meanings for different actors, but, in any case, they are more than merely events to listen to those who have always been excluded.

The way the private sector has been addressing consultation and FPIC in Brazil is another prevailing threat to its realisation. Too often, when companies talk about FPIC they are not really using the same framework with respect to Indigenous Peoples or human rights, and not even considering the duty to obtain Indigenous consent (Oxfam Brasil, 2018). FPIC is used as a tool to move ahead with oil, mining, or construction projects that the company wants to pursue, serving as a supposedly legal and legitimate process to grant them the legal status to operate in or affect Indigenous lands, disregarding their impacts, and entailing consultation but not consent. Oliveira (2014), through his excellent analysis, has found that usually States and private companies:

> limit consultation to the discussion on mitigation and compensation measures. They start from the discourse that it would be possible to reconcile antagonistic perceptions regarding the use of territory and natural resources, and elects 'scientific' knowledge as the only one capable of pointing out the impacts of the project and its respective 'technical' solutions. The predominance of one or other vision will depend on the correlation of force between the social actors involved. Inequality makes it likely that the predominant view will be that of States and companies.
>
> (Oliveira, 2014)

Oliveira also points out that "[e]thnic groups see FPIC as a space for the exercise of autonomy, in which cultural diversity and traditional knowledge are respected. They argue that it is the peoples' final decision on the use of their territories and natural resources" (Oliveira, 2014).

Indigenous protocols for consultations

As a reaction to the non-implementation of consultation and FPIC, and the threats made against them, Indigenous Peoples have been designing their own protocols for consultations and consent, and this can be seen as a good and positive impact that brings some hope. These processes of writing and creating their own protocols, as well as the need to protect their lands and culture by themselves, have in the end empowered them, considering FUNAI, slowness, ineffectiveness and lack of funding.

The case of the construction of the São Luiz do Tapajós dam is well addressed in the work of Oliveira (2016). He mentions that, fortunately, the Munduruku managed to stop the government from considering the consultation process it had carried out to be valid. They claimed that this process was arbitrary, and it did not respect international standards, adopt a culturally-sensitive perspective, or guarantee effective participation. Instead, the Munduruku managed to elaborate their own consultation and consent protocol, setting their cultural standards, explaining how they are organised, how their decisions are made, and how they should be contacted.

Other Indigenous Peoples in Brazil have also written their protocols (Ministério Público Federal, n.d.). For instance, Indigenous Peoples who live (or used to live) by the Xingu river wrote a joint protocol, and one of them, the Juruna, wrote their own separately. The Krenak people in Resplendor, Minas Gerais, built their own instrument collectively between May and August 2017, accompanied by the Federal Prosecutor's Office in Minas Gerais (Ministério Público Federal Krenak, 2017). They live in seven different communities, namely Krenak, Naknenuk, Nakrehé, Takruk, Watu, Atoran and Borum Erehé. Like the Munduruku, the Krenak defined how and when the consultations should take place, who were to be the representatives of the different communities, and other matters. The first contact for State or private entities, for instance, has to be made with the chiefs and representatives of their associations through FUNAI. This means that FUNAI shall contact all of them when there is an issue that will affect their lives, land, and culture. Then, Krenak chiefs and representatives shall decide whether or not all the Krenak should take the question into account or if each village can decide separately. None of these chiefs can speak on behalf of all the Krenak. In fact, they decide everything collectively until they reach a consensus, and if this is not reached the decision is made by the majority.

Conclusions

This chapter has brought to light several examples of what happens when international consultation and FPIC standards remain unfilled, as in the case of Brazil. This chapter has based its analysis on the national legal framework and a broad spectrum of documents from national and international multilateral organisations. As a result, it is clear that these violations are usually perpetrated by corporations, or other private entities, within the framework of joint public-private infrastructure projects. Frequently, the Brazilian government has unfortunately taken the side of private companies to the detriment of the protection of Indigenous Peoples and other groups. This also raises the question of how the law has been used to benefit the most powerful in an asymmetrical relationship.

Among the cases analysed, Raposa Serra do Sol and Belo Monte draw particular attention due to the disrespect for international and domestic human rights norms. Despite the judicial decisions that pose threats to their human rights, Indigenous Peoples have mobilised. Similar to the experience of Indigenous Peoples in other Latin American countries, they have managed to produce their own protocols of consultation and FPIC as a resistance tool, supported not only by institutions like FUNAI and the Federal Prosecutor's Office but also other Indigenous, human rights and environmental organisations.

Undoubtedly, the Brazilian government has failed to protect Indigenous Peoples (as well as other ethnic communities, such as the Afro-Descendants), despite the mandatory legal framework that requires the government to do so. Not only have these communities not been consulted, they have not been able to give their consent either, thus providing a clear understanding that consultation and FPIC are far from being adequately implemented in the country. In fact, the current context of intolerance, harassment, and increasing violence in Brazil shows there is a real risk that the current institutional human rights protection framework, constructed as a result of Indigenous, and other groups' struggles, will be destroyed. This means that the very existence of Indigenous Peoples is at stake.

Notes

1 Victoria Tauli-Corpuz (2016b, para. 92), the UN Special Rapporteur on Indigenous rights, also acknowledges the impacts of this crisis on Indigenous communities.
2 Professor and anthropologist Manuela Carneiro Da Cunha in 1981 stated that Indigenous Peoples were no longer used as a workforce, they were substitutes "for cattle, building dams or exploring minerals" (Carneiro Da Cunha, 2012, p. 114). In 2017, the same author stated that "land conflicts are endemic" referring to the rights of Indigenous Peoples being opposed by the mining industry and other actors interested in infrastructure construction, during the 1988 Constituent Assembly (Carneiro Da Cunha et al., 2017, p. 404).
3 See further on this issue in the chapter by Cantú Rivera in this volume.

References

Advocacia Geral da União. (2012). Portaria No. 303, 16 July 2012. Retrieved from http://www.agu.gov.br/atos/detalhe/596939.
Advocacia Geral da União. (2017). Parecer Normativo No. 001/2017/GAB/CGU/AGU, 19 July 2017. Retrieved from http://www.agu.gov.br/atos/detalhe/1552758.
Anaya, J. (2009). Report on the situation of human rights of indigenous peoples in Brazil. UN Doc. A/HRC/12/34/Add.2, 26 August 2009.

Articulação dos Povos Indígenas do Brasil & Mobilização Nacional Indígena. (2018). Documento final do acampamento terra livre 2018. Retrieved from https://mobiliza caonacionalindigena.wordpress.com/2018/04/26/documento-final-do-acampamento-terra-livre-2018-o-nosso-clamor-contra-o-genocidio-dos-nossos-povos/.

Branford, S. (2018, April 30). 3,000 indigenous people gather in Brasilia to protest ruralist agenda. Retrieved from https://news.mongabay.com/2018/04/3000-indigen ous-people-gather-in-brasilia-to-protest-ruralist-agenda/.

Branford, S. & Torres, M. (2017, January 5). The end of a People: Amazon dam destroys sacred Munduruku "Heaven". Retrieved from https://news.mongabay.com/2017/01/the-end-of-a-people-amazon-dam-destroys-sacred-munduruku-heaven/.

Brazilian Constitution. (1988). Constituição da República Federativa do Brasil de 1988. Retrieved from http://www.planalto.gov.br/ccivil_03/constituicao/constituicao.

Business & Human Rights Resource Centre. (2018). Centro afirma que cresceram ataques a defensores de direitos humanos; Brasil, México e Colômbia são mais perigosos. Retrieved from https://www.business-humanrights.org/pt/centro-afirma -que-cresceram-ataques-a-defensores-de-direitos-humanos-brasil-m%C3%A9xico-e-col%C3%B4mbia-s%C3%A3o-mais-perigosos.

Carneiro Da Cunha, M. (2012). *Índios no Brasil: história, direitos e cidadania.* Sao Paulo: Claro Enigma.

Carneiro Da Cunha, M., Caixeta, R., Campbell, J.M., Fausto, C., Kelly, J.A., Lomnitz, C., … Vilaça, A. (2017). Indigenous peoples boxed in by Brazil's political crisis. *HAU: Journal of Ethnographic Theory*, 7(2), 403–426. Retrieved from https://www.journals.uchicago.edu/doi/pdfplus/10.14318/hau7.2.033.

Duprat, D. (2016). A convenção 169 da OIT e o Direito à consulta prévia livre e informada. Retrieved from http://reporterbrasil.org.br/2016/08/a-convencao-169-da -oit-e-o-direito-a-consulta-previa-livre-e-informada/.

FUNAI. (2016). Nota técnica da Funai sobre a Portaria n° 303/12 da AGU. Retrieved from http://www.funai.gov.br/index.php/comunicacao/notas/2336-nota-tecnica-da -funai-sobre-a-portaria-n-303-12-da-agu.

FUNAI. (n.d.). *Índios no Brasil – Quem são.* Retrieved from http://www.funai.gov.br/index.php/indios-no-brasil/quem-sao.

Garzón, B.R., Yamada, E.M., & Oliveira, R. (2016). Direito à consulta e consentimento de povos indígenas, quilombolas e comunidades tradicionais. São Paulo: Rede de Cooperação Amazônica – RCA. Retrieved from http://www.dplf.org/sites/default/files/direito_a_consultaprevia_no_brasil_dplf-rca-3.pdf.

Governo do Estado do Pará. (2018). Decree (Decreto) No. 2,061, 2 May 2018. Retrieved from https://www.sistemas.pa.gov.br/sisleis/legislacao/4079.

IBGE. (2018). Censo 2010: População indígena é de 896,9 mil, tem 305 etnias e fala 274 idiomas. Retrieved from https://censo2010.ibge.gov.br/noticias-censo?busca=1& id=3&idnoticia=2194&t=censo-2010-poblacao-indigena-896-9-mil-tem-305-etnia s-fala-274&view=noticia.

Instituto Socioambiental. (2008). Consulta livre, prévia e informada sobre medidas legislativas. Retrieved from https://www.socioambiental.org/pt-br/especial/consulta -livre-previa-e-informada-na-convencao-169-da-oit.

Instituto Socioambiental. (n.d.). O Dever de Consulta Prévia do Estado Brasileiro aos Povos Indígenas. Retrieved from https://pib.socioambiental.org/files/file/PIB_institu cional/Dever_da_Consulta_Previa_aos_Povos_Indigenas.pdf.

IACtHR. (2018). *Pueblo Indígena Xucuru y sus miembros v Brasil*, Judgement of February 5 2018 (Excepciones Preliminares). Inter-Am. Ct. H.R., (Ser. C) No. 346.

Justiça Federal. (2012). Justiça Federal Seção Judiciária do Estado de Mato Grosso 2a Vara. Precautionary Decision, 26 March 2012. Retrieved from https://processual.trf1.jus.br/consultaProcessual/processo.php.

Justiça Federal. (2014). Tribunal Regional Federal da Primeira Região. Decision of the Public Civil Action No. 14123–14148.2013.4.01.3600, 13 September 2014. Retrieved from http://www.prpa.mpf.mp.br/news/2014/arquivos/liminar.consulta.pre301via.pdf.

Justiça Federal. (2015a). Tribunal Regional Federal da Primeira Região. Decision of the Public Civil Action No. 3883–3898.2012.4.01.3902, 15 June 2015. Retrieved from http://www.prpa.mpf.mp.br/news/2015/arquivos/Sentenca%20uhe.TAPAJOS.pdf.

Justiça Federal. (2015b). Justiça Federal de Primeira Instância Maranhão, 8a vara. Preliminary Injunction, 26 June 2015. Retrieved from https://processual.trf1.jus.br/consultaProcessual/processo.php.

Ministério Público Federal. (2017). MPF defende obrigatoriedade de consultar povos afetados por empreendimentos hidrelétricos na Amazônia. Retrieved from http://www.mpf.mp.br/pa/sala-de-imprensa/noticias-pa/mpf-defende-obrigatoriedade-de-consultar-povos-afetados-por-empreendimentos-hidreletricos-na-amazonia.

Ministério Público Federal Krenak. (2017). Protocolo de Consulta Prévia do povo Krenak. Retrieved from http://www.mpf.mp.br/atuacao-tematica/ccr6/documento s-e-publicacoes/protocolo-de-consulta-dos-povos-indigenas/docs/ProtocoloConsulta KRENAK_.pdf.

Ministério Público Federal, Defensoria Pública do Estado do Pará, Ministério Público do Estado do Pará, & Defensoria Pública Da União. (2018). Recomendação Conjunta No. 007/2018. Retrieved from http://www.mpf.mp.br/pa/sala-de-imprensa/documentos/2018/recomendacao-decreto-consulta-previa.

Ministério Público Federal. (2018). Nota técnica No. 02/2018–6CCR. Retrieved from http://www.mpf.mp.br/atuacao-tematica/ccr6/dados-da-atuacao/atos-do-colegiado/nota-tecnica/2018/NT02_2018.pdf.

Ministério Público Federal. (n.d.). Protocolo de Consulta Prévia dos Povos Indígenas. Retrieved from http://www.mpf.mp.br/atuacao-tematica/ccr6/documentos-e-publica coes/protocolo-de-consulta-dos-povos-indigenas.

Oliveira, R. (2014). Consulta prévia: um instrumento em disputa Compreenda as disputas em torno da Consulta Prévia, um instrumento estratégico para a garantia de direitos de indígenas e de outros povos tradicionais, 16 December. Retrieved from https://fase.org.br/pt/informe-se/artigos/consulta-previa-um-instrumento-em-disputa/.

Oliveira, R. (2016). A ambição dos Pariwat: Consulta prévia e conflito socioambiental (Master's dissertation). Federal University of Pará, Pará.

Oxfam Brasil. (2018). NÃO É NÃO O estado do Consentimento Livre, Prévio e Informado nas políticas corporativas das multinacionais brasileiras. Retrieved from https://www.oxfam.org.br/sites/default/files/publicacoes/informe_nao_e_nao.pdf.

Pontes, F. (2016). Seminário: hidrelétricas na Amazônia, conflitos socioambientais e caminhos alternativos requerimento No. 128/2016 - do deputado Nilto Tatto (PT/SP). Retrieved from https://www2.camara.leg.br/atividade-legislativa/comissoes/comissoes-p ermanentes/cmads/seminarios-e-outros-eventos/eventos-2016/06-12-2016-hidreletricas-na -amazonia-conflitos-socioambientais-e-caminhos-alternativos/notas-taquigraficas-manh a/view.

Presidência da República. (1992). Law (Lei) No 8,437, 30 June 1992. Retrieved from http://www.planalto.gov.br/ccivil_03/Leis/L8437.htm.

RCA. (2018). Contributions of the RCA on the right to Free, Prior and Informed Consultation and Consent, to the thematic study of the UN Expert Mechanism on the Rights of Indigenous Peoples. Retrieved from https://www.ohchr.org/Docum ents/Issues/IPeoples/EMRIP/FPIC/AmazonCooperationNetwork_EN.pdf.

Rodríguez-Garavito, C. (2011). Ethnicity.gov: Global Governance, Indigenous Peoples, and the Right to Prior Consultation in Social Minefields. *Indiana Journal of Global Legal Studies*, 18, 263–305.

Rodríguez-Garavito, C. & Baquero Díaz, C. (2018). The Right to Free, Prior and Informed Consultation in Colombia: Advances and Setbacks. Retrieved from http s://www.ohchr.org/Documents/Issues/IPeoples/EMRIP/FPIC/GaravitoAndDiaz.pdf.

Sarraf, M. (2018). Quilombolas e indígenas repudiam decreto do governo do Pará sobre consulta prévia. Retrieved from http://amazoniareal.com.br/quilombolas-e-in digenas-repudiam-decreto-do-governo-do-para-sobre-consulta-previa/.

Supremo Tribunal Federal. (2017). STF decide que Mato Grosso não tem direito a indenização por demarcação de terras indígenas. Retrieved from http://www.stf.jus. br/portal/cms/verNoticiaDetalhe.asp?idConteudo=352624.

Tauli-Corpuz, V. (2016a). End of Mission Statement. Retrieved from https://nacoesu nidas.org/wp-content/uploads/2016/03/SR-on-IPs-end-of-mission-statement-Brazil-1 7-03-2016-final.pdf.

Tauli-Corpuz, V. (2016b). Report of the Special Rapporteur on the rights of indigenous peoples on her mission to Brazil. UN Doc. A/HRC/33/42/Add.1, 8 August 2016.

Valente, J. (2018). Inter-American Court Condemns Brazil for Violating Indigenous Rights. Retrieved from http://agenciabrasil.ebc.com.br/en/direitos-humanos/noticia/ 2018-03/inter-american-court-condemns-brazil-violating-indigenous-rights.

Yamada, E. (2017). Direitos humanos e povos indígenas no Brasil: Relatório da relatoria de direitos humanos e povos indígenas e plataforma de direitos humanos – Descha Brasil. Brasília: Terra de Direitos. Retrieved from http://www.plataformadh.org.br/ 2017/05/04/2017-relatorio-da-relatoria-de-direitos-humanos-e-povos-indigenas/.

Yamada, E. & Villares, L. (2010). Julgamento da Terra Indígena Raposa Serra do Sol: todo dia era dia de índio. *Revista Direito GV*, 6(1), 145–157.

Part V
Rethinking prior consultation

16 Implementation of the right to prior consultation of Indigenous Peoples in Guatemala

Lucía Xiloj

Introduction

This chapter focuses on how the prior consultation of Indigenous Peoples is devised in Guatemala, how it has been shaped by national jurisprudence, and the elements which need to be determined for its effective implementation as a right. In this country, it has been legal rulings which have contributed to the interpretation, implications, and implementation of prior consultation. This is due to the fact that several Indigenous communities have denounced the violation of their right to prior consultation through constitutional channels. Therefore, this chapter adopts a legal methodology by analysing different judgements made by the Guatemalan Constitutional Court, which, with reference to international law standards, have ruled over various cases where prior consultation has been omitted.

The chapter comprises six sections. The first section briefly describes the situation of Indigenous Peoples in Guatemala and the legal recognition of their rights. The next section outlines how prior consultation and Free, Prior and Informed Consent (FPIC) have been defined in the Guatemalan context. The third section discusses how the judicial argument regarding the right to prior consultation came into being. The fourth section analyses the different forms of prior consultation of Indigenous Peoples in Guatemala, and the various judgements of the Constitutional Court. The fifth section contains an explanation of specific initiatives for the implementation of the right to prior consultation by means of the Congress of the Republic's ordinary statutes, regulations, and guidance from the Executive Branch. Finally, the concluding remarks are based on the current interpretation of this right and the challenges which are still to be faced.

Indigenous Peoples in Guatemala

In the 1990s, a series of Indigenous rights were recognised in Guatemala. In 1995, the Agreement on Identity and Rights of Indigenous Peoples (AIRIP) was signed, which reaffirms the identity of the Mayan people, as well as that of the Garifuna and Xinca peoples. It also recognises the right to participation of Indigenous Peoples in the decision-making of the country's political

DOI: 10.4324/9781351042109-22

life (UNESCO, 1996, p. 75). On 5 May 1996, Guatemala ratified the ILO Convention No.169 Concerning Indigenous and Tribal Peoples in Independent Countries of 1989 (ILO 169). Nevertheless, before its approval by the Congress of the Republic, a consultative opinion was sought from the Constitutional Court (hereinafter, also CC), in order for it to pass judgement on the constitutionality of the Convention's content.

On 18 May 1995, the Court eventually advised:

> It can be affirmed that the Convention [...], in its entirety, does not contravene the Constitution, since it does not regulate any matter that conflicts with fundamental law but which, on the contrary, deals with aspects that have been considered constitutionally destined to be applied by means of ordinary legislation.
>
> (Corte de Constitucionalidad, 1995, pp. 6–7)

Therefore, two important pillars were put into place to support the demands of Indigenous Peoples' rights which would later be complemented by others, including the United Nations Declaration on the Rights of Indigenous Peoples (UNDRIP), which likewise became an instrument for legal interpretation and which no longer refers solely to prior consultation but also to FPIC.

The Right to Prior Consultation and Free, Prior and Informed Consent

As declared by FLACSO-ANDES (2012), "[r]ecognition of the rights of Indigenous peoples within the international regulatory framework arises as a need to demonstrate the systematic exclusion of these peoples in the history of humanity" (p. 9). In the preliminary discussions for the approval of the ILO 169, it was said that the objective of the prior consultation was to provide Indigenous Peoples with real power to influence decisions which could affect them. (Fundación Konrad Adenauer, 2015, pp. 41–44). Prior consultation is conceived as a fundamental and collective right which lays the foundation for exercising other rights related to the very existence of Indigenous Peoples; that is why it cannot be considered a simple, routine right or a mere requirement for the approval of administrative and legislative measures. In more recent years, it has become necessary to avoid understanding prior consultation as just a right or a goal, and a discussion has been established on the need to obtain FPIC in order to guarantee the full exercise of Indigenous Peoples' rights.

The IACHR (Inter-American Commission on Human Rights) has stated that: "There is therefore a national duty to consult and, in specific cases, to obtain the consent of Indigenous peoples in relation to plans and projects of development, investment or the exploitation of natural resources in ancestral lands [...]" (Comisión Interamericana de Derechos Humanos, 2010, para. 290). For its part, the UN Food and Agriculture Organization (2016) has pointed out that:

FPIC is a specific right which belongs to indigenous peoples and is acknowledged in the UNDRIP [United Nations Declaration on the Rights of Indigenous Peoples]. It allows them to grant or deny consent to a project which might affect them or their lands. Once consent has been given, they can retract it any time.

(p. 6)

This progressive interpretation of human rights is reaffirmed with the adoption of the American Declaration on the Rights of Indigenous Peoples (ADRIP) in 2016. Its article 23.2, relating to the participation of Indigenous Peoples and the contributions of the Indigenous legal and organisational systems, rules that States will hold prior consultations and cooperate in good faith with the Indigenous Peoples concerned through the bodies representing them, before adopting and applying legislative and administrative measures which may affect them, in order to obtain their FPIC (Organización de los Estados Americanos, 2016).

A brief background of the implementation of prior consultation in Guatemala

In 2005, prior consultation governed by Guatemalan laws began to be put into practice by Indigenous Peoples themselves. The first case is related to the authorisation of the Marlin mining project in the towns of San Miguel Ixtahuacán and Sipacapa in San Marcos.

The Sipacapa Municipal Council, supported by articles 6 and 15 of ILO 169, decided to convene a so-called "consultation in good faith", so that its residents could declare themselves in favour of or against the mining activity (Corte de Constitucionalidad, 2008a, p. 1). In view of this decision, the company owning the mining project brought two constitutional actions: 1) the writ of *Amparo* (*juicio de amparo*), denouncing a threat to the rights they had acquired and the violation of constitutional principles; and, 2) a petition of unconstitutionality for the same acts.

The first resolution was about unconstitutionality, with the CC pointing out that:

The right of indigenous peoples to be consulted is unquestionable; however, it considered that [...] Convention [ILO] 169 lacks precision with regard to the appropriate procedure that 'representative bodies' should carry out [...], for which reason it deemed suitable any consultative method which allows the opinions of the communities' members to be gathered accurately [...]; which supposes that the consultation must be prior to the application of the measure.

(Corte de Constitucionalidad, 2007, p. 8)

Hence, the CC considered that a prior consultation process by means of casting votes constituted a suitable method of participation for gathering the opinions of the concerned communities in Guatemala (Corte de

Constitucionalidad, 2007, pp. 8–9). In this judgement there is also evidence of the existence of a multiple legal framework that regulates the right to prior consultation: ILO 169 on the one hand, and Guatemala's domestic laws on the other (i.e., the Municipal Code, and the Urban and Rural Development Councils Act). However, as further analysis shows, the prior consultations regulated by these national mechanisms are different.

On the basis of those considerations, the Constitutional Court asserted in another decision that: "[p]opular consultations constitute important instruments of popular expression, by means of which various constitutionally recognised rights are made effective, such as those of freedom of action and of expression, as well as the right to manifest" (Corte de Constitucionalidad, 2007, p. 13).

The CC also determined that regulation relating to popular consultations is fairly broad and not very precise on procedures. Therefore it considered urging Guatemala's Congress to put into effect appropriate legal reforms and to adopt a prior consultation law that clearly regulates how consultations should be carried out (Corte de Constitucionalidad, 2007, p. 13)

The writ of *Amparo* was eventually dismissed, utilising the majority of the arguments in the aforementioned unconstitutionality judgement. However, the CC specified certain aspects about competence for calling municipal prior consultations, asserting the following: "[…] [t]he Sipacapa Municipal Council, […], on making the announcement and issuing the ruling through the acts in question has not assumed duties and so causes no grievance to the rights of the postulant" (Corte de Constitucionalidad, 2008a, p. 8). This means that the Sipacapa Municipal Council had not exceeded its competences, and it had thus legitimately called the above-mentioned consultation in good faith.

These two judgements thus reflect the complexity of implementing the right to prior consultation of Indigenous Peoples, the lack of precision as regards who should call these processes, the need to clarify the types of consultations regulated by Guatemalan legislation, and the primary implications of its results.

Forms of Consultation for Indigenous Peoples in Guatemala

Guatemala's Constitution and ordinary legislation regulate the different types of citizen consultation in general. However, for the purposes of this article, only the types of consultation associated with Indigenous Peoples will be mentioned.

First, reference will be made to the so-called "community consultations" (*consultas comunitarias*) which are recognised at the international level in the right of self-determination of Indigenous Peoples and are mainly expressed at the Inter-American level through the right to participate in government (see article 23 of the American Convention on Human Rights). These consultations are carried out in the actual communities without intervention from any State entity. Second, the "municipal consultations" will be dealt with, which in turn have different methods, and which have been used to gain opinions about extractive projects. Finally, there will be an analysis of judgements which refer to the right to consultation recognised in ILO 169. These types of consultation have different scopes, and different authorities are required to oversee their compliance.

Community consultations

As the UNDP (2006) has noted, "[f]or Indigenous peoples, consultation is linked to trust, and they aim to maintain balance amongst themselves and with their territory" (p. 119). When a measure is likely to affect them, consultation is carried out by means of community assemblies. According to the UNDP (2017):

> [c]ommunity Assemblies (*Ri nima comon chomanik-Nim chomanik*) are the maximum expression for the community organisation for dealing with matters of general benefit and great importance, in order to make decisions collectively. In Community Assemblies there are no restrictions on those who participate, as they can be old or young people, men, women, girls and boys.
>
> (p. 62)

This type of consultation should not be confused with the community consultations regulated in the Urban and Rural Councils Development Act, Decree No.11–2002, and in the Municipal Code, Decree No.12–2002, both issued by the Congress of the Republic, and introduced in 2002. In its article 26, the Urban and Rural Development Councils Act establishes: "In accordance with the law which governs consultation for indigenous peoples, consultations for the Maya, Xinca and Garifuna peoples, on development measures put forward by the Executive Body and which directly affect these peoples, can be conducted through their representatives in the Development Councils."

However, as has already been noted, there is no intervention of State agencies in community consultations, which are carried out through community assemblies and directed by their own authorities, these being chosen in accordance with their own methods of appointment or election. In the case of the Urban and Rural Development Councils System, it has been conceived as outside the practices and usual organisation of Indigenous Peoples.

A study by the UN Development Programme (UNDP) can clarify this situation:

> [i]n more general terms, participation in community decisions can be divided into two kinds: a) inside the community, and b) outside the community. The first type is related to decision making of significance in community assemblies and the meetings of authorities who help to

resolve the problems which continually arise. The external type encompasses those decisions carried out by authorities representing their communities before State institutions such as Municipalities, Departments and Offices, but also by means of structures determined by laws like the Development Councils.

(UNDP, 2017, p. 12)

Although they have not been given judicial recognition (Corte de Constitucionalidad, 2013a, p. 13), these community consultations have reinforced the political and social organisation of Indigenous Peoples. As Araujo (2013) clearly points out in his research on consultations in the municipality of Huehuetenango:

Community consultations are gradually beginning to establish themselves as an interesting forum for democratic political participation very similar [...] to a democratic community or consensual model that is governed by at least four fundamental principles for peoples native to America and Africa: 1. The priority of duties towards the community over individual rights. This is a factor of membership and a condition for rights; 2. Service engenders obligation and is directed at the common good, requiring the participation of all; 3. The common good is guaranteed by means of procedures and systems of political life which assure participation by all in the public sphere. This implies that by means of the proceedings of participative democracy, the authoritarianism or supremacy of one group over another is prevented, thus being the logic of 'to command by obedience' (*mandar obedeciendo*); 4. The final logic is consensus and therefore decisions need everyone's opinion.

(Araujo, 2013, p. 23)

Importantly, in the context of Guatemala, these consultations have managed to halt the commencement of extractive projects.

Municipal consultations

The aforementioned Municipal Code regulates several participation mechanisms including the consultation of residents, consultation at the request of residents, consultations of a municipality's Indigenous communities, and consultation of a municipality's Indigenous authorities.

Several communities have used consultation on the request of residents as a defence mechanism for their land, which has its foundation in articles 17, 64, and 65 of the aforementioned Municipal Code, and is of particular interest for the purposes of this chapter. These articles establish the procedure for carrying out these consultations, ruling that they can be performed by means of a ballot strictly and specifically designed for the purpose, establishing in the official announcement the matter to be dealt with, and the date and locations where the consultation will be held.

Unlike the right to prior consultation provided by the ILO 169, municipal consultations constitute a uniform mechanism with very specific requirements. Several actions of unconstitutionality have been put forward against rulings approved by Municipal Councils which have convened this type of consultation. For instance, in the consultations held in the municipalities of Nueva Santa Rosa and Santa Rosa de Lima in 2011, the article which referred to the binding nature of the results was declared unconstitutional (Corte de Constitucionalidad, 2012, p. 10).

However, later on the Court started to change its arguments, and therefore left in force the entire content of the rulings issued by other Municipal Councils, dismissing actions of unconstitutionality that were subsequently put forward by Indigenous communities.

In these judgements, the CC declared:

> The results of the consultation will be binding for municipal authorities [...], consequently requiring: a) The issuing of municipal resolutions in the framework of their ability to assert the will of the municipality's population [...] and b) The remittal of the results of the residents' consultation [...] so that in full compliance of the constitutional mandate to guarantee social peace for the municipality's inhabitants [...], they serve as indications when issuing resolutions of their competence regarding the matter consulted[...].
>
> (Corte de Constitucionalidad, 2013b, p. 12)

For its part, the Municipal Code in its article 65 refers to consultations with Indigenous communities or authorities of a municipality, ruling that:

> [w]hen a matter affects, in particular, the rights and interests of the communities of the municipality or of the authorities themselves, the Municipal Council will carry out consultations at the request of the indigenous communities or authorities, including the application of criteria adhering to the indigenous communities' customs and traditions.
>
> (Congreso de la República de Guatemala, 2002, Art. 65)

Likewise, evidence can be found that the management of this type of consultation lies with the Municipal Council, a body outside the normal organisation of Indigenous Peoples. In addition, municipal precepts establish that Indigenous communities must be registered in the Municipality (article 20) of their jurisdiction, which would also have an effect upon this type of consultation. The CC made reference to the issue, stating

> [it is not acceptable] to condition the recognition of community organisations representing indigenous peoples, [...] to their prior enrolment in the local municipality; if so, their access to defence mechanisms of their

fundamental rights would depend on the submission of their particular declarations of social organisation or authorities.

(Corte de Constitucionalidad, 2016, p. 2)

In Guatemala, municipal autonomy is recognised in article 253 of the Constitution, and therefore this judgement confirmed the position held by the municipalities, allowing them within their competence and jurisdiction to encourage land-use management and prioritise their own systems of development.

Prior consultation

Interpretation of the right to prior consultation has been given by the Constitutional Court in judgements on actions undertaken by Indigenous communities due to the violation of the right to participation, to normal social organisation, prior consultation, and to cultural identity. These are fundamental rights that are recognised not only in the Guatemalan Constitution (articles 66 and 67), but also in international mechanisms on the rights of Indigenous Peoples, such as ILO 169, UNDRIP, and – more recently – the ADRIP (articles 19 and 20, and 19, 20, and 23, respectively).

The first lawsuits were rejected by both the Supreme Court of Justice and the Constitutional Court, contending that the Indigenous Peoples had omitted to partake in the administrative process at the time when publications were entered in the Official Journal. The judgement dated 9 January 2008 concerned a writ of *Amparo* for the alleged omission of a prior consultation on the concession of the exploratory metals open-pit mining project called "Marlin" in the municipalities of Sipacapa and San Miguel Ixtahuacán in the department of San Marcos. The concession had been given by the Guatemalan Ministry of Energy and Mines (MEM) to the Montana Exploradora de Guatemala company. Eventually, the CC determined that:

> The concession procedure was observed in accordance with what is established in the Mining Law, which establishes the right to raise objections against anyone who considers themselves affected by an application for mining rights; given that there was no opposition, there was also the opportunity to urge for a reposition [...], and thus make an administrative appeal viable. It should be pointed out conclusively that: a) The operational licence was granted in accordance with the mandate conferred on the MEM; b) That the postulants had the opportunity to assert their arguments in the appropriate way but did not do so.
>
> (Corte de Constitucionalidad, 2008b, p. 7)

This judicial criterion was modified in subsequent judgements, making it clear that the publication of an edict did not supplant the right to prior consultation regulated in ILO 169, as exemplified in the ruling dated 14 September

2015. This was a writ of *Amparo* brought by Indigenous Maya Ixiles authorities against the authorisation granted to the Hidroxil company for a fifty-year contract to use public land to install two hydroelectric plants called "Hidroeléctrica La Vega II" in the municipality of Santa María Nebaj, in the department of El Quiché – lands that are the ancestral property of the Ixiles people. In this case, the CC affirmed that:

> The mere publication [...] of an edict which contains generalities of the application for authorisation does not represent a dialogue process that is culturally appropriate or directed at obtaining agreements. [...] in many cases that practice is not even an efficient mechanism for simple dissemination, due to the limited circulation of some of the newspapers used for this purpose and the country's linguistic diversity.
>
> (Corte de Constitucionalidad, 2015a, p. 36)

In all these judgements, the CC urged the Guatemalan Congress to issue a law to regulate how prior consultations should be carried out. However, this exhortation has not been observed. In the judgement passed on 25 March 2015, the Constitutional Court had to decide over a writ of *Amparo* for the alleged omission of a prior consultation and the omission of notification to the Indigenous Maya Ixiles communities concerned about the authorisation granted to the Transmisora de Energía Renovable company by the Guatemalan MEM to transport electricity. This project was called "Subestaciones Uspantán y Chixoy II y línea de transmisión Uspantán-Chixoy II" and was located in the municipalities of San Juan Cotzal, San Miguel Uspantán and Chicamán, in the department of El Quiché, and in the municipality of San Cristóbal Verapaz, in the department of Alta Verapazthe. Eventually, the CC put forward an indicative way in which a prior consultation could be carried out, which is summarised in 10 stages to follow over a period of six months (Corte de Constitucionalidad, 2015b, p. 4).

These stages are: (i) Transfer of full information about the project by the Ministry of the Environment and Natural Resources (MARN) to the Executive Body's Office for Indigenous Peoples and Interculturalism (hereinafter, the Office): (ii) The Office requests information from the MEM on the administrative procedure for the project's authorisation; (iii) The Office undertakes a summons through all media and communication outlets with news coverage [in the municipality where the project is supposed to be carried out], in both Spanish and the local language; (iv) The Office calls upon, at the very least, the following individuals and institutions with the aim of appointing two designated representatives and two deputies: the MEM and the MARN, the Municipal Council, the local Indigenous communities, the linguistic association's governing board, the Community Development operating in the municipality, the Deputy or Community Mayors who practise locally as such, and the personal legally responsible for giving authorisation for the project's execution. Apart from the figures already mentioned, the following should also be summoned: a representative of the Human Rights Ombudsman, who should have a deputy representative; the municipality's Indigenous mayor; and representatives of the University of San Carlos of Guatemala and the Private Universities incorporated into the Departmental Development Council which operates in the department (Corte de Constitucionalidad, 2015b, pp. 47–50).

Step (v) would follow once the particular bodies have been accredited. The pre-consultation stage is carried out so that the entities responsible can convey objective, truthful, and relevant information on the project's implications; (vi) the mechanisms for carrying out the consultation are then defined; (viii) commencement of the consultation itself proceeds; (ix) once agreements have been reached, the appropriate government and municipal authorities should define and, where appropriate, authorise the methods and requirements intended for guaranteeing actual execution; and, (x) everything previously relating to this should be carried out within a period of no longer than six months (Corte de Constitucionalidad, 2015b, p. 51).

In accordance with the legal adviser of the Indigenous mayor's office in San Juan Cotzal,

> this ruling has not been followed through, given that the government does not have professionals with the vision and ability to carry out a consultation process. The staff are too positivist and want everything to be regulated already. The planning process then followed and what Cotzal leaders said was approved. After that, an information process was started, but at this stage too it was clear that there was no willingness on the part of the government sector, and they even said that there were details which were not connected to the ruling and therefore they were going to set aside that information..
>
> (Xiloj, 2019)

Finally,

> an implementation plan was approved, which ought to begin with the transfer of information; however, since March 2018, when meetings commenced, government representatives have not explained what the project entails, solely informing about how an Environmental Impact Study (EIS) is approved. For this reason one assumes that they have no interest in the process, due to the fact that the Court did not suspend the project.
>
> (Xiloj, 2019)

Although by the end of 2016 the CC had resolved various related actions to the omission of prior consultation, it was the Oxec case which intensified the debate over how prior consultation should be carried out and the different stances on the issue. For the private sector, consultation ought not to hold back the country's development, and it asked for a law to be created; however, the Indigenous Peoples insisted that it was impossible to standardise a procedure, since different peoples had their own methods of participation and decision-making.

The Oxec case followed a writ of *Amparo* sued by several Q'eqchi Mayan communities of Cahabón, in Alta Verapaz, who alleged the violation of a number of their rights, including prior consultation, after the authorisation of two hydroelectric plants that were already in operation. Provisionally, the CC decided to suspend the projects while it resolved the *Amparo* petition. This produced a media campaign led by the owners of the hydroelectric plants, who alleged a violation of their acquired rights and assets.

On 26 May 2017, the CC delivered a judgement which it described as structural, since it aimed to resolve all cases relating to the right of prior consultation. It determined that:

> [the proposed procedure] should be implemented not only for this case but also for any future matter which may arise in connection with the right of consultation [...], whilst no [domestic] law exists which extends the said right nationwide, the former is to give a structural response to the need for providing a worthy right with full compliance.
>
> (Corte de Constitucionalidad, 2017a, p. 112)

It thus set a period of 12 months for its execution.

Moreover, this decision clarified that the administrative process of licence authorisation in accordance with the Law on Electricity and its regulation do not comply with the requirements for a due application of Indigenous Peoples' right to prior consultation. The CC indeed affirmed that:

> [t]he mere publication of an edict that contains the generalities of the application for the [license's] authorisation in specific, written means of communication does not constitute a process of dialogue that is culturally appropriate and aimed at reaching an agreement [...]. In addition, in the majority of the cases, this practice does not even represent an efficient mechanism of ordinary dissemination due to the limited circulation of the gazettes utilised for these purposes and the linguistic diversity of the country [Guatemala]. [...] It is clear that the dispositions that were provided for the regulation of the above-mentioned application for authorisation omits a specific reference to the duty of consultation to indigenous peoples. And, apart from this fact, the way the procedure was designed does not match the referred prerogative that [consultation] is carried out according to its nature and purpose since it [the procedure] is oriented to obtain two types of MEM initiatives, [i.e.] of opposition or competence; both of them are alien to and also contradict the objective that consultation pursues,

which is, as mentioned, to reach an agreement through dialogue with consent as the way to make a decision [...].

(Corte de Constitucionalidad, 2017a, pp. 65–66)

This is the only judgement that has been eventually executed. However, several communities have made the point that it does not respect the minimum standards laid down by ILO 169, and they did not agree with the process that was implemented, because, according to the MEM, it only took into account communities directly affected.

The judgement in the San Rafael Mining case

For the first time, the judgement in the San Rafael Mining case by the Guatemalan Supreme Court of Justice makes an in-depth analysis of the social, economic, and cultural impacts of an extractive project, and it refers to the need to obtain consent from Indigenous Peoples for this type of activity. Furthermore, it places a duty on the State to grant economic resources to the concerned communities so that they can participate effectively in the consultation process.

In May 2017, a writ of *Amparo* was submitted by an environmental organisation, the Centre for Legal, Environmental, and Social Action in Guatemala (CALAS, for its Spanish acronym), for the violation of the rights to prior consultation and equality of the Xinka people. The Supreme Court of Justice examined it at first instance, declaring that the right to prior consultation of these Indigenous Peoples had indeed been violated (Corte Suprema de Justicia, 2017b).

Several appeal proceedings were brought in opposition to this judgement. Guatemala's private sector, as well as the mining firm, requested that the case should be resolved based on the Oxec and Oxec II rulings, since, according to them, the CC had already set a precedent in a similar case; that is to say, it should order the prior consultation to be carried out without suspending the extractive activities. In September 2018, the CC passed its judgement, giving important guidelines, although without resolving the underlying problem, i.e., it did not state clearly the scope and the implications of the results of the prior consultation (Corte de Constitucionalidad, 2018).

In sum, the Constitutional Court (2018): (i) maintains the suspension of mining activity, while the consultation process is carried out; (ii) orders the MEM not to authorise any more licences without complying with the right to consultation, with the warning that, if it did so, it could incur civil and criminal liability; (iii) does not set a period of time for carrying out the consultation; (iv) makes executing the consultation conditional on two previous studies, both scientific and technical in nature, relating to the protection of archaeological sites located in the area in which the mining projects are taking place, appointing the MEM and the Ministry of Culture and Sport as responsible for the process, and with the revision of the project's area of influence assigned to the MARN; and (v) orders the Ministry of Finance to put in place appropriate funds for the MARN and the MEM to cover the expenses incurred by the involvement of private, specialist bodies unconnected to the dispute, which provided technical and scientific advice to the Xinka people (Corte de Constitucionalidad, 2018, pp. 543–550).

Initiatives for implementing the Right to Prior Consultation

Several attempts have been made via the Executive Body to implement the right of prior consultation by means of regulations and guidelines, which have not been successful due to the lack of participation by Indigenous Peoples. The Constitutional Court (2011) had declared that:

> Being as the right of consultation is one of the rights that helps indigenous peoples, it is clear that the development of legislation regulating such a right, whether by legislative or regulatory means, must be carried out by Guatemala's State Government with coordinated, systematic, and harmonious participation, with the members of those peoples, since it is inconceivable that the right of consultation, in pursuit of the realisation of indigenous peoples' rights, be regulated without the full participation of they themselves.
>
> (p. 9)

However, the Congress of the Republic has not demonstrated any intention of wanting to pass a prior consultation law. In this regard, neither of the two initiatives which were put forward in 2018 have had a response from the Labour Commissions, nor are they a priority in the legislative agenda. The initiatives of the Executive Body and the Legislative Body, in developing the right to prior consultation, are described and analysed below.

Regulations established by the Executive Body

In June 2011, the President of the Republic of Guatemala, Álvaro Colom, convened different sectors and Indigenous Peoples to receive inputs about the Regulation for the Consultation Process of ILO 169, a project that was drawn up by the Ministry of Work and Social Security (MINTRAB).

This initiative was challenged through a writ of *Amparo* that was put forward by several Indigenous individuals. The Constitutional Court eventually determined that:

> Even when the operations of the challenged authority show evident good faith, in the pursuit of a legitimate objective and within the framework of constitutional powers [...]; the adopted procedure limits the scope of the right of consultation which gives assistance to indigenous peoples [...]. To this end, it must be taken into account that, in agreement with article 2 of

the aforementioned Convention, Governments must assume the responsibility for the development of a coordinated and systematic process with the participation of the interested peoples, with a view to protecting the rights of those peoples [...].

(Corte de Constitucionalidad, 2011, pp. 8–9)

This judgement thus made it clear that every initiative which aimed to deal with the right to prior consultation, through a ruling or legal act, must count upon the full and effective participation of Indigenous Peoples.

Operational Guide for the Implementation of Prior Consultation for Indigenous Peoples

On 18 July 2017, the Operational Guide for the Implementation of Prior Consultation for Indigenous Peoples – prepared by MINTRAB and based on the above-mentioned rulings in the Oxec and Oxec II cases – was presented in public (Gobierno de Guatemala, 2017, pp. 3–5).

Once the guide had been produced, dialogues were carried out locally in different areas of the country, "with the aim of compiling inputs and relevant information, they also gave the government the opportunity to get to know the Indigenous peoples' own mechanisms and concepts for consultation in draft form" (Gobierno de Guatemala, 2017, p. 4). However, several of the Indigenous Peoples' representatives declared that there had been disinformation in these dialogues, given that an analysis had been made of progress in the implementation of ILO 169, and not about the guide (Nómada, 2017).

The guide is composed of two fundamental parts. The first contains the prior consultation's basic principles, and the second illustrates the consultative procedure, which is divided into eight stages that in turn set out guidelines which must be observed by the Executive Body's officials. The following are the principles of consultation according to the guide: prior; informed; in good faith; culturally suitable; flexible; and, free (Gobierno de Guatemala, 2017, p. 8).

The stages of the consultation procedure are as follows: a preparatory stage; notifications; creation of a consultation plan; gathering of information; analysis of the information; intercultural dialogue; agreements; and definition of compliance guarantees (Gobierno de Guatemala, 2017, pp. 11–14).

However, this document has been rejected by the Indigenous Peoples, who argue that it is attempting to standardise the way prior consultation is undertaken, and thus disregards the fact that Indigenous Peoples have different ways of carrying out this process (Rivera, 2017).

Legal initiatives before the Congress of the Republic of Guatemala

In 2018, there came before Guatemala's Congress of the Republic two legal initiatives which sought to regulate the right to prior consultation of

Indigenous Peoples. The first, known as No. 5416 and dated 26 February 2018, was submitted by deputy Oliverio García Rodas and heard by a plenary of the Congress of the Republic on the following 13 March.

This initiative, called the Law of Consultation of Indigenous Peoples, compliant with the ILO 169 standards, has 29 articles based on this treaty's contents. It does not take into account other international standards that have already been put forward on this subject, and it rather restricts the content and scope of this right.

By way of example, two articles are cited which completely counteract the right to prior consultation. Article 3, which refers to the consultation's sphere of application, proposes that

[t]he right of prior consultation takes effect when the State of Guatemala carries out the appropriate dialogue procedure of good faith, which has the purpose of gaining information and giving rising to agreements, and listening to points of view even in the case of disagreements [...].
(Congreso de la República de Guatemala, 2018a, p. 3)

However, this limits the real content of the prior consultation, which is to reach an agreement or obtain consent. What is more, it cannot be a simple process of information.

The content of article 4, from the same legal initiative, which refers to measures liable to affect Indigenous Peoples, adds that:

[s]ubject to consultation are those legislative and administrative measures which could be expected to have some direct effect upon the indigenous peoples [...]. The studies, information devices, documents, reports, instruments and decisions that form the record on which the public authority will issue a resolution that will have a direct bearing on the indigenous peoples and communities will not be consulted [...]. Consequently, the final resolution or authorisation, when the possibility of being affected materialises, is the only one which is subject to consultation [...]. Likewise, the construction and maintenance of infrastructure will not be subject to prior consultation, as well as that needed for the provision of public services which are concerned with benefiting the population in general.
(Congreso de la República de Guatemala, 2018a, p. 3)

This article severely limits the content and principles which govern the right to prior consultation, and leaves out the possibility of holding consultations on EIS. Furthermore, it declares that the final decision is the one which will be subject to consultation, although this right presupposes that Indigenous Peoples have a real chance to decide on the final resolution.

The second initiative was put forward by Deputy Amílcar Pop and other members of Congress on 10 May 2018, and is known as Initiative No. 5450.

It was heard in a plenary session on 7 August 2018 and conveyed to the Commission on Indigenous Peoples and the Commission on Constitutional Affairs. It is a bill of eight articles called the Law which Guarantees the Right of Free, Prior and Informed Consultation in Good Faith for Indigenous Peoples (Congreso de la República de Guatemala, 2018b). This bill establishes that the result will be binding, and affirms that the process can be revised whenever it can be shown that there has not been compliance with one or other of the essential requirements in the current administrative law. Yet it fails to regulate what will happen to the results and does not offer procedural certainty (Congreso de la República de Guatemala, 2018b, p. 1).

Finally, neither of the initiatives refers to FPIC, and they are limited to regulating the right to prior consultation as it is understood in ILO 169, when there are other international instruments such as the ADRIP, which declares that the sole purpose of prior consultation is to obtain FPIC in its article 28.3.

Concluding remarks

This chapter has referred to the different forms of prior consultation for Indigenous Peoples which Guatemala's judicial system regulates. In addition, it has analysed how Guatemalan courts have interpreted this right. However, there is a need to move from prior consultation to FPIC as a guarantee for the full participation of Indigenous Peoples in making decisions about matters which may affect them. In this regard, some final reflections are noted below.

Community consultations have been a legitimate practice for Indigenous Peoples. These consultations should not be framed solely as a right according to ILO 169 because they rather correspond to an ancestral principle which has governed the lives of Indigenous communities and came into being *before* the operations of this international instrument. In community consultations, there is no intervention by any State authority and its practice is framed in their rights to self-determination and participation.

In Guatemala, the interpretation of the right to prior consultation is increasingly restrictive. Indigenous Peoples have been bringing the right to prior consultation before the courts for around thirteen years, but the response has been neither uniform nor positive for them. Moreover, the interpretation of this right does not take into account international standards, nor is there any reference to FPIC; this is in spite of the fact that, in accordance with the international principle of progressivity towards human rights, the purpose of prior consultation should be, precisely, to obtain consent via the effective participation by Indigenous Peoples in matters which may affect them.

Another matter is that the issuing of a prior consultation law is not acceptable to the Indigenous Peoples themselves. The government has lost legitimacy and credibility in the eyes of Indigenous Peoples, due to the fact that there is no willingness to comply with the judgements delivered by the

Constitutional Court which order consultations to be carried out, such as in the San Juan Cotzal case. It is essential for the State to show signs of good faith in order to create an atmosphere of confidence so as to enable a closer relationship with Indigenous Peoples and thus reduce the conflicts involving the use of natural resources.

In addition, all initiatives that aim to regulate a method of effecting prior consultation must count on the participation of Indigenous communities and their authorities, right from the project's conception, as the Constitutional Court has stipulated.

References

Araujo, E.Y. (2013). *Las consultas comunitarias en Huehuetenango: Construyendo Democracia*. Guatemala: Cara Parens.

Comisión Interamericana de Derechos Humanos. (2010). Derechos de los pueblos indígenas y tribales sobre sus tierras ancestrales y recursos naturales: normas y jurisprudencia del sistema interamericano de derechos humanos, OEA Doc No. OEA/ Ser.L/V/II.Doc.56/09, 30 December 2009. Retrieved from https://www.oas.org/es/ cidh/indigenas/docs/pdf/Tierras-Ancestrales.ESP.pdf.

Congreso de la República de Guatemala. (2002). Decreto 12-2002, Código Municipal, Retrieved from https://www.sepaz.gob.gt/index.php/component/phocadownload/ca tegory/2-biblioteca?download=3:biblioteca.

Congreso de la República de Guatemala. (2018a). Iniciativa 5416, Iniciativa que dispone aprobar Ley de Consulta a Pueblos Indígenas, Conforme el Convenio 169 de la Organización Internacional del Trabajo-OIT. Retrieved from https://www.con greso.gob.gt/iniciativa-de-ley-detalle/?id=5442.

Congreso de la República de Guatemala. (2018b). Iniciativa 5450, Iniciativa que dispone aprobar Ley que Garantiza el Derecho de Consulta de Buena Fe, Libre, Previa e Informada a Pueblos Indígenas. Retrieved from https://www.congreso.gob. gt/iniciativa-de-ley-detalle/?id=5499.

Corte de Constitucionalidad. (1995). Opinión Consultiva sobre la constitucionalidad del Convenio 169, Expediente 199–195, 18 May.

Corte de Constitucionalidad. (2007). Inconstitucionalidad, Consulta Sipacapa, Expediente 1179–2005, 8 May.

Corte de Constitucionalidad. (2008a). Amparo Consulta Sipacapa, Expedientes Acumulados 1643–2005 y 1654–2005, 28 February.

Corte de Constitucionalidad. (2008b). Amparo por falta de consulta en el municipio de San Miguel Ixtahuacán, Expediente 123–2007, 9 January.

Corte de Constitucionalidad. (2011). Amparo en contra de la Reglamentación de la Consulta Previa, Expediente 1072–2011, 24 November.

Corte de Constitucionalidad. (2012). Inconstitucionalidad, Consultas Municipales de Nueva Santa Rosa y Santa Rosa de Lima, Expedientes Acumulados 2432–2011 y 2481–20011 y Expedientes Acumulados 2433–2011 y 2488–2011, 5 December.

Corte de Constitucionalidad. (2013a). Amparo por falta de Consulta en el Municipio de San Agustín Lanquín, Alta Verapaz, Expediente 4419–2011, 5 February.

Corte de Constitucionalidad. (2013b). Consulta a Vecinos de la Villa de Mataquescuintla, Expedientes Acumulados 4639–2012 y 4646–2012, 4 December.

Corte de Constitucionalidad. (2015a). Amparo por falta de consulta en Nebaj, Quiché, Expedientes Acumulados 4957–2012 y 4958–2012, 14 September.

Corte de Constitucionalidad. (2015b). Amparo por falta de consulta en San Juan Cotzal, Quiché, Expedientes Acumulados 156–2013 y 159–2013, 25 March.

Corte de Constitucionalidad. (2016). Amparo por falta de consulta en el municipio de Sipacapa, San Marcos, Expediente 3753–2014, 13 January.

Corte de Constitucionalidad. (2017a). Amparo por falta de Consulta en Cahabón, Alta Verapaz, Expedientes Acumulados 90–2017, 91–2017 Y 92–2017, 26 May.

Corte de Constitucionalidad. (2018). Amparo por falta de consulta al pueblo Xinka, Expediente 4785–2017, 3 September.

Corte Suprema de Justicia. (2017b). Amparo por falta de Consulta al Pueblo Xinka, Amparo No. 1076–2017, 8 September.

FLACSO-ANDES. (2012). Análisis de la consulta previa, libre e informada en el Ecuador. Retrieved from http://biblio.flacsoandes.edu.ec/catalog/resGet.php?resId= 54028/.

Fundación Konrad Adenauer. (2015). *Los trabajos preparatorios del Convenio No. 169 sobre pueblos indígenas y tribales en países independientes.* Lima: Fundación Konrad Adenauer (FKA).

Gobierno de Guatemala. (2017). *Guía operativa para la implementación de la consulta a pueblos indígenas.* Guatemala.

Nómada. (2017). Por qué estandarizar las consultas indígenas es quitarle poder al pueblo, 18 July. Retrieved from https://nomada.gt/cotidianidad/por-que-estandariza r-las-consultas-indigenas-es-quitarle-poder-al-pueblo/.

Organización de los Estados Americanos. (2016). Declaración Americana sobre los Derechos de los Pueblos Indígenas.

PNUD. (2006). *Cosmovisión Maya, plenitud de la vida.* Guatemala: PNUD.

PNUD. (2017). *Sistema Juridico Maya y Autoridades Indígenas. Una aproximación.* Guatemala: Serviprensa.

Rivera, N. (2017, July 21). Autoridades Ancestrales Rechazaron la Guía y Reglamento de las Consultas. *Prensa Comunitaria.* Retrieved from http://www.prensacomunita ria.org/autoridades-ancestrales-rechazaron-la-guia-y-reglamento-de-las-consultas/.

UNESCO. (1996). Acuerdo de Identidad y Derechos de los Pueblos Indígenas. Retrieved from http://www.lacult.unesco.org/docc/oralidad_08_70-79-anales.pdf.

UN Food and Agriculture Organization. (2016). Consentimiento Libre, Previo e Informado: Manual para profesionales en el terreno. Retrieved from http://www.fao. org/3/a-i6190s.pdf.

Xiloj, L. (2019). Interview with J. S. Sapón (lawyer), 11 January.

17 From consultation to consent

The politics of Indigenous participatory rights in Canada

Martin Papillon and Thierry Rodon

Introduction

In May 2018, the Canadian House of Commons voted in favour of Bill C-262, An Act to ensure that the laws of Canada are in harmony with the United Nations Declaration on the Rights of Indigenous Peoples (UNDRIP). For Romeo Saganash, the Member of Parliament from the Cree Nation of Eeyou Istchee who sponsored the legislation, this was the culmination of tireless work to have the UNDRIP recognised in Canadian law (Saganash, 2017). Delaying tactics from the opposition at the Senate meant the bill ultimately failed to become law, providing yet another illustration of the considerable challenges Indigenous Peoples face in their struggle to have their rights implemented in Canada.

Nowhere is this more apparent than in debates surrounding the participatory rights of Indigenous Peoples concerning decisions affecting their traditional territories. The principle of Free, Prior and Informed Consent (FPIC) is one of the backbones of the UNDRIP, but it is also one of its most contested provisions. As other chapters in this volume suggest, FPIC implementation remains elusive in a number of countries, in no small part because of diverging interpretations of the principle and its implications for States (Tomaselli, 2017; Schilling-Vacaflor, 2017; Leifsen, Gustafsson, Guzmán-Gallegos & Schilling-Vacaflor, 2017; Fontana & Grugel, 2016; Doyle, 2015). Canada is no exception in this respect. While for some indigenous leaders and activists FPIC should be interpreted as a decision-making right, federal and provincial authorities as well as industry actors tend to limit the principle to an obligation to consult in order to seek, but not necessarily obtain, consent (Coates & Flavell, 2016; Papillon & Rodon, 2017a).

If debates over FPIC interpretation and implementation in Canada share many similarities with those encountered in Latin America, they are also shaped by the specific legal and historical context under which they take place. In contrast to the track-record of Canadian mining companies in the Global South, Canada has a well-developed internal regime of Indigenous rights protection. In this chapter, we introduce this domestic regime and discuss its implications for the implementation of the international norm of

DOI: 10.4324/9781351042109-23

FPIC. We argue that while this domestic regime does provide Indigenous Peoples with opportunities to have a voice in decisions over natural resource extraction through consultative mechanisms and negotiated agreements, it also paradoxically serves to constrain how FPIC is interpreted and translated in domestic practices.

Faced with the limits of this domestic regime, Indigenous Peoples in Canada are increasingly mobilising to assert their own interpretations of FPIC. We underline three different modes of Indigenous agency in shaping how FPIC is defined in Canada. First, Indigenous Peoples have adopted *confrontational* strategies and challenged the legitimacy of existing participatory mechanisms through legal actions and civil disobedience. Second, they have also sought to *collaborate* with government authorities to maximise their influence in existing regulatory processes and create mutually satisfactory mechanisms for expressing consent. Finally, Indigenous communities are increasingly *reappropriating* the decision-making process through the creation of their own parallel mechanisms to express consent (or lack thereof).

We begin the chapter with an overview of Canada's evolving position on the UNDRIP and how it intersects with the country's internal Indigenous rights regime. We then discuss the limits of existing mechanisms in fostering Indigenous participation in the context of land and resources management, and then conclude with a discussion of Indigenous strategies to challenge the *status quo*.[1]

Canada's ambiguous commitment to the international Indigenous rights regime

While Canada is generally considered a world leader in promoting human rights, it was until recently reluctant to commit to international Indigenous rights mechanisms (Grover, 2015; Lightfoot, 2016). Canada is not a signatory to the ILO Convention No.169 Concerning Indigenous and Tribal Peoples in Independent Countries of 1989 (ILO 169) and, while it is a member of the Organisation of American States, it has not ratified the American Convention on Human Rights and has shown little interest in the American Declaration on the Rights of Indigenous Peoples of 2016. The Canadian government also initially refused to endorse the UNDRIP, largely out of fear the Declaration's wording in some of the sections concerning FPIC could be interpreted as a veto on government decision-making and regulatory processes (Canada, 2007).

Canada's vote against the Declaration in 2007 stood in sharp contrast with the very active role Indigenous Peoples from that country played in the drafting of the text (Henderson, 2008; Lightfoot 2016). Following both internal and international pressures, the Canadian government eventually gave the UNDRIP qualified support in 2010. It nonetheless maintained its caveats with regards to the provisions of the Declaration pertaining to FPIC, arguing again that FPIC, "when used as a veto", was not consistent with Canada's constitutional framework (Canada, 2010).

Eager to break with the previous administration's confrontational approach with Indigenous Peoples, the newly elected Liberal government announced in May 2016 that it was committed to the full implementation of the UNDRIP, without caveats (Canada, 2016). In a statement to the United Nations Permanent Forum on Indigenous Issues, the Minister of Indigenous Affairs took extra steps to note that this endorsement includes FPIC. The same statement nonetheless specifies that FPIC will be interpreted in continuity with Canada's existing regime of Indigenous rights (Canada, 2016). The federal government further clarified its position in 2017 with a policy statement titled *Principles respecting the Government of Canada's relationship with Indigenous Peoples* (Canada, 2017a). Among other pledges, the government announces in this statement that it will fulfil its commitment to implementing UNDRIP "through the review of laws and policies, as well as other collaborative initiatives and actions" (Canada, 2017a). The document also addresses FPIC directly, stating that "meaningful engagement with Indigenous Peoples *aims to secure* their FPIC when Canada proposes to take actions which impact them and their rights on their lands, territories, and resources" [our emphasis] (Canada, 2017a). It adds that "to this end, the Government of Canada will look for opportunities to build processes and approaches *aimed at securing* consent, as well as creative and innovative mechanisms that will help build deeper collaboration, consensus, and new ways of working together" [our emphasis] (Canada, 2017a).

The adoption in Bill C-262, the previously mentioned UNDRIP Implementation Act (Canada, House of Commons 2018), would have further reinforced Canada's commitment to the Declaration, but its failure to reach enough support at the Senate illustrates the polarizing nature of debates surrounding international Indigenous rights in Canada. Once again, the interpretation of FPIC as a veto was a central preoccupation for the bill's opponents.

The abovementioned 2017 policy statement is, therefore, Canada's clearest commitment to date on FPIC. As the above quotes suggest, the government agrees that existing mechanisms for engaging Indigenous Peoples may be insufficient and that "new approaches" are needed. At the same time, the wording of the statement suggests obtaining FPIC is interpreted as an aspirational goal rather than as a firm obligation. New processes should be "aimed at securing" consent rather than simply establishing consent.

While this position may be consistent with the UNDRIP according to some Canadian commentators (Coates & Flavell, 2016; Newman, 2017), it also falls short of what many Indigenous leaders consider the crux of FPIC, that is the capacity to freely decide on the future of their lands and communities. For Inuit leader Natan Obed, "...FPIC entitles Indigenous Peoples to effectively determine how decisions are made and what the outcome of decisions that affect us is. (...) It is not merely a right to be involved in such processes" (Obed, 2016). This latter understanding of FPIC underscores the root of FPIC in the principle of self-determination (Doyle, 2015) as well as the importance for Indigenous Peoples to have a say both in the *outcome* of decisions and in the *process* through which they are to express or withhold their consent (Papillon & Rodon, 2017a).

Finally, while we focus on the federal government in this paper, it is also important to note that provincial governments play a significant role in land and natural resource management in the Canadian federation and have therefore a key role in establishing mechanisms for implementing FPIC. Provinces have so far been careful in statements concerning the UNDRIP and FPIC. Alberta, British Columbia, and Ontario have issued statements on the UNDRIP, but they have been less forthcoming on its implications for existing participatory mechanisms (Government of Alberta, 2016).

From principle to practice: consultation and consent under Canadian law

While Canada has adopted a cautious approach to its international commitments concerning FPIC, it paradoxically has a well-developed internal regime of Indigenous rights protection. As Lightfoot (2016) argues, it is in fact partly out of concern for inconsistencies with its own Indigenous rights regime that Canada initially refused to endorse the UNDRIP.

It is also worth noting that the notion of Indigenous consent has deep historical roots in Canada. The British Crown legitimised its assertion of sovereignty on the land through the negotiation of treaties and other agreements with Indigenous nations (Borrows, 2017; Russell, 2017). In doing so, it recognised their status as autonomous political entities, as well as their specific capacity to engage in mutually consenting relationships. While this initial nation-to-nation relation was progressively replaced with a model of *de facto* sovereignty, the principle of mutual consent still shapes to this day how Indigenous Peoples define their status in relation to the Canadian federation.

The rights of Indigenous Peoples have been formally recognised in the Canadian Constitution since 1982 (Canada, 1982). While the constitutional text itself doesn't specify the substance of those rights, the Supreme Court of Canada has developed a robust jurisprudence to that effect. The Court has affirmed on numerous occasions Indigenous rights to fishing, hunting, and access to lands for cultural and economic purposes. It has also established that the Crown (the executive branches of the federal and provincial governments) has a duty to consult Indigenous Peoples and accommodate their interests whenever their asserted or established Aboriginal rights may be affected by government conduct.[2] This duty, the Court has argued, is derived from the original relationship between British authorities and the first peoples of the land. Its aim is to "reconcile the assertion of Crown sovereignty with the pre-existing rights of Aboriginal Peoples" through mutual accommodation (Supreme Court of Canada, 2004, para. 34).

The Supreme Court makes it clear that the duty to consult does not constitute an Indigenous veto on government decision-making processes (Supreme Court of Canada, 2004, paras. 42 and 48). It has nonetheless recognised that, in some instances, Indigenous Peoples should be empowered to consent to activities that have an impact on their rights. The strongest wording to that effect is found in the *Tsilhqot'in* decision (Supreme Court of Canada, 2014) and concerns infringement on a recognised Aboriginal title. It is worth quoting at length (our emphasis):

> Where Aboriginal title is unproven, the Crown owes a procedural duty imposed by the honour of the Crown to consult and, if appropriate, accommodate the unproven Aboriginal interest. (...) By contrast, where title has been established, governments and others seeking to use the land must obtain the consent of the Aboriginal title holders. If the Aboriginal group does not consent to the use, the government must establish that the proposed incursion on the land is justified under s. 35 of the *Constitution Act, 1982*.
>
> (Supreme Court of Canada, 2014, para. 76)

While the duty to consult establishes strong legal foundations in Canadian law for Indigenous participation in land and resources management, Canadian courts have so far refused to explicitly consider the UNDRIP and other international documents in their interpretations of the domestic norm. As a result, while there are certainly parallels with international debates, the domestic doctrine has evolved largely out of *sui generis* sources (Newman, 2017).

The duty to consult, as it is currently defined by Canadian courts, is also a far cry from Indigenous expectations that they should be decision-makers when their rights are affected. As the above quote suggests, the duty to obtain consent is limited both in its scope (to cases where the Aboriginal title is recognised) and in its nature (governments can justify infringement for compelling public interest). For now, at least, when the Aboriginal title is not established, only consultations are mandatory.

Beyond these general principles, the Court has thus far defined the modalities of the duty to consult on a case by case basis, using a spectrum approach under which the scope of the Crown's obligation depends on the nature of the project and its potential impact on asserted or established Aboriginal rights. The result is a high degree of uncertainty as there are few criteria to clearly establish the level of required consultation and accommodation measures.[3] The extension of the duty to consult in the territory of consent adds to this uncertainty. What is an adequate consultation in this context? Who should be consulted and when? If consent is required, how should it be expressed?

Existing participatory mechanisms and their limits

In light of this uncertainty, federal and provincial authorities have adopted a minimalist interpretation of their constitutional duty. There is, for now, no distinctive government body specifically responsible for Indigenous consultation and no legislative framework to support federal Indigenous consultation processes. With the exception of specific regimes established through modern land claims agreements in Northern Canada (Rodon, 2017), it is therefore mostly through existing public engagement mechanisms under environmental and social Impact Assessment processes (IA) that Indigenous Peoples are consulted in Canada.

Canadian IAs have proved very controversial as mechanisms to implement Indigenous participatory rights. For one, IA consultations are often ill-adapted to Indigenous realities. They are generally very formal in nature and focused on evidence-based science. Testimonies reporting personal experiences or those grounded in traditional knowledge are not necessarily valued in the same way as expert reports (Dokis, 2015; Damman & Bruce, 2012; Udofia et al., 2017).

Second, and perhaps more significant, while Indigenous Peoples can voice their concerns through consultation hearings, and potentially have an impact on the design of a project, or even its approval by regulatory agencies, there is no guarantee their views will be taken into consideration when it is time to authorise a project. In most cases (again, there are exceptions under some modern treaties), regulatory bodies have no obligation to follow the recommendations of the IA review process. In such cases, IA hearings contribute to legitimising government authority, while in effect limiting Indigenous influence on the decision-making process itself (Papillon & Rodon, 2017a).

An expert panel recently tasked by the Canadian government to make recommendations for a renewed federal environmental assessment process heard many complaints from Indigenous groups concerning the existing IA model:

> EA (environmental assessment) processes are viewed as being based on flawed planning, misinformation, mischaracterisation of Indigenous knowledge and Aboriginal and treaty rights, and opaque decision-making. (...) Instead of advancing reconciliation, EA processes have increased the potential for conflict, increased the capacity burden on under-resourced Indigenous groups and minimised Indigenous concerns and jurisdiction.
>
> (Canada, 2017b, para. 2.3)

In its 2014 report on Canada, the UN Special Rapporteur on Indigenous rights similarly noted the frustrations of Indigenous leaders concerning the mechanisms put in place by federal and provincial authorities to consult Indigenous communities. The Rapporteur noted that these processes "are generally inadequate, not designed to address Aboriginal and treaty rights, and usually take place at a stage when project proposals have already been developed" (UNSR, 2014, p. 20). The lack of a consistent framework for the implementation of the duty to consult, the Rapporteur notes, "is contributing to an atmosphere of contentiousness and mistrust that is conducive neither to beneficial economic development nor social peace" (UNSR, 2014, p. 20).

Indigenous Peoples are not alone in critiquing existing consultation processes. Conflicts over the interpretation of Indigenous rights and the absence of legitimate mechanisms to establish Indigenous support for projects are also proving costly for the resource extraction sector (Coates & Flavell, 2016). Project proponents have therefore developed their own mechanisms for establishing legal clarity and minimising the risks of Indigenous legal and political challenges. Impact and Benefit Agreements (IBAs) are one such mechanism (Papillon & Rodon, 2017b; O'Faircheallaigh, 2016).

IBAs are private agreements negotiated with Indigenous communities to seek their support for a project, in exchange for some benefits and impact mitigation measures. They have become commonplace in Canada, as industry actors seek to compensate for the inadequacies of State-driven regulatory processes while minimising conflicts and maximising their benefits for a proposed resource extraction project. For Indigenous communities, IBAs can be attractive as they allow for direct engagement with project proponents in order to minimise its negative impact and to maximise its potential benefits. More importantly, IBAs are a practical recognition, by private interests, of Indigenous Peoples' right to have a say in the future of their traditional lands (O'Faircheallaigh, 2016).

That being said, IBAs also have significant drawbacks. For one, there is generally an asymmetry in resources, capacity, and knowledge between the proponent and the Indigenous community. In the absence of government regulation to mediate this power imbalance, it is often reflected in the content of these agreements (Szablowski, 2010; Caine & Krogman, 2010). IBAs are also often negotiated behind closed doors through interposed lawyers, with limited community input (Papillon & Rodon, 2017b).

Perhaps more significantly, IBA negotiations are generally premised on the assumption that the project will be approved. Our own analysis suggests Indigenous communities sometimes sign IBAs less because they support the project, than because they have little choice, knowing fully that the project is likely to go ahead even if they oppose it (Papillon & Rodon, 2017b). The focus of negotiations is therefore less on the value of the project itself or its impacts, than on the content of the compensation package Indigenous communities can bargain for in exchange for their consent. Under such circumstances, it is hard to conclude that consent is truly free and informed.

The limits of IBAs as a mechanism to secure Indigenous consent became apparent in the ongoing debates surrounding the construction of the Trans Mountain pipeline, designed to carry heavy crude oil from Alberta to the port of Vancouver on the Pacific Coast. Kinder Morgan, the company originally behind the project, negotiated IBAs with more than 40 Indigenous communities situated along the pipeline's route. While it leveraged these agreements as a show of community consent during environmental assessment hearings, a number of Indigenous nations affected by the project also challenged the legitimacy of the federal authorisation process in court, arguing they were not adequately consulted and never supported the project. The ensuing uncertainty led to the proponent's withdrawal from the project, which was eventually bought by the federal government following pressure from Alberta and the oil industry (Harris, 2018).[4]

Despite Canada's comparatively positive record on the protection of Indigenous rights, there is therefore still an important gap between the principle of Indigenous consent and existing practices. Part of the challenge lies with the diverging interpretation of the norm, as mobilised by different actors. For governments, Indigenous consent historically served as a means to establish the legitimacy of Crown sovereignty. Contemporary consultation under IA processes is approached from a similar logic: the goal is to inform, and legitimise, Crown decisions; it is not, as Indigenous Peoples would like it, conceived as a way to share decision-making authority. Industry actors also increasingly recognise the necessity of seeking Indigenous support for their projects through negotiated agreements, but they do so for strategic purposes, with a purely economic goal in mind. This results in a narrowing of FPIC as either a passive voice in decision-making (under existing IA processes) or a bargaining tool in IBA negotiations.

Indigenous strategies for asserting FPIC

Faced with the inherent limits of existing mechanisms for translating FPIC in practice, Indigenous Peoples have adopted various strategies. We briefly discuss three of these approaches: confrontation, collaboration and what we call "reappropriation". These approaches are not mutually exclusive. In fact, they are often intertwined in complex conflicts. Each strategy nonetheless obeys a distinctive logic and has different aims, costs, and potential rewards for Indigenous Peoples.

The first and arguably most common strategy is to challenge the authorisation of a project through legal and political mobilisation. It can sometimes be the only option when regulatory authorities approve a project while ignoring Indigenous opposition. The constitutional protection afforded to Aboriginal rights and the duty to consult provide the foundations for most court challenges. Of course, there is no guarantee of success for Indigenous Peoples using the court system. While some have succeeded in quashing a permit for exploration or a project authorisation obtained without sufficient consultation (Supreme Court of Canada, 2017b), others have seen their hope dashed (Supreme Court of Canada, 2017a). However, even without a guarantee of success, legal action can be effective in delaying a project and instilling costly financial uncertainty for the project proponent, which may then reconsider its priorities. The already discussed case of the Trans Mountain pipeline is a good example.

Courts nonetheless have their limits as a site for mobilising FPIC. As mentioned, Canadian tribunals have been reluctant to consider international norms in domestic Indigenous rights cases. At best, the court can clarify, expand, or contract governments' obligations under the doctrine of the duty to consult, but it is unlikely to radically reshape the existing legal order. For Indigenous Peoples, going to court also implies at least tacit acceptance of the existing norms under Canadian law. In order to have a chance at winning their case, they have limited choice but to argue for more and better consultation, not greater control over the decision-making process itself.

Some Indigenous organisations and activists are consequently reluctant to engage in legal activism and favour more direct confrontation strategies, such as protests, occupation, and blockades, as well as public relations campaigns. FPIC then becomes a mobilising focal point to attract media attention to Canada's failings in upholding its international human rights commitments (CBC News, 2018).

Confrontational strategies can be effective in blocking or delaying a project, but they do not offer concrete alternatives to existing decision-making practices. A second type of approach to asserting FPIC is to collaborate with governmental authorities to reform existing mechanisms. This approach can be successful when governments are either compelled or willing to recognise Indigenous Peoples' role in decision-making beyond consultative processes. Governments can do so as a result of negotiated agreements, such as land claims settlements, or simply because they see their interest in minimising conflicts and legal uncertainties over resource extraction (Papillon & Rodon, 2017a).

The best example of FPIC implementation through a collaborative process in the Canadian context comes from the Northwest Territories (NWT). In essence, the NWT collaborative consent model establishes guidelines for negotiating protocols with Indigenous Peoples under which the latter are involved in determining both the process for reaching a decision and the decision-making itself. This model is based on the principle that Indigenous and non-Indigenous governments each have their respective authority on the land, and therefore have a mutual interest in developing a common process for reaching decisions over land use planning, regulations, and project authorisation (Simms, Phare, Brandes & Miltenberger, 2018).

A collaborative strategy can be effective in adapting existing decision-making mechanisms and creating a process under which Indigenous Peoples have an enhanced role in decision-making without going through the court system or legislative change. It is only viable, however, when both partners see value in sharing authority. Of course, this mutual interest rarely comes out of sheer governmental generosity. In the NWT, Indigenous Peoples benefit from a strong legal basis for asserting their authority, thanks to modern treaties which include self-government and co-management rights. Elsewhere in the country, the incentives for governments to share their decision-making authority are not as strong.

A third approach to translating FPIC into practice is for Indigenous Peoples to assert their rights through the creation of their own decision-making mechanisms over resource extraction projects. In doing so, they re-interpret the norm through their own prism. This process of reappropriation is sometimes used in conjunction with oppositional strategies, as communities simultaneously reject government-driven and industry-driven mechanisms, and assert the legitimacy of their own, alternative, decision-making processes. In many cases, however, Indigenous-led mechanisms are simply driven by a desire to establish a fair, relevant and legitimate process to express their FPIC (Gibson et al., 2018). As with the collaborative approach, the key is to establish a level playing field under which other parties have an interest in recognising, or are compelled to recognise, Indigenous decision-making authority.

One such example is the Cree Nation of Eeyou Istchee, in Northern Quebec. Through years of litigation and a series of negotiated settlements, the Cree have successfully established their role as decision-making actors on their traditional lands (Rodon, 2015; 2017). They have formalised this authority in a policy statement, the Cree Nation Mining Policy, which defines the conditions under which they are willing to support mining projects and negotiate IBAs. According to former Cree Grand Chief Matthew Coon-Come, the goal of the policy is to establish the principle of Cree consent:

> It is clear that no mining development will occur within Eeyou Istchee unless there are agreements with our communities. Those agreements will need to address a wide range of social, economic and environmental concerns on the part of our communities. Through these agreements, we will ensure that mining development is in keeping with our traditional approach to sustainable development.
>
> (Cree Nation Government, 2010, p. 2)

Project proponents who refuse to collaborate with the Cree run the risk of facing years of legal uncertainty, political pressure, and negative publicity. For example, a uranium mine proposal initially authorised by the federal and provincial regulatory bodies eventually lost its permit following a Cree campaign to have its decision to reject the project respected (Bourgeois, 2017). While the Cree of Eeyou Istchee have rejected a number of mining projects, they have also accepted others and have become partners in a number of resource development projects on their lands.

The Squamish Nation offers a different model of re-appropriation through the creation of a community-driven impact assessment process, this time in the context of a liquefied gas terminal project on its traditional territory. Through the Squamish Process, which ran parallel to both provincial and federal environmental assessment processes, the community was "able to develop its own criteria for deciding on the project, based on community hearings, impact studies and dialogue with the project proponent" (Bruce & Hume, 2015, p. 5). Most significantly, the project proponent, Woodfibre LNG, agreed to fund the process and respect the outcome through a legally binding agreement (Squamish Nation, 2016). The Squamish Nation eventually authorised the project with a series of conditions, which the proponent has, at the time of writing, agreed to follow.

The Squamish Process was successful because the proponent agreed to collaborate and respect the outcome. Woodfibre LNG saw its interest in collaborating with the Squamish Nation. In doing so, it avoided the risk of protracted legal battles over insufficient consultation mechanisms and the related costs, delay, and negative publicity around the project (Bruce & Hume, 2015).

By asserting control over the process through which they express their consent, the Cree and Squamish have sought to redefine FPIC as a matter of self-government rather than as a right to participate in State-led decision-making. In a sense, this is most consistent with the genealogy of FPIC as an expression of Indigenous peoples' right to self-determination (Doyle, 2015). If this kind of reappropriation potentially shapes how FPIC translates into practice, the success of such a strategy ultimately depends on recognition by third parties. While some Indigenous communities engage in unilateral decision-making in the absence of any dialogue with State and corporate actors, these two cases suggest that a more effective way to shape FPIC implementation is to create processes that benefit from some level of recognition and legitimacy by third parties through negotiated agreements or through power politics.

Conclusions

While FPIC is increasingly part of Canada's normative landscape, it is still a contested norm. Its meaning, scope, and operationalisation are all debated in the political and legal arenas by actors who seek to establish their own vision of the norm and their own mechanisms to translate it in practice. Of course, this political process is not taking place in a vacuum. Settler colonial policy legacies and Canada's pre-existing regime of Indigenous rights protection, however imperfect it might be, profoundly shape how FPIC is apprehended in the Canadian context. Jurisprudential interpretations of constitutionally protected Aboriginal and treaty rights offer a backdrop against which FPIC is negotiated and implemented.

As elsewhere (Shilling-Vacaflor, 2017), the language of consultation and consent is appropriated by Canadian government and industry actors to limit the normative and practical implications of FPIC. By engaging in existing federal and provincial IA mechanisms, Indigenous Peoples agree to an operationalisation of FPIC understood as a participatory right of consultation, within decision-making structures they do not control. While IBAs create a process under which Indigenous consent is explicitly recognised by project proponents, they can also result in a rarefied form of consent premised on a cost-benefit logic that does not necessarily reflect community concerns and ontological views of land and natural resources development.

In light of this limited landscape, Indigenous Peoples are asserting their own understanding of FPIC. The Indigenous agency in this context can take various forms, depending on the legal, political, and economic context, from oppositional strategies to collaboration and what we have defined as reappropriation. Through these strategies, Indigenous Peoples both give sub-stance to the norm and establish concrete mechanisms to implement it. In the absence of a clear legal framework or procedure to seek Indigenous consent, it is therefore through politics that Indigenous Peoples operationalise FPIC in Canada.

Notes

1 The political-legal analysis of FPIC implementation in Canada presented in this chapter draws from discourse analysis and fieldwork conducted by the authors, including interviews and archival work conducted in 2016 and 2017.
2 The expression "Aboriginal rights" is used for consistency with the text of section 35(1) of the Constitution Act, 1982. For a detailed analysis of the jurisprudence on the duty to consult, see Newman (2014; 2017) and Imai (2017).
3 For contrasting cases on the adequacy of consultation procedures, see the decisions of the Supreme Court of Canada in the cases *Clyde River* (Supreme Court of Canada, 2017b) and *Chippewas* (Supreme Court of Canada, 2017a), as well as the recent decision by the Federal Court of Appeal of Canada (2018) in the *Tsleil-Waututh Nation* case.
4 The project's initial approval by the regulatory authorities was overturned by the Federal Court of Appeal in light of what it considered a breach of the federal gov-ernment's duty to consult. The Court ordered the federal government to engage in new, "enhanced" consultations with those Indigenous groups that were most affec-ted by the project. See the *Tsleil-Waututh* decision (Federal Court of Appeal of Canada, 2018), as well as Papillon (2018) for a brief comment on the case.

References

Bill C-262. (2018). An Act to ensure that the laws of Canada are in harmony with the United Nations Declaration on the Rights of Indigenous Peoples. 3rd reading, 30 May 2018, 42nd Parliament, 1st session. Retrieved from www.parl.ca/Docum entViewer/en/42-1/bill/C-262/third-reading.

Borrows, J. (2017). Canada's Colonial Constitution. In J. Borrows & M. Coyle (Eds.) *The Right Relationship: Reimagining the Implementation of Historical Treaties.* Toronto: UTP.

Bourgeois, S. (2017). *Comprendre la construction du moratoire administratif sur l'ex-ploration/l'exploitation uranifère: L'influence des coalitions allochtones et auto-chtones* (Master's thesis). Department of Political Science, Université Laval, Quebec, Canada.

Bruce, A. & Hume, E. (2015). The Squamish Nation Process: Getting to Consent. *Ratcliff & Company.* Retrieved from http://www.ratcliff.com/publications/squam ish-nation-assessment-process-getting-consent.

Caine, K. & Krogman, N. (2010). Powerful or Just Plain Power-Full? A Power Ana-lysis of Impact and Benefit Agreements in Canada's North. *Organization and Environment,* 23(1), 76–98.

Canada. (1982). The Constitution Acts, 1867 to 1982. Retrieved from https://laws-lois. justice.gc.ca/pdf/const_e.pdf.

Canada. (2007). Statement by Ambassador McNee to the General Assembly on the Declaration on the Rights of Indigenous Peoples. Retrieved from https://www.cana

da.ca/en/news/archive/2007/09/statement-ambassador-menee-general-assembly-decla ration-rights-Indigenous-peoples.html.

Canada. (2010). Statement of Support on the United Nations Declaration on the Rights of Indigenous Peoples. Retrieved from http://www.aadnc-aandc.gc.ca/eng/ 1309374239861/1309374546142.

Canada. (2016). Fully Adopting UNDRIP: Minister Bennett's Speech at the United Nations. Retrieved from http://www.northernpublicaffairs.ca/index/fully-adoptin g-undrip-minister-bennetts-speech/.

Canada. (2017a). Principles Respecting the Government of Canada's Relationship with Indigenous Peoples. Retrieved from www.justice.gc.ca/eng/csj-sjc/principles-p rincipes.html.

Canada. (2017b). Building Common Ground: A New Vision for Impact Assessment in Canada. Retrieved from https://www.canada.ca/en/services/environment/conserva tion/assessments/environmental-reviews/environmental-assessment-processes/building-common-ground.html.

Canada, House of Commons (2018). Bill C-262, An Act to ensure that the laws of Canada are in harmony with the United Nations Declaration on the Rights of Indigenous Peoples, Parliament of Canada, 1st Session, 42nd Parliament, May 30, 2018. Retrieved from https://www.parl.ca/DocumentViewer/en/42-1/bill/C-262/third-reading.

CBC. (2018). Indigenous Groups Lead Protest Against Kinder Morgan's Trans Mountain Pipeline Plan. *CBC*. Retrieved from https://www.cbc.ca/.

Coates, K. & Flavell, B. (2016). *Understanding FPIC. From assertion and assumption on FPIC to a new model for Indigenous engagement on resource development.* Ottawa: Macdonald-Laurier Institute.

Cree Nation Government. (2010). Cree Nation Mining Policy, 2010–2017. Retrieved from http://www.gcc.ca/pdf/ENV000000014.pdf.

Damman, D. & Bruce, L. (2012). CEAA 2012: The New Reality for Federal Environmental Assessment. *Influents*, 7, 81–83.

Dokis, C. (2015). *Where the Rivers Meet*. Vancouver: UBC Press.

Doyle, C. M. (2015). *Indigenous Peoples, Title to Territory, Rights and Resources. The Transformative Role of Free Prior and Informed Consent.* London: Routledge.

Federal Court of Appeal of Canada. (2018). *Tsleil-Waututh Nation v Canada* (Attorney General), 2018 FCA 153 (Can).

Fontana, L. & Grugel J. (2016). The Politics of Indigenous Participation Through 'Free Prior Informed Consent': Reflections from the Bolivian Case. *World Development*, 77, 249–261.

Gibson, G., Hoogeveen, D., & MacDonald, A. (2018). Impact Assessment in the Arctic: Emerging Practices of Indigenous-Led Review. *Gwich'in Council International.* Retrieved from https://gwichincouncil.com/sites/default/files/Firelight% 20Gwich%27in%20Indigenous%20led%20review_FINAL_web_0.pdf.

Gover, K. (2015). Settler–State Political Theory, 'CANZUS' and the UN Declaration on the Rights of Indigenous Peoples. *European Journal of International Law*, 26(2), 345–373.

Government of Alberta. (2016). Renewing the Relationship: Alberta's implementation of the principles of the United Nations Declaration on the Rights of Indigenous Peoples. Retrieved from http://Indigenous.alberta.ca/UN-Declaration.cfm.

Harris, K. (2018). Liberals to buy Trans Mountain Pipeline for $4.5B to Ensure Expansion is Built. *CBC*. Retrieved from https://www.cbc.ca/.

Henderson, J. (2008). *Indigenous Diplomacy and the Rights of Peoples: Achieving UN Recognition*. Saskatoon: Purich Publishing.

Imai, S. (2017). Consult, Consent, and Veto: International Norms and Canadian Treaties. In J. Borrows, & M. Coyle (Eds.), *The Right Relationship: Reimagining the Implementation of Historical Treaties* (124–146). Toronto: UTP.

Leifsen, E., Gustafsson, M., Guzmán-Gallegos, M., & Schilling-Vacaflor, A. (2017). New Mechanisms of Participation in Extractive Governance: Between Technologies of Governance and resistance work. *Third World Quarterly*, 38(5), 1043–1057.

Lightfoot, S. (2016). *Global Indigenous Politics. A Subtle Revolution*. New York: Routledge.

Newman, D. (2014). *Revisiting the Duty to Consult Aboriginal Peoples*. Saskatoon: Purich Publishing.

Newman, D. (2017). *Political Rhetoric Meets Legal Reality. How to Move Forward on Free, Prior and Informed Consent in Canada*. Ottawa: Macdonald-Laurier Institute.

O'Faircheallaigh, C. (2016). *Negotiations in the Indigenous World: Aboriginal Peoples and the Extractive Industry in Australia and Canada*. New York: Routledge.

Obed, N. (2016). Free, Prior and Informed Consent and the Future of Inuit Self-Determination. *Northern and Public Affairs*, 4(2). Retrieved from http://www.north ernpublicaffairs.ca/index/volume-4-issue-2/free-prior-informed-consent-and-the-futur e-of-inuit-self-determination/.

Papillon, M. (2018). Collaborative nation-to-nation decision-making is the way forward. *Policy Options Politiques*, 5 September. Retrieved from http://policyoptions. irpp.org/magazines/collaborative-nation-nation-decision-making-way-forward/.

Papillon, M. & Rodon, T. (2017a). Indigenous Consent and Natural Resource Extraction: Foundations for a Made-in-Canada Approach. *IRPP insights*, 16. Retrieved from http://irpp.org/wp-content/uploads/2017/07/insight-no16.pdf.

Papillon, M. & Rodon, T. (2017b). Proponent-Indigenous agreements and the implementation of the right to free, prior, and informed consent in Canada. *Environmental Impact Assessment Review*, 62(2), 216–224.

Rodon, T. (2015). From Nouveau-Québec to Nunavik and Eeyou Istchee: The Political Economy of Northern Québec. *Northern Review*, 38, 93–112.

Rodon, T. (2017). Institutional Development and Resource Development: The Case of Canada's Indigenous Peoples. *Canadian Journal of Development Studies / Revue Canadienne D'études Du Développement*, 39(1), 119–136.

Russell, P. (2017). *Canada's Odyssey. A Country Based on Incomplete Conquest*. Toronto: UTP.

Saganash, R. (2017). United Nations Declaration on the Rights of Indigenous Peoples. Canada Parliament. House of Commons. Edited Hansard 148(245). 42nd Parliament, 1st session. Retrieved from http://www.ourcommons.ca/DocumentViewer/en/ 42-1/house/sitting-245/hansard#9866873/.

Schilling-Vacaflor, A. (2017). Who Controls the Territory and the Resources? Free, Prior and Informed Consent (FPIC) as a Contested Human Rights Practice in Bolivia. *Human Rights Quarterly*, 38(5), 1058–1074.

Simms, R., Phare, M., Brandes, O., & Miltenberger, M. (2018). Collaborative consent as a path to realizing UNDRIP. *Policy Options Politiques*, 11 January. Retrieved from http://policyoptions.irpp.org/fr/magazines/january-2018/collaborative-consent-a s-a-path-to-realizing-undrip/.

Squamish Nation. (2016). Squamish Nation Process/Woodfibre LNG Project. *Update, 4*. Retrieved from http://www.squamish.net/wp-content/uploads/2016/11/SNNewsletterV326Oct2016-01288844.pdf.

Supreme Court of Canada. (2004). *Haida Nation v British Columbia*, 2004, 3 SCR 511 (Can).

Supreme Court of Canada. (2014). *Tsilhqot'in Nation v British Columbia*, 2014, 2 S.C. R. 256 (Can).

Supreme Court of Canada. (2017a). *Chippewas of the Thames First Nation v Enbridge Pipelines Inc.*, 2017, SCC 41 (Can).

Supreme Court of Canada. (2017b). *Clyde River (Hamlet) v Petroleum Geo-Services Inc.*, 2017, SCC 40 (Can).

Szablowski, D. (2010). Operationalizing Free, Prior, and Informed Consent in the Extractive Industry Sector? Examining the Challenges of a Negotiated Model of Justice. *Canadian Journal of Development Studies*, 30(1–2), 111–130.

Tomaselli, A. (2017). The Right to Political Participation of Indigenous Peoples: A Holistic Approach. *International Journal on Minority and Group Rights*, 24(4), 390–427.

Udofia, A., Noble, B., & Poelzer, G. (2017). Meaningful and Efficient? Enduring Challenges to Aboriginal Participation in Environmental Assessment. *Environmental Impact Assessment Review*, 65, 164–174.

UNSR. (2014). Report of the Special Rapporteur on the Rights of Indigenous Peoples, James Anaya. Addendum. The Situation of Indigenous Peoples in Canada. UN Doc. A/HRC/27/52/Add.2, 4 July 2014.

Part VI

Lessons learned

18 From the implementation gap to Indigenous empowerment

Prior consultation in Latin America

Claire Wright and Alexandra Tomaselli

Introduction

Latin America is a region of great cultural and ethnic diversity, home to over 45 million members of Indigenous Peoples (UNDP, n.d.). Thanks to the mobilisation of their organisations globally (Brysk, 2000), and their contestation of norms, Indigenous Peoples have gained significant recognition and protection of their rights in international (ILO 169, UNDRIP) and regional instruments (ADRIP). However, national normative frameworks and, particularly, government practice fall far short of these standards. This inconsistency was identified by the first Special Rapporteur on the Rights of Indigenous Peoples, Rodolfo Stavenhagen, who referred to the "implementation gap" (Economic and Social Council, 2006, para. 5), a much-used concept by scholars working on Indigenous rights, to describe the continuing difficulties in realising Indigenous rights at the domestic level (Espinoza & Ignacio, 2015).

The general aim of this study has been to draw comparative lessons regarding the dimensions and nature of the implementation gap in the case of Indigenous Peoples' right to consultation and FPIC, together with its multiple causes and consequences for the protection of other individual and collective rights and, particularly, their lands, identities, and ways of life. The different chapters included in the volume deal with the analysis of specific issues and/or countries and are grouped together in terms of what they can tell us about processes to define, administrate, institutionalise, avoid, and re-think prior consultation.

Several questions were asked in the introduction in order to guide this collective analysis of the implementation gap, including the following:

- What does the application of international standards on consultation and FPIC mean for different actors at international, national, and local levels?
- What success stories can be told and what factors lie behind these successes?
- Conversely, what happens when these standards remain unfulfilled or are only partially applied? What lies behind these failures?

DOI: 10.4324/9781351042109-25

- And finally, even when consultation processes fall short of international standards, may they have other, positive impacts? As a result of their absence or limitations, can they contribute to the empowerment of Indigenous Peoples?

The next pages will first highlight and discuss the main findings and contributions of the different chapters and sections of the book, before providing answers to these questions, in a transversal reading of the text. Finally, a discussion is opened on the need to shift from a "top-down" to a "bottom-up" approach to the enjoyment of the right to prior consultation and FPIC in Latin America; specifically, by approaching the issue from the perspective of Indigenous empowerment.

Significant findings

Defining prior consultation

The first section of this volume narrated processes to define meanings and standards on prior consultation and FPIC, at both the international and regional levels. The first chapter charted the active participation of Indigenous organisations in multilateral negotiations over the creation of international documents – particularly the United Nations Declaration on the Rights of Indigenous Peoples (UNDRIP) – but their somewhat limited impact on the scope of participation, consultation, and consent in the texts finally adopted by States (see the chapter by Del Castillo). The impact of extractive industries and development projects on Indigenous rights was analysed in the next chapter, which looked to establish the responsibility of private enterprises with regard to consultation and FPIC within the Business and Human Rights framework, referring to both social licences and sustainability as useful approaches (see the chapter by Cantú). The transformative role of the Colombian Constitutional Court within Latin America was discussed in the next chapter, with special attention being paid to the proposal of binding consent rather than the controversial veto power, which may create conflicts with both the State and other minorities (see the chapter by Herrera; and, further, Leydet, 2019). On the same issue of defining consultation and FPIC, the final chapter of this first section pointed out the limits of Peru's Law of Prior Consultation (which is generally considered to be pioneering in Latin America), arguing that consultations should be understood as an exercise in self-determination, autonomy and resistance, with processes controlled by Indigenous Peoples themselves (see the chapter by Doyle).

Administrating prior consultation

The next section brought together a series of experiences of prior consultations in practice. These processes, in general terms, present a series of flaws and irregularities with respect to international and regional standards. The first chapter in this section found that Environmental Impact Assessments (EIAs) – which should in theory help to protect Indigenous Peoples' interests – in fact account for many deficiencies in consultations over the extraction of natural resources in Bolivia, given that they fail to identify all impacts and reflect the government's pro-extraction bias (see the chapter by Schilling Vacaflor). In an in-depth analysis of a consultation with the Sikuani in Colombia's Orinoquía, the next chapter discovered that the company and government authorities monopolised the process, ignoring the Indigenous Peoples' world vision and offering them few opportunities to participate (see the chapter by Calle). For its part, the chapter on consultation in Peru highlighted that, although processes are clearly deficient, they have opened up an opportunity for new and unexpected spaces for recognition politics in the form of social contestation (see the chapter by Flemmer). The next chapter referred to the case of Chile, where, although a series of consultations have been carried out as of 2009, they have failed to provide substantial results and have actually undermined the State's credibility in pursuing its international and national obligations vis-à-vis Indigenous Peoples (see the chapter by Tomaselli). Similarly, in the Mexican case, the analysis showed that the series of deficiencies regarding consultation processes in practice reflects, precisely, the lack of commitment to Indigenous rights on the part of the government in more general terms (see the chapter by Monterrubio).

Institutionalising prior consultation

The next section of the volume aimed to discuss experiences in which laws or mechanisms to define prior consultation have been created. Following Peru's consultation law of 2011, there has been significant activity in this sense in the region. For instance, in the case of Paraguay (see the chapter by Villalba), a prior consultation mechanism was created as this book was going to press (late December 2018) and in Panama a law was adopted in 2016. The issue of the institutional framework at domestic level is particularly important for two main reasons: first, because the normative framework itself is a highly contested field (see further below); and second, because without a solid consultation processes, and concrete follow-up actions, the resulting laws or mechanisms may lack legitimacy in the eyes of Indigenous Peoples. In the case of Costa Rica, it was the executive branch which decided to create a mechanism, given the inefficiency of the legislative branch in legislating on Indigenous rights. In order to do so, a consultation was carried out on the consultation mechanism, which took over 24 months and gained the FPIC of the vast majority of Costa Rica's Indigenous Peoples (see the chapter by Vega). In the case of Honduras, international actors were a key source of pressure to create the law on consultation (which is still pending), and even though there was a process of consultation it lacked legitimacy in the eyes of several Indigenous organisations. Furthermore, while the executive was presenting its proposal for a consultation law before congress, another proposal was presented in parallel by political opposition (see the chapter by Barreña), which – we would add – seriously undermined the government's project.

Avoiding prior consultation

The fourth section of this book brought together experiences where mechanisms and processes of prior consultation have been systematically avoided, in a context that is unfavourable to Indigenous rights more broadly. In Ecuador, although Indigenous Peoples were involved in creating the favourable normative framework established in the 2008 Constitution, these rights have been largely unobserved and there is no prior consultation law or mechanism. The case of the Yasuní is illustrative in this sense, given that consultation was avoided in order to prioritise the neo-extractivist project, opening up a considerable public debate on the issue (see the chapter by San Lucas). In Argentina there is no legal recognition of consultation at federal level (despite State ratifications of ILO 169 and UNDRIP) and likewise prior consultation remains elusive in the context of land disputes. This does not mean, however, the Indigenous Peoples remain passive, as in the case of Salinas Grandes-Laguna de Guayatayoc, where the Kolla and Atacama Indigenous Peoples have mobilised – reaching Argentina's Supreme Court (see the chapter by Rosti). The final chapter in this section dealt with the case of Brazil, highlighting how, in a context of threats to Indigenous and human rights in general, the government has chosen to avoid prior consultation processes (see the chapter by Mello).

Re-thinking prior consultation

The final section of the volume included two chapters which invite us to re-think how prior consultation and FPIC are conceived and carried out. In the case of Guatemala (see the chapter by Xiloj), the practice of community consultations offers a reminder that legitimacy over participatory practices must always come "from below"; that is to say from Indigenous Peoples themselves rather than international standards per se. The case study on Canada provides a contrast with the other cases included in the volume, given that it falls outside the historical, cultural, and legal context of Latin America and is the only State included in the volume not to have signed ILO 169 and which initially voted against the UNDRIP, although it revised its position in late 2010 (Government of Canada, 2010). Furthermore, there are two distinguishing features of the Canadian experience which set it apart from the Latin American one: the continuing presence of old treaties and agreements from colonial times; and the response of Indigenous Peoples, which has oscillated between confrontation, collaboration, and reappropriation (see the chapter by Papillon and Rodon). However, it also shares several features with the majority of countries included here (which will be discussed further below), reflecting that the implementation of the right to prior consultation and FPIC of Indigenous Peoples face similar challenges even in very diverse parts of the world.

Inside the implementation gap

Having summarised the main findings of each chapter, our attention now turns to the questions established in the introduction. A series of replies are offered in the pages that follow, based on a transversal reading of the different chapters within the volume.

What does the application of international standards on consultation and FPIC mean for different actors at international, national, and local levels?

One of the defining characteristics of consultation and FPIC is that their meanings continue to be contested. For their part, international and regional standards are generally clear (see the chapter by Doyle), but need to be further fine-tuned on certain points (see the chapter by Cantú). For instance, definitions of consent are undoubtedly a weak spot where interpretations have been less consistent (see the chapters by Herrera and Cantú). Likewise, the role of businesses is generally unclear in international standards, and has not been systematically approached (see the chapter by Cantú). This issue is particularly important, remembering that consultation and FPIC are particularly – although not exclusively – linked to the activities of (mainly) international firms in Indigenous territories (see the chapters by Schilling-Vacaflor, Flemmer, and Calle). One of the main reasons for the continuing debate over definitions at the international level is, undoubtedly, the omission of certain Indigenous demands in processes to adopt instruments such as ILO 169 and UNDRIP (see the chapter by Del Castillo). For their part, both regional (IACtHR) and national courts have shown considerable innovation when interpreting these rights, particularly the Colombian Constitutional Court and its proposal of binding consent (see the chapter by Herrera).

Although many Latin American States have publicly reaffirmed their commitments to prior consultation through jurisprudence (see the chapters by Herrera, Doyle, Vega, and Xiloj), legal frameworks and constitutional reforms (see the chapters by Doyle, Flemmer, Calle, Tomaselli, and Mello), and the creation of protocols and mechanisms (see the chapters by Monterrubio, Vega, Barreña, and Villalba), consultation in practice tends to be understood as a box-ticking exercise, little more than a pre-requisite before giving the go-ahead to a project (see, particularly, the chapters by Schilling-Vacaflor, Calle, Flemmer, Tomaselli, and Monterrubio). In this sense, the generalised lack of acceptance of consent as the ultimate objective of consultation processes is evident in many cases (see the chapters by

Schilling-Vacaflor, Flemmer, Calle, Tomaselli, Monterrubio), although there are some noticeable exceptions, such as the Process to Create a Prior Consultation Mechanism in Costa Rica (see the chapter by Vega) or the consultation over a proposed hydrocarbon project in Tres Islas, Peru (see the chapter by Flemmer).

For their part – and although this volume cannot claim to speak on their behalf – it is significant that Indigenous Peoples are also involved in processes to define prior consultation and FPIC. In this sense it is important to remember that the impulse for the recognition of these rights at international level came from Indigenous movements themselves (see the chapter by Del Castillo), although both international and regional standards place the responsibility and locus of consultation processes clearly with the State (see the chapter by Cantú). In the cases studied here, Indigenous Peoples have shown different relationships with and understandings of both consultation and consent: in many cases, consultation processes are viewed as a rather narrow opportunity for participation or negotiating compensation (see the chapter by Calle); for many organisations, consultation is seen as a useful tool in their repertoire of protest by stopping projects through legal proceedings (see the chapters by Rosti, Mello, and Xiloj); in others cases, claims have been made based on the lack of consultation and consent to garner public support for their demands (see the chapters by Flemmer and San Lucas); and – increasingly – consultation has been re-signified by Indigenous peoples themselves, asserting the framework of their self-determination (see the chapters by Xiloj, Doyle, Tomaselli, Rosti, Mello, and Papillon and Rodon). In this sense, *refusing* to participate in prior consultations has actually become a successful way of stopping projects in Peru (see the chapter by Flemmer) and this apparent rejection may be best understood as a reappropriation of the right (on the case of Mexico, see the chapter by Monterrubio, and Wright, 2018). Again, on this point there is a degree of contention as the Peruvian Constitutional Court has ruled that Peruvian consultation is not so much a *right* as an *obligation* for Indigenous Peoples (see the chapter by Doyle).

What success stories can be told and what factors lie behind these successes?

This is another question that was – somewhat optimistically – asked at the outset of this volume and the answer is – rather unfortunately – quite simple: few success stories can be told regarding prior consultation when implemented through political-administrative processes in Latin America. Indeed, in the following section we outline the myriad failures of consultations that very often pay lip service to international standards or, at the very best, reflect serious technical difficulties despite actors' best efforts. However, there may be room for optimism, for two main reasons:

First of all, it is clear that prior consultation and FPIC is firmly on the public and government agenda in Latin America. Despite their evident flaws, governments in the region have at least carried out processes under the nomenclature "prior" or "indigenous" consultation (see the chapters by Doyle, Schilling-Vacaflor, Calle, Flemmer, Tomaselli, Monterrubio, Vega and Barreña); courts have on many occasions ruled to protect the right to consultation, with reference to international law (see, particularly, the chapters by Herrera, Doyle, Mello and Vega); and there has been a recent push to create laws, regulations, methodologies, and mechanisms to institutionalise prior consultation (see the chapters by Doyle, Flemmer, Tomaselli, Monterrubio Vega, Barreña and Villalba). Although they are by no means free from criticism, some relative success stories in the region in this sense are the jurisprudence of the Colombian Constitutional Court (see the chapter by Herrera), Peru's pioneering Law of Consultation (see the chapters by Doyle and Flemmer), the 24-month long consultation process on the Prior Consultation Mechanism in Costa Rica in which 22 of the country's 24 Indigenous Peoples gave their consent (see the chapter by Vega) and specific consultation processes such as Tres Islas in Peru (see the chapter by Flemmer).

Second of all, in general terms, Indigenous Peoples continue to look to prior consultation and FPIC as a means to make claims and defend their rights in different spheres, mainly through domestic courts (see the chapters by Mello, Vega, Rosti and Xiloj). Likewise, they are increasingly re-asserting prior consultation from a bottom-up perspective, that is to say they have begun to create their own community consultations and protocols within the framework of their self-determination (see, particularly the chapter by Xiloj, this volume, but also the chapters by Doyle, Tomaselli, Rosti, Mello, Villalba, Doyle, and Papillon and Rodon). The fact that Indigenous Peoples do not reject consultation outright may be considered a (mini) success story.

Conversely, what happens when these standards remain unfilled or are only partially applied? What lies behind these failures?

It would be no exaggeration to say that – in general terms – prior consultation processes in practice fall short of international standards on the matter. In several cases, consultations have been systematically avoided (see the chapters by Rosti, Mello and San Lucas). And wherever so-called consultations do take place they are fraught with difficulties and inconsistencies, including: the (short) timeframe; few opportunities for effective participation; the repetition of issues; a lack of transparency; the selection of leaders who are not representative of/legitimate within communities; the omission of documents; false declarations; the lack of proper information; the lack of mutual trust; no conclusion or follow up; threats to terminate consultations without reaching agreements; a lack of an intercultural perspective; and – in general terms – a lack of good faith (see, particularly, the chapters by Tomaselli and Monterrubio). Special mention must also be made of the thorny issue of EIAs, which – rather than providing much-needed protection for the welfare of Indigenous Peoples and their environment – tend to be based on biased information that reflects the economic interests and pro-extraction bias of national governments (see the chapters by Schilling-Vacaflor and Papillon and Rodon). Another problem is the prevalence of private agreements between firms and Indigenous Peoples in which discussions focus on compensation rather than the possibility of having a real say in how the project is developed (see the chapters by Cantú Rivera and Calle), undermining the possibility of meaningful consultations. Finally, the threat of violence surrounds many consultation processes, either in the form of State repression of Indigenous leaders who protest over irregularities (see the chapter by Doyle) or armed groups (see the chapter by Calle).

Rather than technical difficulties, what would appear to be behind these failures are variables of a political nature (see the chapter by Doyle), given that the extent to which laws, jurisprudence, and mechanisms make a real difference to processes in practice remains to be seen (see further below). Indeed, asymmetries of power and vested interest are reflected within many processes of consultation with Indigenous Peoples (see the chapters by Calle and Flemmer), in which State authorities and companies set the agenda and terms of the process, rather than Indigenous Peoples themselves. This underlying difficulty is a reminder of the different conceptualisations of prior consultation and FPIC, and how, for many governments, the aim of consultation is to "tick the box" (see the chapters by Flemmer and Tomaselli) or, even, obtain information rather than offer Indigenous Peoples a real opportunity to reject a project or to take part in decisions regarding administrative or legislative measures that may affect them (on this issue, see the chapter by Monterrubio).

And finally, even when consultation processes fall short of international standards, may they have other, positive impacts? As a result of their absence or limitations, can they contribute to the empowerment of Indigenous Peoples?

In a somewhat paradoxical way, absent, inappropriate or "pseudo" consultation processes can help empower Indigenous Peoples, specifically, by opening up the public debate on the issue of their rights and bringing different parties including government authorities, Indigenous organisations and – in the context of natural resource extraction – corporations to the negotiating table. To use the terminology proposed by Flemmer in this volume, this may lead to entering into the politics of recognition "through the back door". By showing that their right to consultation and/or FPIC has been infringed in public forums, the media, or domestic courts, Indigenous Peoples can garner considerable interest in and support for their plight, and gain political advantage over governments and corporations who are responsible for the shortcomings with regard to international and/or national standards and who – ultimately – lose legitimacy as a result (see the chapter by Tomaselli).

Other implications for policy and research

Having summarised the main contributions of the different chapters and sections, the following paragraphs open up the discussion regarding the implementation gap in relation to prior consultation and FPIC, in the light of some more inductive findings and observations.

On the implementation gap

The chapters included in this volume leave us with an important conceptual question: how useful is the "implementation gap" as an approach to understanding Indigenous rights? This approach starts from "above", in the world of legal frameworks and norms at the international level, trickling down through national laws and practices. At first glance it would appear to be very useful indeed, as much scholarly attention has been paid to international standards and the need for them to be fulfilled at national level.[1] However, while the IACtHR has confirmed the importance of culturally appropriate consultation processes, by placing the onus on the State to carry out consultations, it turns them into a State-orientated rather than Indigenous-orientated practice. This clearly has to do with the fact that international instruments such as ILO 169 and UNDRIP are signed by State parties, but essentially it limits how prior consultation and FPIC may be understood. In this sense, it is important to recall that, although Indigenous organisations actively participated in the processes to create these norms, many of their contributions were excluded from the final texts (see the chapter by Del Castillo). Furthermore, for Indigenous Peoples consultation is an ancestral principle and not just a right enshrined in international law (see the chapter by Xiloj).

Another question that has arisen is to what extent international standards are necessary and useful for (relatively) good or poor practice at national level. For instance, States that have a sceptical attitude towards international Indigenous rights law may actually have a rich domestic practice, particularly thanks to mobilisation and norm contestation by Indigenous Peoples, as is the case in Canada (see the chapter by Rodon and Papillon). Although it has not been approached systematically in this volume, Panama is a particularly interesting case, given that the State has yet to ratify ILO 169 but that has not stopped it from promulgating, in 2016, the second prior consultation law in Latin America (Asamblea Nacional de Panamá, 2016). The same logic is true vice versa, in the sense that virtually all Latin American countries have ratified ILO 169, but that in itself does not appear to have made much difference for practical outcomes, as is often the case with international human rights law. Brazil, Ecuador, and Argentina are all cases in point of where the ratification of international law has – so far – had a very reduced impact at the domestic level, in terms of domestic practice (see the chapters by Mello, San Lucas and Rosti).

Another crucial issue, from this top-down approach, is whether it is a help or a hindrance to create specific laws or regulations on prior consultation. At first glance it may seem logical and desirable to create legal frameworks, and developments in Peru (see the chapter by Doyle), Panama, Costa Rica (see the chapter by Vega), Honduras (see Barreña), and Chile (see the chapter by Tomaselli) can be seen as progress in that sense. However, a specific consultation law or regulation can actually limit the scope of the right, as is the case in Peru (see the chapter by Doyle). Likewise, although in all of the processes mentioned above there were efforts to carry out consultations on the mechanisms developed – following the principle of "consultation on consultation"– nevertheless several Indigenous organisations have criticised both the process and the final results, in each case. There may be other routes to a successful institutionalistion of prior consultation, including the support of jurisprudence (as is the case in Colombia, see the chapter by Herrera) together with historical, informal institutions and practice (see, particularly, the chapters by Villalba, Xiloj, and Papillon and Rodon). Nevertheless, it appears that a lack of clear, domestic legal regulation – beyond the ratification of international treaties, basic constitutional recognition etc. – may also make it easier for authorities to evade their responsibilities. This is clearly the case in Argentina (see the chapter by Rosti), Brazil (see the chapter by Mello), Ecuador (see the chapter by San Lucas) and Mexico (see the chapter by Monterrubio), where there is a noticeable absence of secondary regulation at federal level. In any case, it is important for government authorities to be aware of their responsibilities and take steps to guarantee this right; indeed, in practice Indigenous Peoples often have to remind them (see the chapter by Monterrubio).

Again, it is worth noting that political will and awareness seem to have a greater impact on prior consultation at grassroots level than the institutional route taken (see the chapter by Doyle). That is not to say that jurisprudence and law cannot be used usefully as mechanisms by Indigenous Peoples to claim their rights, but rather that they do not seem to determine the success of prior consultation in practice *per se*. Indeed, there is clearly no unique route to closing the implementation gap from law to practice. Depending on the case, the executive, legislative, and judicial branches may all be foes or allies in this process (on this issue, see for instance the chapters by Vega and Mello). What is important is the commitment different government authorities have to fulfilling rights to consultation and FPIC, and Indigenous rights in general beyond other, competing agendas.

On Indigenous empowerment

It is not uncommon for academic studies to close with a new question, and this volume is no exception. Indeed, having reviewed the different experiences in both Latin America and Canada, the most pressing question seems to be: *how can prior consultation and FPIC be of real use to Indigenous Peoples?* Is it when prior consultations are done by the book, fulfilling international standards and offering an opportunity for co-management in all aspects, including EIAs? Or is it – perversely – through absent or failed consultation processes, which give Indigenous Peoples the opportunity to criticise and make claims against the State at national and international level? Or, is it in the reappropriation of consultations, either in the form of community consultations or the increasingly common practice of Indigenous organisations writing their own protocols to reassert their control over what appears to have become a State-led enterprise? The evidence offered on these particular issues could usefully be broadened with future research, taking into account the importance of Indigenous empowerment rather than the – ever evasive – fulfilment of legal standards.

Final remarks

All things considered, the impact of Indigenous Peoples' right to prior consultation and FPIC in Latin America in legal, political, and symbolic terms cannot be underestimated. Indeed, the issue is firmly on all sorts of agendas throughout the region. However, it is undoubtedly a contested field in which meanings, intentions, goals, and agendas clash, and in which consent is – suspiciously – elusive. Ultimately, States would do well to act in good faith and truly negotiate with Indigenous Peoples, in line with Kingsbury's (2000, p. 22) relational approach of self-determination. Even better, they must give Indigenous Peoples the freedom to truly decide on the matters which affect them, which is nothing more and nothing less than the underlying end of prior consultation. As things stand, consultation in practice may just be serving to re-inforce the centuries-old practice of the "permitted Indian" (Hale, 2004; see also the chapter by Calle), in which the State dictates the role that Indigenous peoples must play, including in processes that – ironically – are established with the aim of defending their rights.

Note

1 See, for instance, studies by Doyle (2015), MacInnes, Colchester, and Whitmore (2017), Tomaselli (2016) and Leydet (2019), among others.

References

Asamblea Nacional de Panamá. (2016). Ley No. 37 que establece la consulta y consentimiento previo, libre e informado a los pueblos indígenas. Published in Gaceta Oficial No. 28090-A, 5 August 2016. Retrieved from https://www.gacetaoficial.gob.pa/pdfTemp/28090_A/GacetaNo_28090a_20160805.pdf.

Brysk, A. (2000). *From Tribal Village to Global Village: Indian Rights and International Relations in Latin America.* Stanford: Stanford University Press.

Doyle, C.M. (2015). *Indigenous Peoples, Title to Territory, Rights, and Resources. The Transformative Role of Free, Prior and Informed Consent.* London: Routledge.

Economic and Social Council. (2006). Human Rights and Indigenous Issues Report of the Special Rapporteur on the Situation of Human Rights and Fundamental Freedoms of Indigenous People, Mr. Rodolfo Stavenhagen. UN Doc. E/CN.4/2006/78, 16 February 2006.

Espinoza, M. & Ignacio, M. (2015). Unimplemented Recognition. An Assessment of the Rights of Indigenous Peoples in Latin America. *Revista mexicana de ciencias políticas y sociales*, 60(224), 251–277.

Government of Canada. (2010). Canada's Statement of Support on the United Nations Declaration on the Rights of Indigenous Peoples. Retrieved from http://www.aadnc-aandc.gc.ca/eng/1309374239861/1309374546142.

Hale, C. (2004). Rethinking Indigenous Politics in the Era of the 'Indio Permitido'. *NCLA Report on the Americas*, 38(2), 16–21.

Kingsbury, B. (2000). Reconstructing Self-Determination: A Relational Approach. In P. Aikio and M. Scheinin (Eds.). *Operationalizing the Rights of Indigenous Peoples to Self-Determination* (pp. 19–38). Saarijärvi: Institute for Human Rights, Åbo Akademi University.

Leydet, D. (2019). The Power to Consent: Indigenous Peoples, States, and Development Projects. *University of Toronto Law Journal*, 69(3), 371–403.

MacInnes, A., Colchester, M., & Whitmore, A. (2017). Free, Prior and Informed Consent: How to Rectify the Devastating Consequences of Harmful Mining for Indigenous Peoples. *Perspectives in Ecology and Conservation*, 15(3), 152–160.

Tomaselli, A. (2016). *Indigenous Peoples and their Right to Political Participation. International Law Standards and their application in Latin America.* Baden-Baden: Nomos.

UNDP. (n.d.). Indigenous Peoples. Retrieved from http://www.latinamerica.undp.org/content/rblac/en/home/ourwork/democratic-governance/political-participation-and-inclusion/citizen-democracy-analysis—advocacy-.html.

Wright, C. (2018). El derecho a la consulta de los pueblos indígenas de México. Un balance de su reconocimiento, implementación e instrumentalización. In C. Wright (Ed.) *Participación política indígena en México. Experiencias de gestión comunitaria, participación institucional y consulta previa* (pp. 219–256). Ciudad de México: Universidad de Monterrey/Ítaca.

Index

Printed in the United States
by Baker & Taylor Publisher Services

Printed in the United States
by Baker & Taylor Publisher Services